AF520615

Commercial Agriculture

Commercial Agriculture

Editors

S.K. Saini & Subhash Chandra

G.B. Pant University of Agriculture & Technology
Pantnagar-263 145, Distt. Udham Singh Nagar
(Uttarakhand)

2011

SATISH SERIAL PUBLISHING HOUSE

403, Express Tower, Commercial Complex, Azadpur, Delhi-110033 (India)
Phone : 011-27672852, Fax : 91-11-27672046
E-mail :info@satishserial.com, hkjain1975@yahoo.com
Website : www.satishserial.com

Published by :

SATISH SERIAL PUBLISHING HOUSE
403, Express Tower, Commercial Complex, Azadpur, Delhi-110033 (INDIA)
Phone : 011-27672852 Fax : 91-11-27672046
E-mail : info@satishserial.com, hkjain1975@yahoo.com

ISBN 81-89304-95-X

Typeset at: Harminder *for* Laxmi Art Creations # 9811482328
Printed at: Salasar Imaging Systems, Delhi-35

Dr. B.S. Bisht
Vice-Chancellor

G.B. Pant University of Agriculture & Technology
PANTNAGAR-263145
Distt. Udham Singh Nagar
(Uttarakhand) INDIA

Foreword

Agriculture across the world has witnessed spectacular changes, because of its large scale commercialization and diversification. Cultivation of various crops of new hybrid varieties, increased area under cash crops, large scale adoption of livestock and fisheries, floriculture, persuance of high-tech agriculture, biotechnology, horticulture, advanced post harvest technologies *etc.* have contributed to commercialization of Indian agriculture.

Presently, agriculture has become a global concern and liberalization and has been acting as a driving force for shifting of traditional subsistance agriculture to profitable commercial agariculture. In the modern life style, the individuals' concern is to attain a better living base and agriculture therefore, is being taken in a commercial perspective.

The fundamental differences in agricultual practices of commercial and subsistance agriculture can be observed in devloped and undeveloped countries. The difference in the economic status of developed and undeveloped countries are well evident. Hence, commercialization of agriculture has become essential to improve our national economy. The essential features of commercial agriculture include larger farm size, mechnical farming, assured market, availability of resources locally, credit facility, sustainability, technological support and minimum risk factors. These warrant its status at par with industry. As a result of globalization of agriculture, ensuring quality farm products within reasonable cost is essential for sustaining in the international market.

In the present era, every earning activity with no exception to agriculture is taken in a commercial view. Hence at this juncture, the book **"Commercial Agriculture"** edited by Dr. S.K. Saini, Professor & Head (Agronomy) and

Dr. Subhash Chandra, Chief Scientist (Water Management), G. B. Pant University of Agriculture & Technology, Pantnagar will be a valuable contribution to understand the long-term consequences of commercial agriculture. Authors of different topics of the book have nicely covered the important contents. It is noticeable that the implications of commercial agriculture on society, environment and soil health have been discussed at length. The advanced technologies of commercial agariculture, such as precision inputs managment, farm operations, pest managment, production cost management *etc.* have been covered in a logical and simple manner keeping the national and global prospects in view. I hope that this book will be a valuable base in the scaling up of the agricultural technologies.

I congratulate the authors for their excellent efforts.

(B.S. Bisht)
Vice-Chancellor

Preface

About two decades earlier, the share of agriculture in country's GDP was more than 30 per cent, while presently it has reduced to merely 17 per cent. Industry has gradually outwitted agriculture on this front, consequently agriculture as business carries less importance. Yet, of-late, Indian agriculture received momentum towards commercialization and is now passing through a very crucial phase as the professionalism is booming-up among the cultivators. Bit-by-bit, it is turning from a mere way of life to a business. The economics more precisely profit has become an integral part of agriculture too.

Further, under shrinking land resource, pressure is building-up on this sector to meet the growing challenge of feeding ever increasing population. Issues like quality of the produce, cost effectiveness in crop production, sustaining crop productivity at higher level, protecting the natural resources, competition across the national boundaries *etc.* have become integral part of agriculture with the globalization. Thus, the agriculture henceforth warrants a different kind of form. For addressing all these issues, appropriate agro-technologies alongwith farmer-benign market policies provide a strong foundation. Recently, some fruitful efforts have been made in this direction and many dimensions (protected cultivation, integrated farming, precision agriculture, conservation agriculture) have been added to make agriculture a profit spinning avocation on a sustainable basis. Thus there is strong need to have a publication, which contains information on various aspects of commercial agriculture such as advanced and adoptable agrotechnologies, economics, social and policy aspects. This prompted the authors to bring out a book on "**Commercial Agriculture**", which is a collection and collation of latest knowledge, intended for making available the invaluable information on almost all the facets of commercial farming for its wider dissemination and use for betterment of mankind in India and abroad.

This book contains forty one chapters which have further been grouped into twelve thematic areas (Insight into commercial agriculture, Mechanization for breaking barriers, Production management agro-techniques, Balancing nutrition, Managing water - the vital input, Combating pests, weeds and insects, Resource conservation, Alternate agro-enterprises, Livestock based farming systems, Harnessing advanced technologies, Promoting agro-industry and Socio-economic and climatic facets) for systematic and easy understanding of the subject. Through this book, the authors have made sincere efforts to putforth the upcoming agricultural scenario and management techniques from its commercial view point. It is our belief that this book will prove worthy to policy makers, planners, researchers, teachers, students, development agencies, NGOs, and entrepreneurs.

We are sincerely grateful to the contributors of this book, who have made all possible efforts and extended all cooperation in bringing out this publication.

Editors

List of Contributors

S.No.	Name (s)	Address
1	S. K. Saini	Professor & Head, Director (Centre for Advanced Studies), Department of Agronomy, G.B. Pant University of Agriculture and Technology, Pantnagar - 263 145
2.	Amit Bhatnagar	Jr. Research Officer, Department of Agronomy G.B. Pant University of Agriculture and Technology, Pantnagar - 263 145
3	Shweta	Sr. Research Fellow, Department of Agronomy, G.B. Pant University of Agriculture and Technology, Pantnagar - 263 145
4	V.K. Singh	Associate Professor (Agronomy), G.B. Pant University of Agriculture and Technology, Pantnagar - 263 145
5	Subhash Chandra	(Chief Scientist Water Management, Agronomy), G.B. Pant University of Agriculture and Technology, Pantnagar - 263 145
6	Gurvinder Singh	Jr. Research Officer, Department of Agronomy, G.B. Pant University of Agriculture and Technology, Pantnagar - 263 145
7	T.C. Thakur	ICAR National Professor (Farm Machinery & Power Engg.), G.B. Pant University of Agriculture & Technology, Pantnagar - 263 145
8	B. Gangwar	Project Directorate for Cropping Systems Research, Modipuram, Meerut - 250 110 (U.P.)
9	Anand Prakash	Professor (Agronomy), G.B. Pant University of Agriculture & Technology, Pantnagar - 263 145

10	Badrul Hassan	Professor & Dean, Faculty of Agriculture, Sher-E-Kashmir University of Agriculture and Technology, Jammu
11	Tahir Ahmad Sheikh	Professor & Dean Faculty of Agriculture, Sher-E-Kashmir University of Agriculture and Technology, Jammu
12	M.K. Sinha	Central Research Institute for Jute and Allied Fibres, Indian Council of Agricultural Research, Barrackpore, Kolkata – 700 120, W. B., India
13	Sabyasachi Mitra	Central Research Institute for Jute and Allied Fibres, Indian Council of Agricultural Research, Barrackpore, Kolkata – 700 120, W. B., India
14	T. Ramasubramanian	Central Research Institute for Jute and Allied Fibres, Indian Council of Agricultural Research, Barrackpore, Kolkata – 700 120, W. B., India
15	B. S. Mahapatra	Central Research Institute for Jute and Allied Fibres, Indian Council of Agricultural Research, Barrackpore, Kolkata – 700 120, W. B., India
16	A.S. Chandel	Ex-Professor (Agronomy), G.B.Pant University of Agriculture and Technology, Pantnagar – 263 145
17	H. Chowdhury	Central Research Institute for Jute and Allied Fibres, Indian Council of Agricultural Research, Barrackpore, Kolkata – 700 120, W. B., India
18	D.K. Singh	Senior Research Officer (Agronomy), G.B. Pant University of Agriculture and Technology, Pantnagar – 263 145
19	Anchal DassScientist	SS, Division of Agronomy, IARI, Pusa, New Delhi- 110 012
20	Renu Pandey	Division of Plant Physiology, Indian Agricultural Research Institute, New Delhi – 110012
21	Vanita Jain	Division of Plant Physiology, Indian Agricultural Research Institute, New Delhi – 110012
22	K. P. Singh	Division of Plant Physiology, Indian Agricultural Research Institute, New Delhi – 110012
23	S.S.L. Tripathi	Professor (Agronomy), G.B. Pant University of Agriculture & Technology, Pantnagar – 263 145
24	S.N. Tiwari	Professor (Entomology), G.B. Pant University of Agriculture & Technology, Pantnagar – 263 145

25	Sumitra Arora	Sr. Scientist, National Centre for Integrated Pest Management, LBS Building, Pusa Campus, New Delhi-110 012
26	A K Kanojia	National Centre for Integrated Pest Management, Lal Bahadur Shastri Bhawan, IARI Campus, PUSA, New Delhi-110 012
27	V. Pratap Singh	Professor, Dept of Agronomy, G.B. Pant University of Agriculture & Technology, Pantnagar - 263 145
28	Neeta Tripathi	Sr. Reserach Fellow, G.B. Pant University of Agriculture and Technology, Pantnagar - 263 145
29	P.K. Singh	Senior Research Officer (Irrigation & Drainage Engineering), G.B. Pant University of Agriculture & Technology, Pantnagar - 263 145
30	J.P. Tiwari	Dean, College of Agriculture, G.B. Pant University of Agriculture & Technology, Pantnagar - 263 145
31	B.C. Saini	Ex- Professor & PI, Tea Research Centre, Soil Science Department, College of Agriculture, G.B. Pant University of Agriculture & Technology, Pantnagar - 263 145
32	Ajit Kumar Kapoor	Assistant Professor (SS) (Horticulture), G.B. Pant University of Agriculture and Technology, Pantnagar - 263 145
33	K.P.S. Kushwaha	SRO (Plant Pathology), G.B. Pant University of Agriculture & Technology, Pantnagar - 263 145
34	M.S. Gill	Project Director, Project Directorate for Cropping Systems Research, Modipuram, Meerut
35	Sanjay Kumar	Associate Professor (Livestock Production & Management), G.B. Pant University of Agriculture & Technology, Pantnagar - 263 145
36	R.K. Sharma	Professor (Veterinary & Animal Science), G.B. Pant University of Agriculture and Technology, Pantnagar - 263 145
37	D.V. Singh	Professor (LPM), G.B. Pant University of Agriculture & Technology, Pantnagar - 263 145
38	S. K. Singh	Department of Livestock Production and Management, G.B. Pant University of Agriculture & Technology, Pantnagar - 263 145
39	Jyoti Palod	Department of Livestock Production and Management, G.B. Pant University of Agriculture & Technology, Pantnagar - 263 145

40	Rajeew Kumar	Jr. Research Officer, Department of Agronomy, G.B. Pant University of Agriculture and Technology, Pantnagar – 263 145
41	Vineeta	Sr. Reserach Fellow, Department of Agronomy, G.B. Pant University of Agriculture and Technology, Pantnagar – 263 145
42	Ramphool Puniya	Sr. Reserach Fellow, Department of Agronomy, G.B. Pant University of Agriculture and Technology, Pantnagar – 263 145
43	Virendra Singh	Professor (Forestry & Hill Agriculture), Hill Campus, Ranichauri – 249 199 (Uttarakhand), G.B. Pant University of Agriculture and Technology, Pantnagar – 263 145
44	V.P.S. Arora	Professor (Agricultural Economics), G.B. Pant University of Agriculture and Technology, Pantnagar – 263 145
45	R.S. Verma	Principal Scientist (Agronomy), Division of Crop Production, Indian Institute of Sugarcane Research, Lucknow-226 002 (U.P.)
46	Jagdish Kumar	Professor & Head (Agricultural Economics), G.B. Pant University of Agriculture and Technology, Pantnagar – 263 145
47	O.P.S. Bana	Professor (Agronomy), G.B. Pant University of Agriculture and Technology, Pantnagar – 263 145
48	A.S. Nain	Senior Research Officer (Agrometeorology), G.B. Pant University of Agriculture and Technology, Pantnagar – 263 145

□□□

Contents

Promoting Agro-Industry

Socio-Economic and Climatic Facets

Insight into Commercial Agriculture

1

Commercial Agriculture — An Overview

S.K. SAINI, AMIT BHATNAGAR AND SHWETA

During the last few decades Indian agriculture has undergone spectacular changes because of large-scale commercialization and diversification. Cultivation of new crops and varieties, increase in the share of area under cash crops, large-scale adoption of livestock and fisheries, pursuance of hi-tech agriculture, biotechnology, horticulture, processing, etc. contributed to commercialization of Indian agriculture. Moreover, liberalization/ globalization acted as driving force for conversion of traditional subsistence agriculture into profitable commercial agriculture.

Subsistence agriculture is the self-sufficient farming system in which farmers grow enough food to feed their family and to pay taxes or federal dues. The typical subsistence farm has a range of crops and animals needed by the family to eat during the year. Planting decisions are made with an eye towards what the family will need during the coming year(s), rather than market prices.

Subsistence grain-growing agriculture first emerged during the Neolithic Revolution when humans began to settle in the Nile, Euphrates, and Indus River Valleys. It was the dominant mode of production in the world until recently, when market-based capitalism became widespread. Roughly, 30,000 years ago hunter-gatherer behaviour fed 6 million people, 3,000 years ago primitive agriculture fed 60 million people, 300 years ago intensive agriculture fed 600 million people and today commercial agriculture is to feed over 6000 million people.

Historical perspective

Before the advent of the British rule, crops such as cotton, tobacco and sugarcane were grown extensively since land revenue had to be paid mostly in cash and the prices of these crops were relatively much higher. Even during the British rule, the situation did not change much. Though the primary concern of the rulers from then onwards was the expansion of trade, some of the policies in pursuit of this objective introduced market forces into agriculture. In the process, land was rendered marketable in principle as the British vested the property rights on land with the individual farmers for the first time. This, coupled with the growth in population and infrastructural investments in irrigation, communication and transport, resulted in rise in land value. Besides, expanding trade opportunities in agricultural commodity also brought forth inflow of finance from rent-seeking urban traders and money lenders to agriculture. This set off the emergence of a different outlook for farm enterprise from an enterprise that provided a source of livelihood to one that had the potential of a commercial venture.

The All India Rural Credit Survey (1951-52) report threw up a fair idea as to the extent of commercialization with its regional perspective along with its correlates such as cropping pattern, value of gross produce per unit of land, distribution of holdings, land rent paid in cash and kind, wages paid in cash and kind, borrowings, etc.

Impressive growth in yield levels was noticed with the onset of green revolution during mid-sixties in widespread use of modern inputs such as HYV seeds, water, fertilizers and pesticides. As a result, there was a spurt in marketable surplus. The growth in agricultural output also owed it to favorable public policy on prices, marketing and credit, besides widespread adoption of new agricultural technology. The Commission for Agricultural Costs and Prices (CACP), (the erstwhile Agricultural Price Commission) set up in 1966 played an important role in recommending remunerative prices for various crops. Regulation of markets has also been given due importance during the post Green Revolution period. On the credit front too, several innovations had been introduced to augment the working capital of the cultivators as well as to enhance their investment capabilities. Commercialization was also accompanied by diversification of agriculture, outside the crop sector by way of subsidiary enterprises in animal husbandry, poultry, fisheries, sericulture, etc. There has been acceleration in the commercialization, growth and diversification of agriculture since 1980s, especially during 1990s. The positive feature of this phase of commercialization is the coverage of even small and marginal farmers and backward regions.

Concept of commercialization

The production of crops for sale, crops intended for widespread distribution (supermarkets), and any non-food crops such as cotton and tobacco is considered as commercial agriculture. Commercial agriculture also includes livestock production, fishery, poultry and agro-forestry etc. but does not include crops grown for household consumption (backyard garden or from a vegetable garden or fruit tree's)

Commercial farming relies on large arable land and sufficiently advanced technology such as hybrid seeds, fertilizers, pesticides, etc.. It specializes and focuses on one particular crop due to economies of scale. This is further augmented by high levels of technology, which significantly reduces the risk of poor harvests. Commercial farming emphasizes on capital formation, scientific progress and technological development. Due to the expensive nature of capital formation and implementation of technological processes, the landowners of such farms are often large agricultural corporations, especially in developing countries. Commercial farming, in terms of some of its processes, is conceptually not very different from large industrial enterprises.

Thus, conceptually, commercialization can either take the form of product commercialization which can occur on the output side through increased share of marketed surplus or factor commercialization which can occur on the input side through increased use of purchased inputs. In the former case, the surplus can be had in the form of additional amount, over and above self-consumption, or in the form of a cash crop which is allotted a higher proportion of the cultivated area owing to favorable economics. An expected increase of production and income motivate the farm-firms to enter into the exchange economy and become more commercialized. In the latter case, commercialization may be understood as adoption of modern inputs such as seed, fertilizers, pesticides, irrigation and mechanical power, most of which are purchased from market. Thus, at the farm level, the extent of commercialization can be measured either as a ratio of value of market arrivals to total value of agricultural production or as a ratio of value of inputs purchased from the market to total value of agricultural production. Considering different facets of commercial agriculture, some of the ratios that can serve as indicators of the degree of commercialization are, viz., ratio of market arrivals to production, share of non-food grain cash crops and food grain cash crops in gross cropped area as well as value of output, ratio of purchased inputs to value of all inputs (as also output) and share of agricultural exports in GDP (Gross domestic product) in agriculture. The above indicators, however, cannot separately portray the complex phenomenon of commercialization. They cannot capture, for instance, degree of diversification, introduction of new products, extent of involvement of corporate in agri-business and changes in the attitudes of the farmers. In recent times, the corporate seem to be coming into agriculture in a big way and the agriculture-industry linkages are assuming new dimensions (Satyasai and Viswanathan, 1997). Nonetheless, these indicators may give fairly good idea about the degree of commercialization.

Extent of commercialization

Commercialization on product side includes increase in marketable/marketed output relative to production, changes in product-mix, within crop sector in favour of commercial crops, both in terms of area and production, diversification towards different sub-sectors of agriculture, viz., livestock, poultry, fisheries, forestry, advent of high-tech agriculture, diversification towards secondary sector, i.e. agro-industry. These are discussed in the following paragraphs.

I) Change in cropping pattern

"From food crops to non-food crops and low value crops to high value crops"

Historically, food grain crops had dominated Indian agriculture although cultivation of commercial crops like cotton, sugarcane, tobacco, etc., was prevalent from olden days. At the beginning of the planning era, food-grain crops occupied 73.5 per cent of the gross cropped area of which cereals and pulses accounted for 59.2 per cent and 14.3 per cent, respectively. Cereals being the staple food for the whole population and since after independence, the country embarked upon self-sufficiency in food production, the cultivation of cereals by the majority of the small farmers could not be undertaken on commercial lines. As such, during the 1950's major proportion of the gross cropped area (59.2 per cent) was occupied by cereals, which were grown for subsistence. Pulses, however, were not grown for self consumption alone, and, hence, can be considered commercial crops from the beginning. Rest of the gross cropped area was partly occupied by other crop-groups (17.0 per cent) such as oilseeds, fiber crops, fruits and vegetables, narcotics, beverages, spices and condiments, natural rubber, sugarcane, etc., which were traditionally considered as commercial crops as they were cultivated for the market and partly (9.3 per cent) by other miscellaneous crops like fodders and forages.

In globalization, era farmers have shifted from cultivation of food to non-food crops like cotton, tobacco, sugarcane etc., which have low risk and fetch high price in the market. Moreover, low value high volume crops such as wheat, maize, pearl millet etc., have been replaced by high value low volume crops like medicinal and aromatic plants, spices etc. especially in rainfed region of our country to get more income.

II) Diversification

Diversification in agriculture can be broadly defined as producing increased number of remunerative agricultural commodities. In essence, diversification to commercial crops/commodities becomes an essential strategy that can increase income levels in agriculture, reduce risks of crop failures and earn foreign exchange. Further, diversification can be designed to help poverty alleviation, employment generation and environmental conservation. In Indian agriculture, diversification has occurred across sub-sectors, viz., crop sector, livestock sector, forestry, fishery, etc. as well as within each of these sectors. In more recent times, diversification towards high-tech, innovative enterprises in the primary sector such as floriculture, horticulture and towards secondary (manufacturing) sector such as agro/food processing and rural non-farm sector have been gaining momentum.

Diversification towards secondary sector

Of late, diversification into secondary and tertiary sectors has been taking place in the rural economy. Rural non-farm sector, including, agro-processing has been promoted as this sector holds promise for absorption of surplus labour from agriculture and for improving living standards of the rural poor. Overtime, there has been an expansion, accompanied by diversification, in the consumption basket. Consumption of agro-processed commodities has assumed importance among different income groups (Radhakrishna and Ravi, 1990). Coinciding with such changes in consumption pattern, there has been an accelerated growth in the output of several agro-based industries in recent times. During last decade, corporatisation in the agricultural sector has emerged in the form of establishing teak farms, orchards, plantations, floriculture, etc. The cost of land, maintenance etc., are met through public subscription of shares with promises of attractive returns on the investments to the investors. A wide variety of schemes, including schemes for fixed deposits, are being floated. Most of the companies are engaged in developing wastelands.

III) Hi-tech projects

In the recent years, there has been a spurt in hi-tech projects in agriculture and allied sectors. India ranks first in the world in fruit and second in vegetable production. The exports of processed fruits and vegetables have been increasing steadily during the recent years. India has vast potential for development of aquaculture, both in inland and marine fisheries. Other emerging commercial ventures are in floriculture which is mainly confined to cut flowers such as rose, carnation, orchid and wax flowers. Production of hybrid seeds of high value ornamental, indoor foliage plants as well as vegetables is other potential areas. Mushroom production in India is also gaining importance.

Consequences of commercialization

1. *Food and nutritional security*: The food crops have attained a commercial status and have been substituted for other uneconomical crops, the best example being the case of large scale cultivation of rice and wheat in North-Western states. This had a positive influence on our food security. Other important changes in agricultural sector that had positive influence on our nutritional security were increasing share of livestock and fisheries, on one hand, and increase in the share of fruits and vegetables in our food basket on the other. These developments had the influence of improving the availability of animal protein as well as fruits and vegetables in the Indian diet (Satyasai and Viswanathan, 1996).
2. *Increase in household income and change in consumption pattern*: Due to cultivation of remunerative crops along with cereals, higher yields and easy sale of farm produce, the income of farmers have increased which has led to the tendency of farmers to include vegetables in other wise common roti/ chaval- dal diet.
3. *Increased job opportunities*: The commercialization of agriculture has created year-round job opportunities not only for land owners but also for the landless and educated youth.
4. *Environmental threat*: Commercialization, however, has share of liabilities too. Some of the ills associated with commercial agriculture are as follows:

- Huge consumption of water, energy and industrial chemicals
- Increased land, water and environment pollution
- Deterioration in nutritional quality, if crops are bred for cosmetic and shipping characteristics
- Loss of biodiversity
- Deterioration of soil quality
- Deterioration of land/soil due to accelerated erosion and runoff losses and salinity

- Food and fodder toxicity
- Increased use of fossil fuels for agrichemicals and farm machines
- Destruction of natural predators/enemies

Factors influencing commercialization

1. Population

Demographic change is a key determinant of commercialization facilitating or impeding the process depending on the availability of resources. In a situation where expansion of the cultivated area is still possible, and if the marginal labour productivity exceeds the marginal subsistence requirements, population growth may enable in increasing the marketable surplus on one hand. On the other hand, due to perceived food security risks, population growth might lead to reduced volume of marketed surplus in relative or even in absolute terms in regions with deficient market linkages. In the case of Indian agriculture, there is ample evidence to prove that the growth of population had been a driving force to bring about marginal expansion in cultivated area but nevertheless a substantial increase in productivity especially after the seventies was witnessed.

Need for food security was the primary motive for diversification, i.e., to include crops which ensured minimum food is even in the face of adverse weather.

2. Technological factors

Technological change is another factor which influences commercialization. Although increased commercialization can occur without technological change in agriculture but technological change without increased commercialization is unlikely because the increased use of purchased inputs and specialization are inherent elements of technological innovations in agricultural production. With the advent of new technology, the cash requirements of farm households increased on account of increased use of purchased inputs necessitating the cultivation of cash crops on a large scale. As a result, the proportion of area under these crops, especially wheat, increased overtime.

3. Infrastructure

Infrastructural support is an important pre-requisite for commercialization, which , at present, does not commensurate with the production potentials available across states. For instance, concentration of production of fruits and vegetables and infrastructural facilities including processing capacity were observed to be mismatched across states (Viswanathan and Satyasai, 1997). Policies have to focus on aspects that foster commercialization by facilitating an open domestic and international trade environment, improving hard and soft infrastructure for opening up new market opportunities and ensuring legal security. The benefit of commercialization and specialization, to a greater extent, depend upon infrastructure and both have a 'push and pull' relationship. Lack of roads, cold storage facilities, transport (including refrigerated) facilities, air cargo, pre-cooling, etc. Inadequate post harvest facilities relating to processing, grading, packing and erratic supply of electricity etc. are major impediments. Delay in electric connection are yet another important constraint.

4. Government policy

Commercialization in a country also depends on the existing global economic environment and trade relations. Public policies like agricultural price policy, market interventions to make available food to the poor and national priorities such as food security, creation of irrigation infrastructure, wasteland development etc., have influenced the commercialization of crops.

5. Price factor

Relative prices of crops have been the most important factor in determining the crop mix. Coarse cereals have been replaced by more lucrative crops like oilseeds because of their price advantage.

6. Information technology

Information technology can provide some of the most innovative tools/usages in the field of agricultural marketing in general and export trade in particular. If these tools are in conformity with the domestic trade policies and international technological sophistication, the technology can be effectively utilized for planning, analysis and policy making with the help of computer based information management systems, forecasting, modeling and other quantitative techniques.

7. Marketing constraints

Inefficient agricultural marketing system has been a major impediment in the growth of marketed surplus and realization of remunerative prices by the farmers. In spite of best efforts by the state, weaknesses still persist. Some of the maladies that affect the agricultural marketing are presence of unduly large number of middlemen, long marketing channels, malpractices by the intermediaries, information gap due to lack of free and effective flow of information.

8. Input supply

Supply of inputs in adequate quantity and time is essential for success of commercial projects. For example, for establishing plantations successfully, supply of quality seedlings is a prerequisite. In vast tracts of our country which are rain-fed, water is a major constraint for undertaking land based commercial ventures.

9. Institutional constraints

The present level of exploitation of agro based hi-tech potential in the country has been observed to be very low. The supply of credit is inadequate to meet the demand not withstanding the significant role played by NABARD and credit institutions. Lack of awareness among entrepreneurs, bank personnel and other implementing authorities about the relevance of hi-tech and absence of effective coordination among different agencies involved are the major constraints.

The existing provisions of Land Ceiling Act in the country impede large scale cultivation. Most hi-tech and commercial ventures require large extent of area under cultivation to achieve break-even point.

10. Socio-economic constraints

The socio-economic profile of our agriculturalists is not always conducive for adoption of new technology. Green revolution technology being scale neutral could enlist the participation of most of the farmers, sooner or later, irrespective of farm size. However, the new commercial agricultural ventures are capital intensive, require in-depth knowledge and involve high degree of risk and hence, are not amenable for individual farmer's many of them whom are illiterate.

Vision to promote commercial agriculture

- Establishing commercial farming schools to demonstrate cultivation of profitable cash crops and train young farmers in advanced methods to raise productivity.
- Establishing integrated horticulture estates to cultivate profitable vegetable and fruit crops linked to professional processing and marketing by private or public sector agencies.
- Establishing integrated sericulture projects in which all the operations from mulberry cultivation to silk spinning and weaving are carried out within clusters of villages and the products are professionally marketed and exported.

- Establishing intensive aquaculture estates consisting of quarter acre intensive production ponds leased out to small farmers and landless workers with centralized technical support, feed plant, processing, storage and marketing facilities.
- Establishing scientifically run soil labs in every district to test soils for macro and micro-nutrients and prescribe measures to restore soil fertility.
- Encouraging the private sectors to acquire or lease degraded, uncultivable waste lands and to utilize advanced technologies to reclaim them for intensive horticulture and farm forestry.
- Development of Argo-eco tourism on waste land.
- Development of integrated mega food parks.
- Revamping the curriculum of agricultural graduates to impart practical skills in commercial farming and to encourage graduates to take up scientific farming and agri-business ventures.
- Widely publicizing achievements in the agri-business sector to generate awareness.

References

Radhakrishna, R. and Ravi, C. 1990. Food demand projection for India. Centre for Economic and Social Studies : Hyderabad.

Satyasai, K.J.S and Viswanathan, K.U. 1996. Diversification of indian agriculture and food security. *Ind. J. of Agrl. Econ.*, Vol. 51(4) : 674-679.

Viswanathan, K.U. and Satyasai, K.J.S. 1997. Fruits and vegetables production trends and role of linkages. *Ind. J. of Agrl. Econ.*, 52 (3).

Satyasai, K.J.S and Viswanathan, K.U. 1997. Commercialization and diversification of Indian agriculture. Published by National Bank for Agriculture and Rural Development, Department of Economic Analysis and Research, Jeevan Seva Complex (Annexe), S.V. Road, Santacruz (W). Mumbai - 400 054.

□□□

2

Green Revolution — Approach, Implication and Impact

V.K. SINGH

Agriculture is the greatest living, private sector industry of India providing livelihood to over 600 million persons. The development of agriculture system is, therefore, must for the development of the country. In the 20th century, massive public investments in modern scientific research for agriculture led to several breakthroughs in the industrial countries. Modern plant improvement techniques, improved agronomy, development of inorganic fertilizers and pesticides and expansion of irrigated area fueled these advances. Most industrial countries achieved sustained food surpluses by the second half of the 20th century. These advances were much slower in reaching developing countries. By the mid 1960's hunger and malnutrition were wide spread especially in Asia. Back to back occurrence of droughts in India fractured the backbone of agriculture. In response, Rockefeller and Ford foundation took the lead in establishing an international agricultural research system (IARS) to help transfer and adapt scientific advances to the conditions in developing countries. The first investments were made for the research on rice and wheat, the two of the most important food crops of developing nations.

To achieve higher yields for rice and wheat, scientists needed to develop plants that had short and stiffer straw to support the weight of heavier head of grains. They also needed to develop varieties which could be more responsive to nutrients and water, having high harvest index and resistance to biotic and abiotic stresses. Borrowing from rice breeding work undertaken in China, Japan and Taiwan, IRRI developed semi dwarf varieties of rice that met most of these requirements. Similar achievements were made for wheat after Norman Borlaug crossed japanese semi dwarf varieties with maxican wheat at CIMMYT in Mexico. The new palnt type was spectacular and has high yielding potential. In 1968, USAID administrator Willian S. Gaud coined the term 'Green Revolution' to describe this phenomenal growth in agriculture. Although the term green revolution originally described developments for rice and wheat, yet HYVs have since been developed for other major food crops of developing countries including sorghum, maize, millets, cassava etc.

Approaches of green revolution

- Change in plant architecture having high harvest index and lodging resistance through intensive breeding programme.
- Change in physiological rhythm by incorporating insensitiveness to photoperiodism and synchronization in growth and reproductive phases.
- Improved crop management technology.
- Increased use of water, fertilizer and pesticides.
- Mechanization.
- Government support.

Implication of green revolution

The adoption of HYVs occurred quickly. By 1970, about 20% of wheat and 30% of rice area in developing countries was planted under HYVs. Early rice & wheat modern varities were rapidly adopted in tropical and subtropical region with good irrigation system or reliable rainfall. The English wheat, which took nearly 1000 years to increase the yield from 0.5 to 2.0 t/ha, climbed to 6 t/ha only in 40 years. Implication of green revolution in India was so conspicuous that rice and wheat yields during 2006-07 increased by 2.7 and 6.8 times, respectively as compared to base year of 1960-61 (Table 1).

These changes more than doubled the cereal production in Asia between 1970 and 1995 besides, increase in population by 60%. Latin America also experienced significant gains as well. Poor infrastructure, high transport costs, limited investment in irrigation and marketing policies penalized the farmers of much of African countries.

Table 1. Production and yield of cereals and pulses

Year	Population (crore)	Production (m tonne)			Yield (q/ha)		
		Cereals	Pulses	Rice	Wheat	Rice	Wheat
1960-61	43.9	69.32	12.70	34.58	11.00	10.13	8.51
1970-71	54.8	96.60	11.82	42.20	23.83	11.23	13.10
1980-81	68.5	118.96	10.63	53.63	36.31	13.36	16.30
1990-91	84.6	162.10	14.10	74.29	55.14	13.40	22.81
2000-01	102.7	196.00	13.42	84.98	69.68	20.80	27.60
2006-07	114.0	197.7	14.11	92.76	74.89	21.27	26.71

Singh and Chandra (1999) computed the regression function to examine the effect of important inputs on crop productivity for Punjab. They gave the best fit regression equation for fertilizers, irrigation and farm power.

a) *Fertilizer* : (Linear response)
$Y=1160.4+13.972\ X_f$
Where Y = state average food grain yield (kg/ha)
X_f = State average fertilizer consumption

b) *Irrigation* : (Exponential response)
$Y = 278.32\ e^{0.0262xi}$
Where Xi = Average food grain area under irrigation (%)

c) Farm power (Quadratic response)
$^*Y= 534.41 + 1924.0\ Xp - 297.53\ Xp^2$
Where, Xp = State average availability of power (k W/ha)
*1=Farm power showed decreasing effect beyond 3.24 kW/ha

Regression coefficients for different inputs computed for Punjab and Madhya Pradesh conditions have been embodied in Table 2.

Table 2. Regression coefficient for production inputs

Input	Regression coefficient	
	Punjab	Madhya Pradesh
Fertilizer	0.23	0.52
Irrigation	0.35	0.38
Farm power	0.45	0.05

Source: Singh and Chandra (2000), Singh (2000)

Impact of green revolution

a) Positive impacts

1. Gains from modern varieties (MVs) were larger in the 1980s than in the preceding two decades-despite popular perceptions that the Green Revolution was effectively over by this time. The Green Revolution is best understood not as a one-time jump in production, occurring in the late 1960s, but rather as a long-term increase in the trend growth rate of productivity. This was because successive generations of MVs were developed, each contributing gains over previous generations.
2. The MVs contribution to yield growth was higher in the late than in the early Green Revolution period, accounting for almost 50% of yield growth and 40% of production growth for all developing countries. This indicates that in the late Green Revolution period, production gains were more dependent on MVs than in the early period and that MV contributions were greater in the late period.
3. Contrary to some views of the Green Revolution, the rate of MV releases has actually increased since the 1960s.
4. More than 35% of MVs released and adopted were based on crosses made in IARCs; 15% of MVs crossed by the national agricultural research systems of developing countries (NARS).
5. IARC (International Agriculture Research Centre) research complemented NARS breeding by providing improved germplasm. International breeding efforts increased the productivity of national programmes. Because of this IARC-NARS complementarity, the existence of the international centres actually stimulated national investment in NARS research.
6. Although input use intensified in the late Green Revolution period, productivity gains from MVs allowed food production to increase dramatically with only modest increase in area planted to food crops-and with relatively slow growth in the use of inputs such as fertilizer and irrigation.
7. Assessments of the impacts of IARC and NARS programmes showed very high benefit : cost ratio for IARC programmes and for most NARS programmes.
8. Productivity growth associated with MVs had important consequences. Increased food production has contributed to lower food prices globally. Average caloric intake has risen as a result of lower food prices-with corresponding gains in health and life expectancy.

9. Virtually all consumers in the world have benefited from lower food prices. Many farm families also benefited from research-driven productivity gains-most clearly those whose productivity rose more than prices fell, but also those who produce much of their own food.
10. The absolute number of poor people fell from 1.15 billion in 1975 to 825 million in 1995.
11. Real per capita income almost doubled in Asia between 1970 and 1995.
12. Increase in irrigated area after independence increased the input use efficiency.

b) Negative impacts

A revolution of this magnitude was bound to create some problems of its own. Critics charged that took place:

1. Green revolution resulted in environmental degradation and increased income inequality. Inequitable asset distribution worsened absolute poverty.
2. Owners of the large farms were the main adopters of the new technologies because of their better access to improved seeds, irrigation water, fertilizer and credit.
3. Green revolution encouraged unnecessary mechanization thereby pushing down rural wages and unemployment.
4. It spread only in irrigated and high potential rainfed areas and many villages or regions without access to sufficient water were left out. Although evidences suggest that even in these cases, villagers obtained indirect benefits through increased employment, irrigation opportunities and cheaper food.
5. Excessive and inappropriate use of fertilizers and pesticides has polluted water ways, poisoned agricultural workers and killed beneficial insects and other wild life.
6. Faulty irrigation practices have led to salt buildup in the field.
7. Heavy dependence on a few major cereal varieties had led to loss of biodiversity on farms. Some of these outcomes were inevitable as millions of largely illiterate farmers began to use modern inputs for the first time.
8. Inadequate extension and training, input pricing and subsidy policies directed towards monoculture without due consideration of environment which led to continuous decrease in productivity of cereals.

References

Balishter, Gupta, V.K. and Singh R. 1991. Impact of mechanization on employment and farm productivity. *Productivity* **32** (3) : 484-489

Bayliss-Smith, T.B. and Wanmali, S. 1984. The green revolution at micro scale. [In] understanding Green Revolutions. Cambridge University press.

Hazell, P. and Haddad, L. 2001. Agricultural research and poverty reduction. 2020 vision discussion paper 34, Washington, D.C. International Food Policy Research Institute.

IEG, 1975. Technological changes and distribution of gains in Indian Agriculture. The Institute of Economic Growth, Delhi

Sheehy, John 2006. Genomics to pave way for new green revolution in rice. *Indian Express.* August 6, 2006

Singh, G. and Chandra, S. 2000. Analytical approach to growth dynamics of agricultural inputs and their effect in increasing productivity in Madhya Pradesh. *Agricultural Situation in India.* Pp. 723-731.

Swaminathan, M.S. 1983. Science and conquest of hunger, concept, Delhi. P 409.

Tribe, D. 1994. Feeding and greening the world ; The role of International Agricultural Research, Wallingford, U.K. CAB International.

□□□

3

Commercial Agriculture in Farmer's Perspective

SUBHASH CHANDRA, GURVINDER SINGH AND ANCHAL DASS

India can be considered as a rural country, as more than 70% of its population live in village and survive on agriculture. Agriculture has been back bone of Indian economy since ever. Rural prosperity can be improved only by improving the level of agriculture in these areas. Agriculture in India has undergone various changes. In the ancient time, its activities were mainly confined to the domestic and local needs. With growing population and increase in living standard, agriculture turned has from domestic/ subsistence to gradually commercial.

What is Agriculture?

Agriculture can be defined as: Intentional planting of crops and raising of domesticated animals.

or

The deliberate modification of the Earth's surface through cultivation of plants and rearing of animals to obtain sustenance or **economic gain.**

The economic gain is one of the principle characterstics of commercial farming.

What is commercial agriculture?

Commercial agriculture is the production of food primarily for sale off the farm.

What is subsistence agriculture?

Subsistence agriculture is the production of food primarily for consumption by the farmer's family. The most fundamental differences in agricultural practices are between those in less developed countries and those in more developed countries

The level or state of agriculture prevailing with farmers is generally, categorized by the farm size. However, it is also affected considerably by the available farm resources, or technologies and prevailing growing conditions and marketing facilities.

Essential features of commercial agriculture

- Farm size (large)
- Market (assured)
- Resource availability (local)
- Financial support (loans/banks)
- Sustainability (stable yield and profit)
- Technology (feasible and suitable)
- Risk factor (minimum)

The essential features of commercial agriculture warrants its status at par with industry. For this, consistently high farm productivity with good quality and negligible risk is essential. Water plays a pivotal role in success of agriculture, because it is a pre –requisite for improving the efficiency of all other inputs both monetary and non-monetary. The increase in food grain production in the county has largely been seen with increasing irrigated area (Fig. 1).

Presently, in our country about 40% area is irrigated. It has been estimated that 40% irrigated area contributes approximately 55% to the national food basket while contribution from the rainfed areas is nearly 45%. Besides low contribution, the uncertainty of crop productivity over the years in rainfed areas is again a matter of concern. The type/nature of commercial agriculture will, therefore, be different under these two situations.

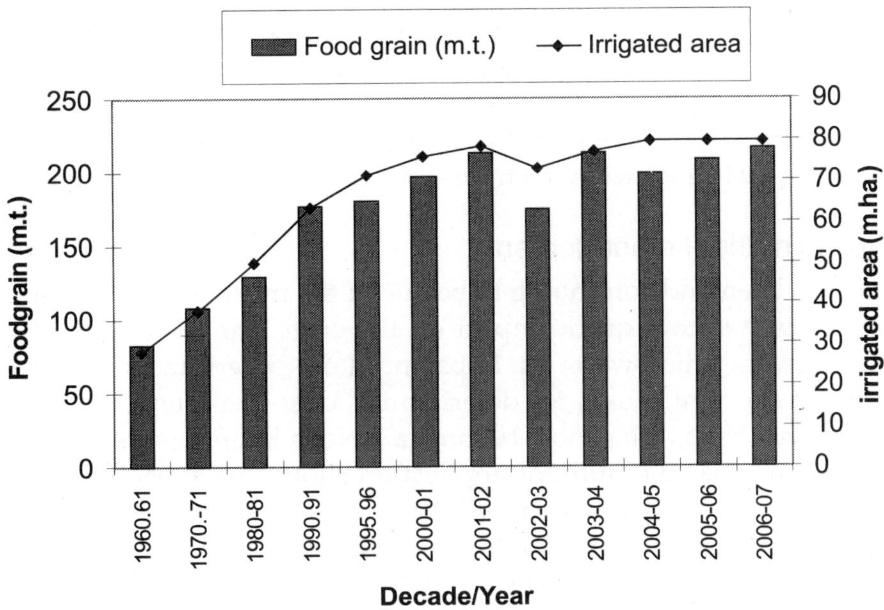

Fig. 1. Irrigation development vis- a- vis total food grain production in India

Major issues in commercial agriculture

In broader sense the commercial agriculture can be compared with industry, where prime focus remains on profitability. Following objectives/ issues have been found closely associated with commercial agriculture.

Profit

Unlike subsistence farming, making more and more profit from per unit area and time is the major aim of commercial farming. For long- term sustainability, of course, it should not be at the cost of environment, soil health and other natural resources. Growing summer rice in *Tarai* region is an important example, wherein farmers aim to get more profit while ignoring excessive use of under-ground water during hot summer months. Farming as such can not be compared with industry as it is a biological system, and greatly affected by climate/growing environment besides crop management level. Studies have shown that making more profit in short term has proven detrimental in the long run. This has resulted into following undesirable practices.

Indiscriminate use of pesticides

Most of the Indian farmers have been very poor in technical know how about the judicious use of pesticides. Mostly, they are guided by the advice of shopkeeper. Farmers always want to see a clean crop; often they do not wait even for the visible effect of the pesticide. A myth has also been observed with the farmers that plant should turn complete healthy within a short time, which tempts them to go for frequent use of pesticides without showing any concern for environment. The farmers due to obvious reasons, generally, do not follow the practices (IPM modules) which minimize the use of synthetic chemicals.

Imbalanced nutrient management

Urea has been and continuing to be one of the most favorite fertilizers of the farmers as it shows quick impact on the crop. The ill- effects of green revolution are not unknown to us. Imbalanced use of fertilizers has been one of the most important causes for deterioration of soil and environment, and unstable yields. Thus, it has now become a serious issue. Invariably at every forum, it is said that the nutrient application should be made based on the soil test values, however, practically it is hardly happening anywhere. Development of deficiency of secondary nutrients in Punjab is one the examples of such irrational practices. Emphasis on regular use of organic manure by the farmer is rare and its use has been restricted to either specific crop or area.

Non-judicious water management

Still in our country, there is no proper water pricing policy. Farmers generally go for excessive application of irrigation water without considering its adverse effects on crop health and nutrient loss. Field / farm leveling is still done manually in most part of the country leading to poor water productivity. It has been reported that land leveling alone can save water upto 30 per cent. Irrigation application criterion has been highly arbitrary; soil moisture and other characteristics are not considered while applying irrigation to a particular crop.

Selection of crop variety

Most of the farmers go by the advice of seed seller or relatives without taking into consideration the suitability of that particular variety for their area. It often results in to failure of that variety.

Farm operation practices

Farmers try to modify the farm operations/ management practices without assessing its long term impact. Although it reduces cost of production and makes it easy to carry out. Use of tanks sprayers with power jet, burning of crop residues, sand mixed application of pesticides, plot to plot irrigation in rice and wheat etc. are few of the farm operations modified by the framers for the sake of the ease of the opertaion.

Agricultural development process in India

In the agriculture development system of our country, basically, three groups of people play major role, viz., farmer, policy maker and researcher. Of these, farming community is the only group directly affected by the ups and downs in the system, because their livelihood depends upon farming. Therefore, his opinion is entirely different about commercial agriculture, and often differs from rest of the two groups. His perceptions about commercial farming have been enlisted below.

- Profit is the major aim/concern
- Least bothered about natural resources especially water and climate
- Always has short term goal/vision
- Resorts to easy going farm operation practices without considering their direct/ indirect ill effects/ consequences, For example, burning of crop residue and use of tractor sprayer with power jet
- Try to copy his neighbourers for more gains
- Cropping pattern is market driven and least concern about the crop diversification and soil health
- Least bothered about common resources and colleague farmers

Policy maker

- National food security is the major concern
- Worried about wrong practices followed by the farmers, climate change and degrading natural resources

Researcher

- Some time unrealistic extrapolation of research findings
- Low scale of research activity
- Often mis-match between the farmers need and the research carried out
- Feasibility of the technology, often not taken into account
- Often recommendation on individual component rather than a complete package

Suggestive measures for better viability of commercial agriculture

- Farmers should not aim maximum yield and profit, rather optimum to have long term sustainability

- Must keep records of the farm activities for future planning and judicious use of inputs and natural resources
- Farmers must work in a cohesive group while maintaining their individual identity to minimize the involvement of the middleman
- Farmers need to be trained to develop or improve their skill to keep pace with the world agriculture
- There is need to develop a national agriculture policy to protect the interests of the farmers
- Local marketing facilities / processing units need to be developed and organized

Conclusion

As most of the agricultural commodities are essential for human beings, therefore, this sector can not be ignored at the cost of industry etc. Efforts must be made to make agriculture a profitable venture to keep intact the interest of the farmers to this risky, complex and labourious activity. A perfect linkage among farmers-researchers-policy makers needs to be established with a concern for one another. Development of sense or feeling about country's food security should be the prime concern and the profit and personal interest should not come in the way of progress of agriculture.

❒❒❒

Mechanization for Breaking Barriers

4

Farm Mechanization in Commercial Agriculture

T.C. THAKUR

India witnessed unprecedented growth in agriculture which has helped the country to graduate from hunger to self-sufficiency in food grain production with an increase from 51 Mt (1950-51) to 231 Mt (2007-08). The technology back-up by the scientists, in the form of *'Green Revolution'* combined with industrial growth, positive policy support, liberal public funding for agricultural R & D, and dedicated work of farmers contributed to phenomenal increase in agricultural production. Application of mechanization in agriculture was equally appreciated by the farmers and today they feel proud to have improved machinery from a traditional country plough to tractor powered rotary tillers, dibblers to pneumatic planters, Persian wheel to drip and micro-sprinkler systems, sickle to combine harvesters and so on.

Indian agriculture has all the three phases of agricultural development i.e. traditional agriculture with low energy, capital and input-output ratio; technologically dynamic agriculture with low capital intensity and market orientation, and technologically dynamic agriculture with high capital intensity and high market integration. Commercial farming, in terms of some of its processes, is conceptually not very different from large industrial enterprises because it deals with industrialized production of crops, livestock, fish and poultry. The country has moved from subsistence agriculture to sustainable, commercial and market driven agriculture. Most of the dairy products, eggs, fruits and vegetables available in supermarkets are produced using the methods of commercial agriculture. Farmers find commercial agriculture more attractive than traditional agriculture because the production is market oriented and it is conducted on the principles governing any business. Commercialization of agriculture in India is critical so as to ensure optimum utilization of scarce resources, to supply raw material to industries and edible goods at low cost; and to create farm surpluses for capital formation and consumer goods. For successful implementation of commercial agriculture, the partnership with producers, industry associations, government agencies, input suppliers, extension specialists and academician is a prerequisite. Many of the agriculture produce such as hybrid seeds, sugarcane, cotton, oil seeds, vegetables etc. are produced through contract farming wherein corporate enters into arrangements with farmers to supply crucial farm and management inputs, and buyback the produce, resulting in a win-win situation for the producer as well as the industry.

Agricultural mechanization refers to interjection of machinery between human and materials handled by them. In agriculture, the materials are soil, water, seed, fertilizers, pesticides, irrigation, agricultural produce and by-products such as food grains, oilseeds, fruits, vegetables, cotton, sugarcane, jute and other cash crops, milk, meat, eggs and fish etc. There is a scope of mechanization in every unit operation of production agriculture, post-harvest processing, storage and marketing. Mechanization has varied connotations that in developed world, it tends to be synonymous to automation but in developing countries like India, mechanization means any improved tool, implement, machinery or gadget that assists in enhancement of workers output, multiplies the human efforts, supplements or substitutes human labour that enables to remove drudgery affecting adversely the efficiency. It also means automation and controls that assure quality of work and produce. Agricultural mechanization in a limited sense refers to production agriculture. Mechanization of agriculture has shown to result in higher productivity (10-35%), saving of seeds and fertilizers (15-20%), saving of time (20-30%), enhanced cropping intensity (1.5-2.5 times), increase in return to farmers (30-50%) and dignity of agricultural profession and better living conditions.

Unlike the large farms of developed countries specializing in production of a particular product at commercial scale, the contract farming is a viable option with regards to small farm size in Indian context. It can be an effective private sector-led mechanism to facilitate the transition to commercial agriculture leading to improved profitability and raising the incomes of small farmers. Due to economies of scale, it becomes more profitable for farmers to specialize and focus on one particular crop. A cafeteria of improved equipment and technologies is available for all types of farms to meet their specific need. However, some of the equipment needed for basic unit operations in production agriculture have been described in this paper.

1. Equipment for soil cultivation

The unit operations required for soil management includes mainly soil loosening, clod disintegration, soil consolidation and compaction, clod sorting, mixing, levelling/smoothing etc. which are needed to achieve desired seedbed for crop establishment. Chiseling and subsoiling are needed for alleviation of subsoil compaction and modification of rootbed soil structure for proper crop growth.

i. Soil loosening/bursting

It is required to burst open the field to obtain granular soil structure seedbed for enhanced infiltration, water retention, air exchange within soil profile, and to provide minimum resistance to root penetration and plant growth. The equipment such as M.B. ploughs, disc ploughs chisel ploughs, subsoilers, ridger ploughs, tines with forward rake angles are commonly used for soil loosening operation (Table 1).

Table 1. Selected tractor drawn/self propelled equipment for different farm operations

S. No.	Equipment	Size, mm	Field capacity, ha/h	Suitability
1. Seedbed Preparation				
1.	M.B. Plough	2x350 3x350	0.20-0.25 0.30-0.35	All regions
2.	Disc Plough	2x350 3x350	0.20-0.25 0.30-0.35	All regions
3.	Disc harrow (16 disc)	2000	0.40-0.50	All regions
4.	Cultivator (9-11 tines)	2500	0.40-0.50	Lighter soils
5.	Rotavator	1500	0.20	All regions
6.	Sweep cultivator	5x500	0.35	Heavy soils
2. Sowing and Planting				
1.	Seed-cum-fertilizer drill	11x225	0.75	All soils with suitable furrow openers
2.	Zero-till drill	11x225	0.75	Direct drilling of wheat, paddy, pulses & oil seeds

Contd...

3.	Strip-till drill	1500	0.25	Direct drilling of wheat, pulses after paddy
4.	Sugarcane cutter planter	2x900	0.25	All regions
5.	Potato planter	2x600	0.25	All regions
3. Weeding and Interculture				
1.	9-11 tyne tiller	2500	0.45	All soils, wider row crops
2.	5 tyned sweep	2500	0.45	All soils, wider row crops
3.	Self-propelled power weeder	500	0.125	All soils winder row crops
4. Plant Protection				
1.	Tractor operated sprayer	5000	0.50	All regions, wider row crops
2.	Stationary Power operated sprayer	5-10 h.p.	0.35	All regions
3.	High clearance self-propelled sprayer	10,000 mm	0.50-1.0	Cotton, potato & other tall crops
5. Harvesting				
1.	Tractor operated VCR	2000	0.40	All regions, rice, wheat, soybean etc.
2.	CIAE soybean reaper	2000	0.40	All regions
3.	Self-propelled VCR	1000	0.25	Paddy, wheat, soybean etc.
4.	Combine harvester	4200	1.5-2.0	Paddy, wheat, gram, soybean
6. Threshing				
1.	Power operated thresher	5-10 h.p.	200-500 kg/h	For single crop, paddy, wheat
2.	CIAE multi-crop thresher	5-10 h.p.	200-500 kg/h	Wheat, paddy, soybean, maize, gram, pigeon pea etc.
3.	High capacity multi-crop thresher	15-20 h.p.	1500-2000 kg/h	Wheat, paddy, soybean, maize, gram, pigeon pea etc.
4.	Tractor PTO operated multi-crops thresher	25-35 h.p.	1500-2500, kg/ha	Wheat, soybean, maize, gram
5.	Paddy thresher	5-15 h.p.	200-1000 kg/h	Paddy
6.	Maize thresher	5-15 h.p.	500-2000 kg/h	Maize
7.	Groundnut thresher	10-15 h.p.	500-1000 kg/h	Groundnut

The top 15-20 cm layers can be disrupted with normal soil cultivation operations but deeper pans must be broken by using special type of equipment such as subsoiler. The conventional curved or straight leg subsoilers (also called chiseler) available in Indian market are not efficient when working at greater depths and require very high specific draft. Beyond a particular depth known as 'critical depth' the subsoiler's leg works like a mole plough forming only a mole channel (slot). The level of the field is also disturbed after operation (Fig. 1a). An improved subsoiler named as 'Pant winged subsoiler with shallow leading tines' has been developed which disturbs over three times more area with only about 20% increase in draft. It also leaves the surface of the field almost level (Fig.1b). The subsoiling is repeated after every 3-4 years to restore the soil structure. The fields with natural compaction, repeated traffic in the fields, tillage operations at the same depth year after year, fields with iron pan and headlands of a field require chiseling and subsoiling at regular intervals. The effect of subsoiling on response of wheat, maize, soybean and sugarcane under the field trials conducted during 2000-2004 at Pantnagar university farm has shown the yield increase of 16-25% depending upon soil types and crops.

a. Conventional

Fig. 1 Subsoilers used for modification of rootbed soil structure

b. Plant winged type

Fig. 1 Subsoilers used for modification of rootbed soil structure

ii. Clod disintegration

This operation is required to produce a granular soil structure in the final seedbed after soil loosening operation. Different types of clod crushers, planker, harrows and tined implements are used for clod disintegration. Clod crushers are effective tools for breaking surface clods. Under high moisture conditions, disintegration is possible through soil cutting or chopping with disc harrow and rotary powered tools such as rotavator.

Plain rollers and land packers are used for breaking of clods and packing of soil for moisture conservation but a spiked clod crusher can give more effective breaking of clods (Fig. 2). Disintegration can also be achieved by throwing the clods against a resistant barrier to break them by impact as in case of a rotavator. A rotavator can prepare the seedbed in one or two passes depending upon soil types and conditions but requires high energy. A 'combined active-passive tillage machine' equipped with a P.T.O. driven rotavator and winged passive tines mounted in the front (Fig.2) prepares the seedbed and rootbed in a single pass, neutralizes the negative draft of active elements and stabilizes the machine during operation. The combined tillage machine prepares a very fine tilth upto 8 cm as seedbed and a coarser tilth with a number of cracks and fissures upto a depth of 20-25 cm for rootbed. The concept of seedbed and rootbed preparation with commonly used equipment such as ploughing (20-25 cm depth) and disc harrowing at 10-15 cm first pass, 8-10 cm second pass and 5-8 cm third pass has saved 40-50% diesel in land preparation in *Tarai region* of Uttarakhand.

a. Pant spiked clod crusher

b. Pant combined active-passive tillage machines

Fig. 2 Equipment for clod disintegration

iii. Soil compaction

Proper compaction of seedbed is needed before sowing operation for conservation of moisture. Wide backward inclined implements such as planker, plain roller, tines and press wheels for compaction at seed depth and others are used for compaction and consolidation of seedbed.

iv. Clod sorting

The last tillage operation performed for seedbed preparation is rearranging and sorting of the smaller aggregates and soil particles, and packing them into large pore spaces for obtaining a well consolidated seedbed. Vertical tines/ pegs such as spike tooth harrow working at correct forward speed performs sorting operation effectively. Vertical tines oscillating perpendicular to the direction of travel are found very effective for sorting of clods.

v. Land smoothing

This operation is required for obtaining a well levelled field for proper operation of sowing machines, efficient utilization of irrigation water, proper drainage and for enhanced nutrients utilization efficiency. For large movement of soil in short hauls different types of levellers and scrapers are used but final smoothing operation is performed with a land plane which is based on the principle of a carpenter plane. Laser land leveller is the recent introduction in India for land smoothing operation (Fig. 3). The first laser leveller was imported to Pantnagar University by Rice-Wheat Consortium from Lahore, Pakistan in the year 2001. There has been an exponential increase in their number from just 4 units (2002) to over 372 units (2007). At present, there are over 925 units operating in the fields and have covered over 0.2 Mha mainly in the Indo-Gangetic Plains (RWC, 2009). This technology is spreading fast with custom hiring provisions in various states. The results of several hundred participatory trials conducted during 2001-06 under collaborative projects among RWC, CIMMYT, IRRI, ICAR and SAUs namely G.B.P.U.A. & T., Pantnagar, PAU, CCSHAU and SUBPUA& T have brought out many benefits

Fig. 3.

such as saving of irrigation cost by 20-30% depending upon crops, increase in nutrient use efficiency (10-15%), enhanced operational efficiency of farm machinery (8-10%), increase in cultivable area (3-6%) by removal of bunds and ridges, yield advantage of 5-25% with higher values in vegetable, and enhanced overall profitability by Rs. 8000 to Rs. 10,000/ha/annum.

2. Equipment for fertilizer application

Most of the seed drills and planters have arrangement for placement of fertilizer in soil at desired depth at the time of sowing/planting. However, for additional dosage of fertilizer at different stages of a crop, fertilizer is required to be applied separately. Depending on the needs of the plants for particular nutrients, the fertilizer is broadcast over the field, placed in the soil at a depth of seed or deeper, or distributed at one or both sides of the rows of crop as close to the plants as may be required for nutrition. Machines for placement of fertilizers deeply into the root zone (20-40 cm depth) while breaking the hard soil pans are the recent development and it is the sole contribution of Pantnagar University in India. The following three new machines have been developed and are being patented under the ICAR National Professor Scheme of the author.

i. Subsoiling and deep placement of fertilizers

The soils which produce low yield due to infertility, hard pan and fixation problems can be improved by subsoiling, and applying plant nutrients around the root zone of crops. A new machine named as *'Pant-ICAR Subsoiler-cum-Differential Rate Fertilizer Applicator'* has been developed (Fig. 4). The machine is used for breaking of hard subsoil pan to a depth of 45-50 cm and simultaneously for depositing varying amounts of fertilizers at two different depths. The shallow leading winged tines of the machine place 80% of fertilizers at 15 to 25 cm depths whereas the central subsoiling tine places the remaining 20% fertilizers at 35 to 45 cm depth. It can also place equal amount of fertilizers at two different depths. This machine disturbs the subsoil within the tractor rear wheels, therefore, it is suitable for controlled field traffic farming. The machine would prove very effective in sodic lands which frequently form hard pan and in rainfed agriculture for moisture conservation to a greater depth of soil profile. It can be operated with a 50 to 75 h.p. tractor depending upon the depths of operation.

The machine has been evaluated extensively by a dozen groups of farmers of U.P. and Uttarakhand for growing of sugarcane since the year 2006. A saving of about 20% fertilizers for the same yield of sugarcane as that of farmer's practice has been obtained. Also, the pooled mean of multi-location trials has revealed an yield increase of over 25% for sugarcane compared to conventional practice. The use of developed machine for subsoiling and deep placement of fertilizer in direct seeded rice has shown less incidence of weeds, stem and shoot borer, vigorous root growth upto 50 cm and yield increase by about 17% over conventional practice of shallow tilling and dry seeding of rice (DSR). The machine would prove to be beneficial for direct seeding of rice in rainfed area as well as drought prone regions for *in-situ* moisture conservation.

a. The developed machine

b. Soil disturbance

Fig. 4 Pant-ICAR Subsoiler-cum-Differential Rate Fertilizer Applicator

ii. Deep soil volume loosener-cum-fertilizer applicator

Levelling of fields with recently introduced laser land leveller is increasing exponentially especially in IGPs. There is a need to develop a new range of soil cultivation machineries which could manipulate soil without inversion and redeposit the disturbed soil almost at the same place so that the level of fields is changed to a minimum. Also, there are row crops such as sugarcane ratoon which requires off-barring (cutting of old roots of plant crop) and placement of fertilizers at both sides of rows for enhanced fertilizer use efficiency. One such machine named as *'Pant-ICAR Deep Soil Volume Loosener-cum-Fertilizer Applicator'* has been recently developed at Pantnagar. This unique machine disturbs soil upto a depth of 30 cm, covers a width of about 120 cm in two rows and deposits fertilizers at 15-25 cm depths with inverted-T openers mounted on two rigid tines spaced at 35-40 cm and cuts old roots of sugarcane ratoon when used as 'sugarcane ratoon manager'. Specially designed paired main tines are positioned immediately behind the tractor rear tyres, avoiding thereby the recompaction of tilled soil. Field trials of the machine in sugarcane ratoon management during 2007-08 have revealed substantial increase in yield. The machine is being tested extensively for inter-row tilling of sugarcane ratoons (75 cm spacing) and also inter-row planting of wheat, moong beans and cowpea in ratoon crop of 90 cm paired rows (30 cm) spacing with very promising results. This technology will be useful for inter-row cultivation and fertilizer application in cotton, castor and others row crops.

iii. Deep soil incorporation of organic fertilizers

The loss of soil organic matter (SOM) leads to deterioration in physical and biological properties of soil. There is great depletion of SOM which stands at 0.2 to 0.4% in different ecosystems, thus there is an urgent need to encourage use of organic nutrients. If SOM is enhanced, the soil becomes a living system and the soil biota helps in enhancing production. Morever, the responses to external inputs would increase, thereby reducing the need for external inputs. Since India is a tropical to sub-tropical country, 2 to 5 t/ha of annual application of organics would be effective.

The production of organic manures and their application to crops are labour intensive processes. Tractor operated manure spreader (TNAU design) is available in the country for uniform surface distribution of FYM in the fields and mixing in the top soil by secondary tillage implements. However, it would be borne in mind that such practice enhances the SOM only at the top soil surface which is already being enriched by several other biomass such as crop stubbles, leaves, roots, weeds and other biota. Also, the top soil is generally dry for most part of the cropping period, thereby making the nutrients unavailable to plants. In addition, the average soil loss of about 16 t/ha/year has been reported under Indian conditions indicating thereby loss of most of the organic fertilizers alongwith fertile soil during rains. There is a need to enrich the subsoil with organic fertilizer which will enhance the microbial activities as the total organics would remain in moist zone of the field without loss due to erosion. A new machine named as *'Pant-ICAR Subsoiler-cum-Vermicompost Applicator'* has been developed at Pantnagar. This tractor operated machine applies 100% organic fertilizers at different depths varying from 20 to 40 cm while performing the subsoiling operation. The field evaluation of this machine in mustard has shown very promising results. The machine is also being evaluated for application of press mud, gypsum, FYM, lime and other soil amendments.

3. Drilling and planting machinery

The objectives of sowing operation is to place seeds and fertilizer at proper depths in rows, maintain plant to plant spacing, cover the seeds with fine soil and provide compaction around the seeds for proper germination and emergence. These objectives differ with type of crops and agro-climatic conditions. A variety of drills and planters (Table 1) equipped with fluted roll, fixed hole, cup feed, inclined plates etc. type metering devices are available. Pneumatic drills for very small seeds (vegetables) to large seeds (beans) as well as transplanter for rice seedlings are being introduced and accepted on a limited scale. Transplanting of seedlings/saplings is also being practiced in fruits, vegetables and agro-forestry where mechanization is needed with efforts to mechanize pit digging by post hole diggers.

Minimum soil disturbance, crop rotation, cover crops, recycling and maintenance of residues on the surface, weed control and use of appropriate tools and implements are the key components of conservation agriculture. Zero-till technology for direct drilling of wheat, pea, lentil, maize in anchored rice stubbles is well accepted in IGPs. For loose straw spread over the surface after combine harvesting of rice, drills with four types of openers i.e. double disc coulter, star wheel punch planter, rotary powered disc and happy seeder are being introduced in IGPs (Fig.5). Happy seeder which works satisfactorily under residue loads of 6-8 t/ha by tilling a narrow strip of 7.5 cm for drilling by openers is gaining popularity amongst farmers.

Furrow irrigated raised bed (FIRB) planting system has opened up new opportunities for replacement of rice with extra short duration pigeon pea and QPM, and for mechanized intercropping of sugarcane with wheat and chickpea, winter maize with potato, beans and vegetable crops, and winter wheat with mint. FIRB technology is gaining popularity amongst farmers as it saves seed (20%) and water (30%), reduces crop lodging, improves grain filling and its quality, and avoids temporary water logging problems.

a. Star wheel b. Dulble disc

c. Rotary powered disc d. Happy seeder

Fig. 5 Openers for planting under loose straw conditions and a happy seeder

4. Equipment for interculture and weed control

The main objectives of interculture and weed control are to improve the soil conditions by reducing evaporation from soil surface by creation of soil mulch, infiltration of rain or surface water, reduce run-off, aeration of root zone, earthing and reduce competition of weeds for nutrients and water. Traditional interculture tools are very slow and unaffordable for large acreage. As an improvement, wheel hoes have been developed with several attachments to perform various interculture jobs (Fig. 6). Tractor mounted cultivators with sweeps are used in row crops. For wider row crops like maize, pigeon pea, groundnut, cotton etc. power weeders have been developed and commercialized (Table 1). The selection of implements depends upon operational and economic factors with an aim to reduce drudgery and achieve timelines of operation.

Wheal Hoe

Power weeding

Fig. 6 Equipment for interculture

5. Plant protection equipment

Sprayers and dusters are used for application of weedicides, insecticides, fungicides as well as growth regulators and micro-nutrients such as zinc, iron, boron, manganese etc. All kinds of equipment i.e. manual, knapsack auxiliary engine operated, power tiller operated, tractor mounted and stationary engine operated are commercially available in the country (Table 1). Aeroblast sprayers have been introduced from Brazil and have found acceptance for cotton and tall trees. Air assisted sprayers and dusters improve the coverage in case of contact type pesticides. Selection of right kind of nozzles in vital in achieving efficacy.

6. Harvesting and threshing equipment

Timely harvesting, threshing and cleaning are basic unit operations required to save the crop from losses. Delayed harvesting leads to over maturity and loss of quality, shattering in the fields, damage due to weather, fire, animals, theft etc. Harvesting operation has been mechanized with the introduction of walking, riding and tractor mounted vertical conveyor reapers (VCRs). Traditional threshing of several crops has been largely replaced with power threshers of high capacity (Table 1). However, combine harvesters are found to be the cheapest method of harvesting and threshing rice, wheat, soybean etc. (Fig. 7). As a result combine harvester has become popular and more than 30 manufacturers are manufacturing all kind of combines i.e. tractor mounted, tractor driven, self-propelled (both wheel type and track type) with 2.5 to 4.5 m cutter bar. For management of straw left behind combines, conventional type of field balers and straw combines (wheat) are being introduced. For harvesting of commercial crops i.e. cotton and sugarcane, demand for cotton pickers and sugarcane harvesters is picking up.

The use of improved farm equipment enhances productivity and profitability of land, labour and inputs such as seeds, fertilizers water and chemicals, and reduces the drudgery of farm workers. The benefits accrued from some of the specialized farm equipment have been mentioned in Table 2.

a. Tractor mounted grain combine

b. Straw (busha) combine

Fig. 7 Combines for harvesting of grain and straw

Table 2 Benefits accrued by using some specialized farm equipment

S. No.	Equipment	Benefits
1.	Laser land leveller	Water saving (15-30%), increase in fertilizer use efficiency (10-15%), increase in cultivator area (3-6%), yield advantage (5-25%)
2.	Rotavator	Fuel saving (50%), saving in time and cost of tilling (65%), better quality seedbed in 1-2 passes
3.	Zero-till drill/minimum till drill/multipurpose tool bar/raised bed planter	Water saving (15% ZT-drill and 30% FIRB), seed and fertilizer saving (25% FIRB), saving in cost of seedbed preparation by Rs.3000/ha (ZT-drill)
4.	Pressurized irrigation	20-30% saving in water
5.	Rotary power weeder	20-30% saving in time and labour
6.	Vertical conveyor reaper (VCR)/grain combine harvester	Timely harvesting and minimization of field losses
7.	Multi-crop thresher	50% saving in labour and time, and 54% saving in cost of threshing
8.	Straw (busha) combine	Recovers 50% straw and also. 70-100 kg grain/ha resulting in average saving of Rs. 1250/ha
9.	Field baler	Recovery of loose straw after combining and sugarcane leaves and making bales (130-150 kg/m^3 density), ease in transportation and low storage space
10.	Straw cutter-cum-spreader (straw manager)	Cuts and spreads the combine harvested straw evenly as surface mulch and helps in sowing by zero-till drill
11.	Improved manual harvester for mango and kinnow	No damage to fruits and higher harvesting capacity

7. Conclusion

Availability of appropriate farm power sources is of paramount importance for mechanization of various farm operations. There is direct correspondence between farm power availability and agricultural productivity. At the present level of mechanization in agriculture and for 200% cropping intensity, an average of 2 kW/ha is considered essential as against current availability of 1.5 kW/ha. Agriculturally advanced states (Punjab and Haryana) have farm power availability of 4 kW/ha. All kinds of equipment are being manufactured in the country for different power sources i.e. human, animal, power tiller and upto 75-80 h.p. tractors. Commercialization of agriculture would need specialized and precision machinery and gadgets. Efforts have been started to adopt appropriate machinery from developed world, However, most of these require high capital cost but can be used on custom-hire-basis or as a part of contract farming by industries.

References

Alam, A. 2006. Future requirements of agricultural machines for mechanizing agriculture. In : *Status of Farm Mechanization in India*. Indian Agricultural Statistics Research Institute (ICAR), New Delhi. Pp 175-195.

Ali, Nabab. 2009. Tools and machines for conservation agriculture to enhance productivity and profitability. *Indian Farming,* 58 (11):38-43.

Arun Kumar and Thakur, T.C. 1999. Effect of subsoiling and tillage methods on wheat crop. *J. Agric. Engg,* 36 (3): 51-54.

Arun Kumar and Thakur, T.C. 2007. An experimental investigation into the performance of conventional and winged type subsoilers. *J. Agric. Engg;* 44 (2) : 8-13.

Rai, Mangala. 2009. Resource Conservation in Indian Agriculture. *Souvenir, 4th World Congress on 'Conservation Agriculture-Innovations for Improving Efficiency, Equity and Environment'* held at New Delhi (India), Feb. 4-7, Pp 1-6.

Thakur, T. C. 2004. Final report of NATP on "Accelerating the Adoption of Resource Conservation Technologies (RCTs) for Farm Level Impact on Sustainability of Rice-Wheat System of the Indo-*Gangetic Plains* (PSR-65)". Department of Farm Machinery and Power Engineering, College of Technology, Pantnagar.

Thakur, T. C. and Arun Kumar 2005. Machinery for zero-till surface managed crop residue systems - progress and prospectus. *In : Abrol, I.P., R.K. Gupta and R.K. Malik (Ed.). Conservation Agriculture-Status and Prospects.* Centre for Advancement of Sustainable Agriculture, New Delhi. Pp. 33-42.

Thakur, T.C. 2008. Soil compaction and response of sugarcane to subsoiling. In : *Proceedings of National Seminar on 'Varietal Planning for Improving Productivity and Sugar Recovery in Sugarcane* held at G.B.P.U.A. & T., Pantnagar 14-15, Pp 88-97.

Thakur, T.C. and Varshaney, B.P. 1993. Comparative performance of spiked and angle iron bar type clod crushers. *J. Agric. Engg.,* 3 (3): 1-7.

□□□

Production Management Agro-Techniques

5

Tools for Identification of Efficient Cropping Systems for Commercialization

B. GANGWAR

The cropping system research is being perused at national level with major focus for identification of location specific efficient systems since 1990. However, on perusal of research publications we find that the data of cropping systems are interpreted in different ways with major focus only on production and net return on hectare basis. In fact, these 2 aspects are not the only indicator of the efficiency of a particular system. Many advances have been made in recent years to identify the efficient system for a particular location. Some of the methodologies, which are of major concern to crop scientists in general and agronomists in particular, are highlighted here.

Indicators of system efficiency

- Productivity
- Production efficiency
- Land use efficiency
- Energy use and its production
- Water use productivity
- Nutrient use productivity

- Profitability
- Economic efficiency
- B: C ratio
- Stability/sustainability
- Employment generation
- Soil fertility trend
- Natural weed and pest management

The refined methodologies

The refined methodologies for assessment of above aspects of cropping systems are as follows :

System productivity

The crop yields-The crop productivity- The system productivity.

- *Crop yield*: Crop yield/unit area generally reported as q/ha or t/ha
- *Crop productivity*: The crop yield /unit area /unit time generally reported as kg/ha/day. It may be calculated on dividing the yield by duration of crop. A crop showing high productivity needs to be preferred over the low productivity for inclusion in a new system.
- *System productivity*: Refers to yield/unit area/ unit time. But, the yield here is the equivalent yield in a single commodity term. The time here is agricultural year (365 days). Therefore, the equivalent yield is calculated than divided by 365 and expressed as kg/ha/day. The equivalent yield should preferably be calculated in terms of *kharif* crop (say paddy in rice based cropping system) using the following formula.

$$PEY = \frac{\Sigma\, Yi \times Pi}{P\,(p)}$$

Where, PEY denotes paddy equivalent yield, Yi= yield of different crops Pi = Price of respective crops, P (p) is the price of paddy.

Inference: Higher the productivity better is the system

Relative production efficiency (RPE)

Generally refers to the capacity of the system for production in relation to existing system and expressed in percentage. The following method is proposed for calculating the RPE.

$$RPE = \frac{(EYD-EYE) \times 100}{EYE}$$

Where EYD denotes the equivalent yield under improved/diversified system while EYE denotes the existing system yield.

Inference

(+) tive figures shows the superiority of the new system over the existing in percentage and considered desirable.

(-) tive figures shows inferiority than the existing system and not desirable for any change.

Any positive figures of more than 20% are considered worth recommending for extension use.

Land use efficiency (LUE)

Refers to the land use during the agricultural year. It can be calculated by using the following formula.

$$LUE\ (\%) = \frac{TND\ (i) \times 100}{365}$$

Where TND (i) denotes the total number of days during which field remain occupied under different crops (i=1....n)

Inference : Higher the land use efficiency denotes the more use of the land in a year. With increase in intensity in cropping system, the land use efficiency generally improves. However, LUE is not a good indicator of system efficiency as higher the land use may not necessarily result in higher output from the system. It depends more on the contribution of individual crop. Therefore, have a limitation.

Energy use and energy production

These are the new considerations for judging the system efficiency. Every input used and outputs obtained are converted into the form of energy. A system more efficient in producing higher output energy and requires less input energy is considered desirable. The standard figures for different inputs as per Binning *et al.* (1983) are available for energy analysis. Similarly, the standard energy conversion figures are available with ICMR on nutritive value of different crops (Gopalan et al, 1978). These figures can be utilized for simple conversion of crop produce into the energy terms (Table-1). The energy of the cropping system is generally expressed in MJ/ha.

Apparent water use productivity (AWUP)

The water is becoming a major concern all over the world. The production even per drop of water is the recent concern keeping in view the future needs and availability. Even decline in water table in certain parts and limited availability of irrigation water calls for more attention on water use productivity. This mainly refers the production or yield / ha cm of water.

The apparent water use productivity (AWUP) may grossly be calculated using the following formula as used by Gangwar et al (2005.)

$$\text{AWUP} = \frac{\text{Equivalent yield of a system / ha}}{\text{Total quantity of water used in ha of land in cm (ha-cm)}}$$

Inference : Higher the AWUP more efficient is the system. This approach is more useful under declining irrigation water conditions.

Apparent nutrient use productivity

The system, which utilizes the nutrient used in different crops more efficiently per unit of nutrient, is considered to be more efficient system. This measure reflects the efficiency of the nutrient use. Apparently it may be calculated on dividing the equivalent yield of the system by the total quantity of nutrients used. However, this measure is an indicator of the trend but not the true measure.

Profitability

Cost of cultivation- Gross return –Net return- Profitability

Pre-requisites

- Crop wise cost of cultivation
- Sale price of produce (S)
- Gross return (S)
- Duration of crop (S)
- Crop profitability (Rs/ha/day) $= \frac{\text{Net return / ha.}}{\text{No. of days field occupied}}$
- System profitability (Rs /ha/day) $= \frac{\text{Net return/ha/year}}{365}$

Inference: Crops and systems, which give higher profitability, are economically beneficial and advisable.

B.C: ratio

This measure shows the return/ benefit on per rupees investment

$$\text{B.C. ratio} = \frac{\text{Gross return}}{\text{Cost of cultivation}}$$ .... Economist point of view

$$\text{B.C. ratio} = \frac{\text{Net return}}{\text{Cost of cultivation}}$$ .... Agronomist point of view

Inference : Wider the ratio more the benefit.

Relative economic efficiency (REE)

This is a comparative measure of economic gains over the existing system. The following method is proposed for calculating the REE.

$$\text{REE (\%)} = \frac{\text{DNR - ENR}}{\text{ENR}}$$

Where, DNR denotes net return obtained under improved/diversified system, while ENR refers to net return in the existing system.

Inference: Higher the economic efficiency -better the system for large scale recommendation.

Stability/Sustainability index (SI)

The trend of yield over the year in systems reflects the stability/ sustainability of a crop/ cropping system. Less fluctuation in yield over the years reflects positive side of stability. The stability index ranges from 0 to 1. The stability index may be calculated using the following formula as adopted by Gangwar *et al.* (2004).

$$SI = \frac{\overline{Y} - sd}{Ymax}$$

Where, SI= stability index, is the average yield over years n, sd is the standard deviation and Y max is the maximum yield obtained in any of the year.

Inference : The value nearing unity shows higher stability reflecting that the system is higher sustainable.

Pre-requisite: Minimum four-year yield data

Employment generation

Refers the additional man days required for the new system. Certain crops specially vegetables requires more man days and considered helpful in employment generation

Intensification of the system adds to the employment generation. But, it adds to the cost. Therefore employment generation needs to be considered in relation to economic benefit and availability of required labour and machinery.

Soil fertility trend

Soil fertility trend over the years is considered to be a good indicator of system efficiency. A system, which maintains or helps in build up of soil fertility with use of recommended doses of nutrients is considered to be efficient system compared to a system which causes imbalances and affects the soil adversely. The initial and final soil fertility status over a minimum period of four years can indicate the system efficiency in respect of soil health.

Table 1. Energy conversion factors for inputs

Particulars	Unit	Energy equivalent(MJ)
Human labour (Adult)	Man-hour	1.96
Diesel	Litre	56.31
Chemical fertilizers		
Nitrogen (N)	kg	60.60
Phosphorus (P)	kg	11.10
Potash (K)	kg	6.70
Plant protection (superior)		
Granular chemical	kg	120
Liquid chemical	ml	0.102
Crop produce (grain)		
Rice	kg	14.70
Wheat	kg	15.70
Vegetable Pea	kg	3.91
Greengram	kg	14.03
Maize	kg	15.10
Mustard	kg	22.72
Pigeonpea	kg	14.07
Soybean	kg	18.14

Source: Gopalan *et al.* (1978) and Binning *et al.* (1983)

Natural weed and pest management

Some systems have suppressing effects on natural weed problems and minimizing pests. A system, which helps in minimizing the major problem, is naturally a better one and considered efficient system over the system which promotes the existing problems. For example, introduction of berseem crop once in 3 years during *rabi* minimizes the risk of natural build up of *Phalaris minor*. Similar are the reports on some of the diseases. Therefore, these aspects have the significance of due consideration for deciding or recommending cropping system for a particular situation.

With due consideration on above aspects one can identify a efficient crop and copping system keeping in view the available resources and emerging needs.

References

Binning, A.S., Pathak, B.S. and Panesar, 1983 .The energy audit of crop production system research report, School of energy studies for agriculture, Panjab Agricultural University, Ludhiana, Punjab (India).

Devasenapathy, P; Ramesh, T. and Gangwar B., 2008. Efficiency Indices for agriculture management research. New India Publishing Agency, Pitam Pura, New Delhi. pp 1-146.

Gangwar, B, Katyal, V. and Anand K.V., 2003. Productivity, stability and efficiency of different cropping systems in Maharashtra . *Indian Journal of Agricultural Sciences* 73 : (9): 471-477

Gangwar, B., Katyal, and Anand, K.V., 2004. Stability and efficiency of cropping systems in Chhatisgarh and Madhya Pradesh. *Indian Journal of Agricultural Sciences*. 74 (10): 521-8.

Gopalan C. Sastri BVR and Balasusbramanium S.C., 1978. Nutritive value of Indian Food pp 66-144. National Institute of Nutrition Indian Council of Medical Research Hyderabad, Andhra Pradesh.

□□□

6

Efficient Cropping Systems for Commercialization in Selected Zones of India

B. GANGWAR

As per an estimate, India would require about 280 million tones of food grains for feeding about 1.3 billion populations by the year 2020. The per capita availability of land which was 0.15 ha in 1999-2000, may reduce to 0.11 ha by 2020. Approximately 90% of total farm holdings are categorized as small and marginal. There is no other option except to increase cropping intensity and enhance productivity. In fact, increase in cropping index will be a key strategy for future gains in crop production. In areas, endowed with adequate irrigation resources, short duration pulses, oilseeds and other high value crops will find their definite niche as sequential or intercrops, rather than replacing the major cereal crops having higher yield potential combined with high stability. Hence, an increased cropping intensity will contribute substantially to additional demands of food and cash crops. However, possibilities have to be explored for their inclusion in the cropping systems in such a manner that they are adjusted within the existing systems without any adverse impact on common food grain crops.

The country is endowed with diverse climate and soil types to support diversified agriculture, which is the major strength to meet food, fodder, fibre and finance requirements of ever increasing population. With the advent of modern agricultural technologies, there is a continuous surge for diversified cropping, primarily on economic considerations. The prevalent cropping systems are the outcome of the technological innovations, household needs, reflection of government policies, and availability of production inputs, market forces and socio-economic compulsion. Yadav and Rao (2001) enlisted 250 double cropping systems of primary and, secondary importance in term of their spread in the country out of which 30 are of primary importance. These are; rice-wheat, rice-rice, rice-gram, rice-mustard, rice-groundnut, rice-sorghum, pearl millet-wheat, pearl millet-gram, pearl millet-mustard, pearl millet-sorghum, cotton-wheat, cotton-gram, cotton-sorghum, cotton-safflower, cotton-groundnut, maize-wheat, maize-gram, sugarcane-wheat, soybean-wheat, soybean-gram, sorghum-sorghum, groundnut-wheat, sorghum-groundnut, groundnut-rice, sorghum-wheat, sorghum-gram, pigeon pea-sorghum, groundnut-groundnut, sorghum-rice and groundnut-sorghum However, among the top ten popular cropping systems in the country (Table 1), cereal-cereal with major share of rice-wheat (9.85 m ha) and rice-rice (5.89 m ha) are the most dominant systems contributing about 52.3 % to the national food basket.

A changed scenario with respect to determinants of food requirement and natural resource use is foreseeable in next two decades. It is imperative that future agricultural systems would be guided not only by the compulsion of improving food security, but also by the concerns for quality of food, environmental protection, resource sustainability and system profitability. Therefore, we may expect a major shift in cereal based agricultural production systems towards systems involving selected pulses, oilseeds, vegetables, fruits and flowers in view of emerging needs and global trade opportunities arising from open economy. Therefore, new potential crops with higher and stable yields and/or profits will continue to find place in to the existing cropping systems of different agro-ecological regions. Further, because of integration of world markets, urbanization and rising personal incomes significant shifts in cropping systems and other farm enterprises may also take place. In fact, there would be more growth in the demand for pulses and milk, relatively slow growth for cereals and decline in per capita demand for root and tuber crops. Therefore, intensification would be essential to cater the emerging need. Majority of predominant systems are cereal based double cropping systems and there is considerable scope to intensify the existing systems using high value short duration crops. Legumes are excellent crops for natural resource management, environmental security, crop diversification and consequently for viable commercial agriculture. Likewise, the importance of oilseeds in crop rotation cannot be ignored. Soybean and groundnut are two important leguminous oilseed crops suited for intensification. Availability of short duration varieties coupled with matching agro technologies has resulted in identification of several remunerative, and more productive cropping systems which have either already shown their promise or have tremendous potential for expansion in new niches and diversification in the existing cropping systems. Series of experiments are being conducted under AICRP on Cropping Systems and elsewhere so to identify need based cropping systems for different agro-climatice zones of India. The efficient alternatives to dominant cropping systems viz rice-wheat, pearl millet and maize-wheat having considerable potential for commercialization in selected zones of India are discussed here for ready reference of researchers, planners and extension workers.

Alternatives to rice-wheat cropping system

The rice-wheat is one of the most important cropping systems in India covering about 10.5 million-hectare area and contributing about 38% food grains to the national food basket. This system is considered as backbone of food grain security and predominates mainly in Punjab, Haryana, U.P., Bihar, West Bengal and parts of M.P., Rajasthan, Gujrat and Orissa states. This system is likely to continue its important role in sustaining self-sufficiency of food grains in future too. The average rice equivalent yield of rice - wheat system was lowest in parts of Madhya Pradesh and Chatishgarh (2.15 t/ha/year) while it was highest in parts of Punjab and Haryana (7.77 t/ha/year). Similar trend was in average system productivity with respective values of 5.9 and 21.3 kg grain/day/ha (Gangwar *et al*). The potential rice equivalent yield (11.8 t/ha/year)of the system could be achievable in Punjab and Haryana. It is closely followed by Jammu (10 t/ha/year), Bihar (10.66/t/ha/year) and Uttar Pradesh including parts of Uttaranchal (10.4 t/ha/years) and it is lowest (7.67 t/ha/year) in state of Himachal Pradesh. The yield gap between actual present yield and potential yield is ranging from 51 to 290% in different states. It is highest (three times) in Madhya Pradesh and parts of Chattisgarh, while lowest in Punjab and Haryana. It clearly shows that management needs to be improved in states where wide gap exists for achieving higher production and productivity. The system production can be increased up to the level ranging from 9.1 to 21.5 t/ha/year depending upon the choice of high value crops for diversification of existing cropping system. Continuous cropping of rice and wheat in sequence during last 3 decades has been causing many second-generation problems like decline in water table, emergence of multi-nutrient deficiencies, formation of hard pan and build up of weeds like *Phalaris minor* in wheat. Moreover, the stagnation in system productivity and profitability is experienced in recent years. As a result, system sustainability is under threat. There is need to identify and consider partial substitution of rice-wheat system through efficient alternatives as diversification approach. Therefore, some of such suitable cropping systems as emerged profitable at various locations (Table 1) under All India Coordinated Research Project on Cropping Systems are as follows.

- In mid high altitude, intermediate zone of J&K, rice-potato-wheat cropping system was found more efficient in terms of rice equivalent yield (13.9 t/ha/year) that was 51% more than the existing of rice-wheat system. Similarly, productivity was also highest (38.2 Kg/ha/day). But, rice-wheat -sorghum + cowpea fodder was more profitable than other systems which gave Rs. 121/day/ha with relative gain of 38% more than the existing rice-wheat system.
- In central plain zone of Punjab, the system involving rice-potato-groundnut was identified to be more productive with highest production (19.9 t/ha/year), productivity for 54.5 kg grain/ha/day and profitability of Rs. 96/ha/day. This system was more productive with relative production efficiency of 82 % and relative economic efficiency of 191% in comparison to rice-wheat system. But, in terms of water use productivity, maize-potato-groundnut and soybean-wheat-pearlmillet were equally good which gave about 140 kg rice equivalent yield /ha cm of water.
- In Bhabar and Tarai zone of Uttaranchal, rice-vegetable pea-rice system was found more productive (14.0 t/ha/year) with highest productivity (38.4 kg rice equivalent yield /ha /day) and profitability of Rs.112 /day /ha.
- In central plain zone of U.P., maize-potato-wheat was more productive (15.8 t/ha/year) with highest productivity of 43.2 kg grain/day/day. But, maize-pea-sunflower was found more profitable which gave highest profitability of Rs. 78/ha/day that was relatively 114% higher than the existing rice-wheat cropping system.
- This system was also more efficient in terms of resource use productivity, which gave 115 kg grain/ha/cm of irrigation wáter and 29 kg grain/kg of NPK used.
- In Vindhayan plain zone of U.P., rice-potato-greengram system was more productive with highest rice equivalent yield (14.3 t/ha/year), productivity (39.2 kg grain /day/ha) and profitability (Rs. 58.2/ha/day). This system was relatively superior to the tune of 69% in terms of relative production efficiency and 80% in terms of relative economic efficiency when compared to the existing rice-wheat system.
- In south alluvial zone of Bihar, rice-potato-onion was found distinctly better with highest system yield of 18.1 t/ha/year in terms of rice equivalent yield, productivity of 49.7 kg/day/year and relative

production efficiency of 135% compared to existing rice-wheat system. But, rice-berseem - maize + cowpea fodder system was more remunerative which gave highest profitability of Rs. 125/ha/day and relative economic efficiency of Rs. 224% compared to existing rice-wheat system

Table 1. Efficient alternatives to rice - wheat cropping system

Cropping Systems	Average Yields (t/ha) Kharif	Rabi	Summer	Total REY (t/ha)	System productivity (Kg/day/ha)	Profitability (Rs/ha/day)	Resource use ability: Nutrient use productivity (kg grain/kg NPK use)	Irrigation water use productivity (kg grain/ha cm water)	Relative production efficiency (%)	Relative economic efficiency (%)
Mid high altitude Intermediate zone of J & K (R.S.Pura)										
Rice-Wheat-(E)	5.07	3.80	-	9.2	25.3	87.4	26.4	53.2	-	-
Rice-Potato-Wheat	4.89	20.83	2.81	13.9	38.2	99.6	25.3	78.2	51.0	14.0
Central Plain zone of Punjab (Ludhiana)										
Rice-wheat -(E)	5.54	4.95	-	11.0	30.0	33.0	28.1	53.4	-	-
Rice-Potato-G.Nut	6.08	21.38	2.61	19.9	54.5	96.1	40.2	78.0	81.7	191.4
Bhabar & Tarai zone of Uttarnchal(Pantnagar)										
Rice-Wheat- (E)	5.70	5.70	-	11.9	32.7	91.5	23.9	101.1	-	-
Rice-Vegpea-Rice	5.65	7.13	4.81	14.0	38.4	111.5	24.7	67.4	17.6	21.9
Central Plain zone of U.P (Kanpur)										
Rice-Wheat (E)	3.98	4.38	-	8.8	24.0	36.4	18.1	42.7	-	-
Maize-Potato-Wheat	3.12	25.84	3.31	15.8	43.2	51.0	18.0	93.5	79.9	39.8
Vindhyan Plain zone of U.P (Varanasi)										
Rice-Wheat- (E)	3.86	4.22	-	8.5	23.2	32.3	17.7	41.4	-	-
Rice-Potato-G.Gram	3.79	22.82	0.93	14.3	39.2	58.2	22.2	50.8	68.7	80.4
South Bihar Alluvial Plain zone of Bihar (Sabour)										
Rice-Wheat- (E)	4.13	3.27	-	7.7	21.1	38.6	21.4	116.7	-	-
Rice-Potato-Onion	4.44	20.36	12.29	18.1	49.7	83.2	25.2	146.2	135.4	115.8
West Central Table Land zone of Orissa (Chiplima)										
Rice-Wheat (E)	3.94	2.87	-	7.1	19.4	37.0	15.9	37.4	-	-
Rice-Tomato-Ladyfinger	4.20	17.82	5.35	17.1	46.8	82.2	15.8	46.8	168.6	211.5
Plain zone of Chattisgarh (Raipur)										
Rice-Gram (E)	5.04	1.77	-	8.6	23.7	90.2	32.0	146.3	-	-
Rice-Tomato	4.13	22.13	-	15.2	41.6	167.8	40.0	166.1	76.0	86.1
Key more Plateau and Satpura hill zone of M.P.(Rewa)										
Rice-Wheat (E)	1.69	4.60	-	6.7	18.4	53.7	16.7	44.3	-	-
BlackGram-Potato+Wheat	0.75	11.7+	-	11.8	32.3	100.8	22.8	169.6	75.6	87.9
Vindhyan Plateau Zone of M.P.(Jabalpur)										
Rice-Wheat (E)	1.67	3.36	-	5.3	14.6	33.4	14.1	35.3	-	-
Soybean-Wheat	1.39	4.51	-	7.1	19.6	62.3	23.8	94.0	33.7	86.5
South Gujarat Heavy Rainfall zone of Gujarat (Navsari)										
Rice-Wheat (E)	4.52	2.18	-	6.9	18.9	48.3	20.9	40.6	-	-
Rice-Sorghum (F)-G .Nut	4.76	19.27	1.21	9.1	24.9	63.3	32.4	51.0	31.7	31.0

- In western central tablel and zone of Orissa, rice-tomato-lady finger sequence was found better with highest rice equivalent yield (17.1t/ha/year), productivity (46.8 kg/grain/ha/day), resulting a gain of 169% over existing system. But, rice-maize (cobs)-lady finger system was more remunerative which gave highest profitability of Rs. 98/ha/day resulting 271% gain over existing rice-wheat system.
- In plain zone of Chhatisgarh, rice-tomato system was found to be better than existing rice-wheat system which gave highest production (15.2t/ha/year) with productivity of 41.6 kg/ha/day and profitability of Rs 168/ha/day. But, in terms of resource use efficiency, rice-berseem system was better which gave highest water use productivity (225 kg grain/ha cm water used) and nutrient use productivity (54 kg grain/Kg of NPK applied). Both the systems are suited for the area and advisable.
- In key more plateau and Satpura hill zone of M.P., the system involving black gram–potato+wheat was better in terms of production (11.8 t/ha/year), productivity (32.3 kg rice grain equivalent /day /ha) and water use productivity (170 kg grain/ha cm of water). But, rice-berseem fodder-berseem seed system was more remunerative giving highest profitability of Rs.103/ha/day and nutrient use productivity of 48 kg grain equivalent/kg of nutrient (NPK) used.
- In Vindhyan plateau zone of M.P., soybean-wheat system was found to be distinctly better in terms of highest rice equivalent yield (7.1 t/ha/year), productivity (19.6 kg grain/ha/day), profitability (Rs. 62/ha/day) and water use productivity (94 kg grain/ha cm of water)
- In south heavy rainfall zone of Gujarat, rice-sorghum fodder-groundnut system was more productive (9.1 t/ha/year REY) with highest productivity (25 grain/day/ha) and profitability (Rs. 63/ha/day) resulting average gain of 31% over existing rice-wheat system.

As per our estimates, the present contribution of rice-wheat system (52 million tons) may be doubled through adoption of improved management practices while on diversification of only 20% area, a equivalent production potential of 117 million tones is achievable with use of improved technologies.

Alternatives to pearlmillet based cropping systems

Pearlmillet (bajra) based cropping systems are predominant systems in arid and part of semi-arid ecosystem of India covering an estimated area of about 6.64 million ha. The major spread of this system is in the low rainfall areas mainly located in Agra region of Uttar Pradesh, parts of Maharashtra,

Rajasthan, Gujrat, Haryana and Madhya Pradesh. The major pearlmillet based systems practiced in our country are: pearlmillet-fallow, pearlmillet-wheat, pearlmillet-mustard and pearlmillet-gram. The pearlmillet-based systems are considered to be more efficient due to low water requirement and being practiced over years in these states. However, in recent years, pearlmiilet attracted very low prices in the market and Government very often procures it at minimum support price. Moreover, the produce of pearlmillet cannot be stored for long time due to fast deterioration in its quality. To over come this problem, the lands that are blocked under pearl millet cultivation can be used to produce some other crops, which may give more returns per unit area and time to the farmers. Moreover, in certain patches of these states where pearl millet is traditionally a dominant crop some irrigation facilities have also been created. As such, the question arises which are the most efficient alternate crops and cropping systems for these areas so as to diversify the existing pearl millet based cropping systems. The estimates for production potential and yield gaps clearly shows that the present production (3.65 t/ha/year) can be doubled (7.32 t/ha/year) using available technologies under the existing systems (Gangwar *et al*) However, through diversification and use of improved package of practices 4 to 5 times increase (19.10 t/ha/year) is possible under assured irrigated conditions. Among states, the maximum potential exists in Rajasthan (up to 28.72 t/ha/year) and in part of Uttar Pradesh (20.45 t/ha/year). Among the series of crops and cropping systems evaluated during a period of 3 to 10 years (1991-2001), the most efficient cropping systems as identified are given in table-2. Considering various aspects of comparative efficiency of different systems, the following conclusions are drawn.

- In North-west zone of Gujrat, pearlmillet-potato-groundnut system was found to be distinctly better with highest system yield of 17606 kg/ha / year in terms of pearlmillet equivalent yield, productivity of 48.2 kg/day/ha, net return of Rs.36046 /ha/year, economic efficiency of Rs.98.8 /ha/day and irrigation water use productivity of 146.7 kg grain/ha cm of water. The synthesized system like castor -groundnut may also be suited for commercialization.

- In western zone of Haryana, pearlmillet-potato-greengram system was better in terms of production and productivity when compared with other systems. But, considering profitability and nutrient use productivity, the pearlmillet-pea-maize fodder and cotton-wheat were equally good with respective to net return of Rs.24625 and Rs.24793/ha/year, economic efficiency of Rs.67.5 and 67.9/ day /ha, nutrient use productivity of 36.9 and 38. 4 kg grain/ kg nutrient and irrigation water use productivity of 165.4 and 148.0 kg grain/ ha cm of water used.

Table 2. Efficient alternatives to pearl millet – wheat systems

Cropping systems	Productivity					Profitability		Resource use ability	
	Average yield (kg/ha)			Total PEY (t/ha/ year)	System productivity(kg/day/ ha)	Net return (Rs/ha/ yr)	Economic efficiency (Rs/ha/day)	Nutrient use productivity (kg grain/kg of nutrient use)	Irrigation water use productivity (kg grain/ ha cm water)
K R S	*K*	R	S						
North west Gujarat (S. K. Nagar)									
P.Millet - Mustard - P.Millet	0.67	1.64	2.52	7.62	20.88	13116	35.93	21.2	84.7
P.Millet - Potato- G.Nut	0.71	20.76	2.76	17.61	48.24	36046	98.76	23.6	146.7
Castor- Pearlmillet	3.04	-	2.42	9.92	27.19	20188	55.31	40.5	141.8
Western Haryana (Hisar)									
P.Millet- Wheat	1.87	4.17	-	7.40	20.27	5505	15.08	18.6	119.3
Soyabean-Wheat - Cowpea(Fodder)	1.26	4.78	18.18	12.51	34.27	20532	56.25	31.1	136.0
P.Millet- Potato- G.Gram	2.37	20.78	7.56	13.75	37.67	15243	41.76	26.3	154.5
P.Millet Pea - Maize (Fodder)	2.27	1.98	21.28	11.91	32.63	24625	67.46	36.9	165.4
Cotton- Wheat	1.71	4.77	-	12.58	34.47	24793	67.93	38.4	148.0
Semi-arid eastern plain zone of Rajasthan (Durgapura)									
Pearlmillet- wheat	1.08	2.77	-	4.75	13.01	2148	5.89	19.8	94.9
Pearlmillet- Onion	1.26	22.01	-	14.62	40.05	35162	96.33	39.5	208.8
Clusterbean- Onion	0.55	27.39	-	18.10	49.58	50767	139.09	58.4	258.5
	1.26	15.70	-	10.80	29.54	23899	65.48	25.4	134.8
South western semi-arid zone of U.P (Bichpuri)									
Pearlmillet-Wheat	2.38	4.11	-	7.83	21.44	7405	20.29	20.6	111.8
Pearlmillet-Wheat- G.gram	2.45	4.06	6.22	9.51	26.06	10238	28.05	20.9	90.6
G.Manure- Potato- Sunflower	31.7	23.09	8.88	12.71	34.82	21114	57.85	28.9	105.8

- In semi-arid eastern plain zone of Rajasthan, clusterbean-onion system was identified to be distinctly better than other systems including existing pearlmillet-wheat system with highest production of 18098 kg/ha/year, productivity of 49.6 kg/ grain/ha/day, net return of Rs.50, 767/ha/ year, economic efficiency of Rs.139.1/ha/day, nutrient use productivity of 58.4 kg grain/kg of nutrient applied and water use productivity of 258.5 kg grain/ha cm of water used.

- In South-western semi-arid zone of U.P. mainly representing Agra region, the green manure-potato-sunflower system was found to be a most suitable and viable choice to replace existing pearlmillet -wheat system with highest production of 12711 kg grains /ha/year, productivity of 34.8 kg grain/ha/ day, net return of Rs.21114/ha/year, economic efficiency of Rs.57.9 /ha/day, nutrient use productivity of 28.9 kg grain/kg nutrient used

An estimated average yield of the existing pearlmillet based systems (3653 kg/ha/year) can be doubled through improved agronomic management. Moreover, a quantum jump to the tune of producing 19104 kg grain /ha/ year in terms of pearlmillet equivalent yield is possible through diversification of pearlmillet based existing systems clearly revealing a huge potential to reap up in future. Such systems are worth considering for commercialization.

Alternatives to maize based cropping systems

The maize based cropping systems are mainly practiced in the states of Himachal, Jammu and Kashmir, parts of Uttar Pradesh, Jharkhand, Punjab and Rajasthan covering an area about 2.73 million hectares. Among maize based systems, maize-wheat occupies the major area (1.86 million ha) mainly in northern India with average yield of 5.08 t/ha as maize equivalent yield of the system that is considerably low. Diversification is considered to be a good alternative to improve system yield with enhanced profitability. But, the question arises that how to diversify the system so as to improve the productivity and profitability on one hand and efficiently use the applied or available inputs/resources and also to obtain other advantages of diversification. Any system that is capable of producing more in terms of grain, money and per unit of resource use over time is considered to be efficient for the specified location. Based on these considerations, the efficient cropping systems as an alternative to the existing maize based systems under irrigated conditions of selected zones (Table 3). as identified are summarized below.

- In mid hills sub-humid zone of Himachal Pradesh, maize -pea-potato and maize-pea frenchbean system were equally good and distinctly better than existing maize-wheat system with maize equivalent yield ranging from 13.7 to 14.3 t/ha/ year, productivity of 37.4 to 39.2 kg/ha/day, net return of Rs. 14484 to 16998 /ha/year, economic efficiency of Rs. 39.7 to 46.6 /day/ha, nutrient use productivity of 29.8 to 34.2 kg grain/ kg of nutrient used and production of 118.8 to 147.5 kg grain/ha cm of water (Table 3).

- In Punjab, both rice-wheat and maize-wheat system are practiced. Comparatively rice-wheat is more productive but at the cost of higher resource use in terms of water giving only 64.9 kg of maize grain equivalent /ha cm of water. The system involving maize-potato-onion was identified to be most suited with highest maize equivalent yield of 31.0 t/ha/year, productivity of 85.0 kg/ha/day, net return of Rs. 73,248/ ha/year, economic efficiency of Rs 201.20/ha/year, nutrient use productivity of 43.7 kg grain/kg nutrient used and water use productivity of 248.1 kg grain /ha cm of water. The other sequences involving maize-potato-groundnut, maize–wheat-green gram were also comparatively better in the order of merit when compared with existing rice-wheat and maize-wheat systems.

Table 3. Efficient alternatives to maize based systems

Cropping Systems	**Productivity**					**Profitability**		**Resource use ability**	
	Average Yields (t/ha)			Total MEY (t/ha/ year)	System productivity (kg/ day/ha)	Net Return (Rs/ha/ year)	Economic efficiency (Rs/ha/ day)	Nutrient use productivity (kg grain/kg of nutrients)	Irrigation water use productivity (kg grain/ha cm water)
K R S	K	R	S						
Mid-hills Sub humid- (Palampur-HP)									
Maize-Wheat	3.61	3.26	-	7.93	21.73	13069	35.81	18.44	105.75
Maize-Pea-Potato	5.05	1.58	11.12	14.31	39.20	14484	39.68	29.81	147.51
Rice-Veg Pea-Potato	3.01	6.56	7.74	11.11	20.66	19073	52.25	25.84	75.59
Central Plain (Ludhiana - Punjab)									
Maize-Wheat	5.95	4.88	-	12.42	34.01	39497	108.21	29.56	151.40
Maize-Potato-onion	4.87	19.89	28.36	31.01	84.97	73426	201.17	43.68	248.11
Maize-Potato-Groundnut	3.39	23.59	2.97	22.12	60.61	58790	161.07	42.14	170.17
Central and North (Ranchi- Jharkhand)									
Maize-Lentil	2.41	1.17	-	5.56	15.22	15661	42.91	23.64	123.47
Soybean-Wheat	2.13	2.31	-	7.19	13.06	13279	36.38	24.39	143.88
Maize-Potato	2.35	18.00	-	10.44	8.57	27425	75.14	21.99	153.57
Humid South eastern plain zone (Banswara, Rajasthan)									
Maize-Wheat	2.63	4.62	-	8.75	23.97	22300	61.10	27.34	109.36
Soybean-wheat-Greengram	1.39	4.63	1.9[illegible]	14.22	38.95	45172	123.76	44.43	131.63
Cotton-Wheat-Greengram	1.84	4.36	0.73	14.46	39.62	45041	123.40	51.65	111.25

- In central and eastern U.P., maize-potato-sunflower system was identified to be the most suited alternative to the rice-wheat and maize-wheat existing systems with highest maize equivalent yield of 15.8 t/ha/year and resource use productivity of 24.8 kg grain/kg of nutrient and 125.0 kg grain/ha cm of water. But in terms of profitability, rice-potato-wheat was better with net return of Rs. 28934/ha/year and economic efficiency of 79. 3 kg /day/ha. The system involving maize-pea-sunflower was also good in the order of merit. Therefore, offering a good choice so as to achieve enhanced productivity and profitability gains with less stress on resource use specially on irrigation water.
- In central and North zone of Jharkhand, maize-potato, soybean-wheat and groundnut-wheat proved to be better systems intend of existing maize-lentil system.
- In humid south eastern zone of Rajasthan, cotton-wheat-greengram and soybean-wheat-greengram systems were equally good and distinctly better than existing maize-wheat system with highest production of 14.2 to 14.5 t/ha year, productivity of 38.0 to 39.6 kg grain/ha/day, net return of 45041 to 45172, economic efficiency of Rs. 123.4/day/ha, nutrient use productivity of 44.4 to 51. 7 kg grain/kg nutrient used while in terms of water use productivity, soybean- wheat-greengram was better with grain out put of 131.6 kg grain/ha cm of water.

The average yield of existing maize based systems (4.94 t/ha/year) can nearly be doubled (9.3. t/ha/year) through improved agronomic management practices. Moreover, average productivity to the tune of 22.4 t/ha/year in term of maize equivalent yield is possible through diversification of maize-based systems clearly revealing the huge gap, which can be realized in future.

Suggestions

- Instead of practicing same system year after year on same price of land it is always desirable to change the crops and cropping systems in certain years so as to break the monotomy of any particular system.
- Choose a crop and cropping system based on expected net return, effect on soil fertility and likely effect on pests including weeds.
- Synthesize the most suitable cropping system involving high potential crops considering all aspects of production system on equivalent yield terms and economic efficiency with comparative emphasis on resource use.
- Choose a most suited variety that fits well in the system. In general, fast growing and short duration cultivars having high production potential are considered to be the better choice.

- Use recommended doses of inputs specially nutrients through organic and inorganic sources for obtaining good yield of crops in the system and also to maintain the soil fertility over time.
- Under the diversified intensive cropping systems the contingent planning in advance is always desirable to minimize risk. Therefore, suitable adjustments are essential under adverse conditions.

Future thrusts

- The emerging needs and prospects in relation to resource availability are considered to be the driving forces for any change or shift in the system. In fact, crops are not the bar for inclusion in any system. The crops should fit well. Inclusion of non-conventional low volume high value crops including medicinal and aromatic crops, spices etc having regional and global prospects needs to be evaluated for diversification of the existing system on comparative advantage basis.
- Instead of random inclusion of a crop in any system, there is need to synthesize high potential cropping systems on considering individual crop relevance in the system and should be evaluated both at on station and on-farm conditions with farmers participatory approach.
- The crop varieties need to be identified for the cropping system as a whole instead of only considering the season.
- Location specific production approaches needs to be prepared and standardized with specific notes on contingent planning too.
- Since, water is likely to be the major scare resource in future, more efficient systems in limited water resource conditions with high productivity needs to be developed.

It may be conclusively stated that lot of potential exists in selected zones of the country and the food grain targets can be achieved through adoption of alternative diversified systems taking into consideration the availability of resources. However, the site-specific resource management would be necessary to achieve the quantum jumps in achieving production targets. The diversified most efficient systems as discussed above.

References

Gangwar, B., Dhoon, S. K. and Pandey, D. K. 2005. Diversification in pearlmillet based cropping systems. *Indian Farming*.55 (2):17-21.

Gangwar, B., Dhoon, S. K. and Pandey, D. K. 2006. Diversification of maize based cropping systems. *Indian Farming*.56 (6)::20-24.

Gangwar, B., Duhoon, S. K. and Pandey, D. K. 2007. Profitable cropping systems for rice-wheat farmers. *Indian Farming*. 59 (5):4 –8.

□□□

7

Non-monetary Inputs Management in Commercial Crop Production

ANAND PRAKASH

Commercial farming is a progression from diversified farming, when the land owner acquires a sufficiently large amount of arable land and sufficiently advanced technology. Commercialization can either take the form of product commercialization which can occur on the output side through increased share of marketed surplus or factor commercialization which can occur on the input side through increased use of purchased inputs. Commercialization on product side includes increase in marketable out put relative to production, changes in product mix, within crop sector in favour of commercialization crops, both in terms of area and production, diversification towards different sub-sectors of agriculture.

Non-monetary inputs do not require any money but they influence the productivity considerably. These inputs are.

- Crop rotation
- Selection of a field
- Planting time
- Planting technique
- Selection of a variety

- Planting depth
- Planting density (Row to row and plant to plant)
- Time of irrigation
- Time of fertilizer application
- Time and method of weed control
- Time of disease control
- Time of harvesting and threshing

Crop rotation

Growing different crops in a succession on a piece of land in a specific period of time with an object to get maximum profit from least investment without impairing the soil fertility.

- Fixed land
- Fixed no. of crops
- Definite order
- Definite duration
- Rotating crops having drastically dissimilar life cycles or cultural conditions so as to break the cycle of the weeds is among the most effecting of all weed management practices.
- Soil type (sandy loam to loam for pulses)
- Low land (for rice cultivation)

Planting time

For all high yielding semi dwarf varieties of wheat recommended for irrigated conditions, adjustment should be made in such a way that the date of sowing is reached after the mean daily temperature has dropped down to 22-23 ^{0}C. Time of planting, the crop can be manipulated in such a way that ecological conditions for germination of weed seeds are not met due to escape mechanism. Growing rice by transplanting method suppresses weeds.

Selection of a variety

Choose the latest recommended varieties for your area and for the required cultural conditions. i.e. irrigated timely sown, irrigated late sown and rainfed condition.

Depth of seeding

Germination depends on large extent to depth of seeding. Sowing very deep or shallow often results poor germination. The poor germination will cause poor stand at the beginning as well as at harvest and ultimately yields will be low. There are following problems if sowing of wheat is either deep or shallow.

Deep

- Emerge later and take a longer time to develop in to a healthy plant
- This delay keeps the plant in the tillering phase for longer time
- In deep sowing tillering buds once delayed remain dormant and inactive and do not produce tillers.

Shallow

- Shallow seeding combined with high available moisture will quickly initiate the formation of crown roots and will expose the tiller buds to micro climate in general and to high temperature in particular.
- At the time of CRI, stage the tiller buds are subjected to high temperature resulting in loss of effective tillers.
- Shallow seed at near the soil surface may not germinate due to lack of soil moisture.

Planting density

Planting density is very important for getting the higher yield. In case of wheat, about 400-500 spikes per square meter area are required.

Time of irrigation

If irrigations are not given at required time then there is drastic reduction in yield under water deficit condition, the growth and development of the crop in checked.

Time of fertilizer application

The time of fertilizer application depends upon type of soil and climatic conditions. The fertilizer should be applied on the basis of soil test values. The phosphatic and potassic fertilizers must be applied as basal while nitrogenous fertilizers should be applied as basal and top-dressed in split doses.

Time and method of weed control

It is always better to use clean seed free from weeds seeds. Timely application of herbicides is some what more important than ensuring proper time in other operations as none of herbicides is completely selective and therefore herbicide application at a wrong time might seriously injure the crop plants. Some suitable alternative rabi crops like berseem and gram should be grown in rotation. Getting the harvesting machines cleaned when they move weed infested field.

Time of disease control

Timely control of disease is important. Use latest resistant varieties. To control loose smut, the seed may be soaked in water for four hours in morning 6 AM to 10 AM and then sun dryied in May/June on cement floor.

Time of harvesting and threshing

Timely harvesting in an important, otherwise the yield will be reduced due to shattering. Timely threshing is also important; otherwise the crop will be damage by rats.

Reference

Lal, Pyare, 2004. Manipulation of non-monetary inputs for weed management in rice. Proceedings of the training course G.B.P.U.Agri. & Tech. Pantnagar. 276-278.

□□□

8

Crop Diversification towards Commercial Agriculture

BADRUL HASSAN AND TAHIR AHMAD SHEIKH

Crop diversification in Indian is generally viewed as a shift from traditionally grown less remunerative crops to more remunerative crops (Hazra, 2003a). Crop diversification has been recognized as an effective strategy for achieving the objectives of food and nutritional security, income and employment generation, poverty alleviation, judicious use of natural resources and ecological management (Reddy and Suresh, 2008). Although fruit cultivation is labor intensive as compared to cultivation of staples, fruits are relatively capital intensive, making them a less advantageous choice for small-holders who tend to have low capital endowments. Continuous adoption of the rice- wheat cropping system has led to the problems of specific weeds, reduced soil fertility in specific root zone with special references to micronutrients and infestation of similar kind of pests which ultimately resulted in declining the efficiency of the system however, inclusion of oilseeds and leguminous crops (green manure or grain crops) in the system improve soil health and increases the production efficiency of the system (Kumar *et al.* 2008). With globalization of market, now crop diversification in agriculture means to increase the total productivity in terms of quality, quantity and monetary values under diverse agro-climatic situations of the country. Crop diversification is now of urgent necessity for the third world developing countries like India, since growing of basic staples such as cereals alone cannot support the economic development not withstanding the need to ensure food security to the people (Satyasai and Viswanathan, 1997).

In agriculture, crop diversification essentially refers to shift from one crop to another crop. But in real sense it is bringing out a desirable change in the existing cropping pattern towards more balanced cropping system to meet ever increasing demand for cereals, pulse, oilseeds, fibre, fodder, fuel etc. and aims to improving soil health and agro-ecosystem (Singh, 2008a). Crop diversification can be approached in two ways. The main form and the commonly understood concept is the addition of more crops to the existing cropping system, which could be referred to as horizontal diversification. For instance, cultivation of field crops in rice fields or growing various types of other crops in uplands have been defined as crop diversification. However, this type of crop diversification means the broadening of the base of the system, simply by adding more crops to the existing cropping system utilizing techniques such as multiple cropping techniques coupled with other efficient management practices. The systems of multiple cropping have been able to increase food production potential to over 30 t/ha, with an increase of the cropping intensity by 400-500 percent. The other type of crop diversification is vertical crop diversification, in which various other downstream activities are undertaken. This could be illustrated by using any crop species, which could be refined to manufactured products, such as fruits, which are canned or manufactured into juices or syrups as the case may be. Vertical crop diversification will reflect the extent and stage of industrialization of the crop. It has to be noted that crop diversification takes into account the economic returns from different crops (Gunasena, 2003). This is very different to the concept of multiple cropping in which the cropping in a given piece of land in a given period is taken into account. Besides the above, some other terminologies are also used to define crop diversification. There are terms such as "crop substitution" and "crop adjustment". It is necessary to indicate here that crop substitution and adjustment are linked to the main concept of crop diversification and are strategies often used to maximize profit of growing varieties of crops. The level of diversification will also be different in various countries. Diversification at farm level will involve growing of several crops for achieving self-sufficiency, but it may be a totally different approach at the national level. Crop diversification at national level will demand more resources and require selection and management of a specific crop or a group of crops sold freshly or value added to achieve higher profits. There are several advantages of crop diversification, which could be listed as follows:

- Comparatively high net return from crops
- Higher net returns per unit of labour
- Optimization of resource use
- Higher land utilization efficiency
- Increased job opportunities

In order to achieve the above benefits the process of diversification should be changed from very simple forms of crop rotations, to intensive systems such as relay cropping and intercropping or specialization by diversifying into various crops, where the output and processing etc., could be different. This process could be similar at farm level and national level.

Crop diversification: an effective strategy for achieving

1. Food security

The estimated food demand of the country by the end of Five Year Plan (2006-07) was 243 mt and further breakup of these estimates indicates that there is need of 103.5 mt rice, 84.3 mt wheat, 34.4 mt coarse grains, 21.5 mt pulses and 9.5 mt edible oils. Besides these 110.7 mt vegetable, 70.5 mt fruits will be needed in near future. The long -term projections (2030) reveals that 36-40 % of total dietary energy demand is to come from other than food grains. Undoubtedly to satisfy this need, the diversification of cereal dominated production system induction of short duration oilseed crops like toria and yellow sarson for oilseed is highly essential.

1.1 Balanced food

After meeting the cereal requirement of burgeoning population of the country, time has come to endow with balanced food demand of the entire Indian malnourished population. As socio-economic status improves, the proportionate consumption of other than cereal product also increases. Conjunction of pulses, oilseeds, horticultural crops and agro-forestry with cropping would ensure the seasonal access to pulses, oilseed, vegetables and fruit. In this way diversified farm would not only meet the food demand but also cater the need of protein, fat, vitamins and mineral required for good health (Singh, 2008*a*).

2. Increasing income on small farm holding

The dominance of small and marginal farmer is one of the major problems of agricultural economy in India. More than 75% of the farm holdings are below 2 ha, and a large portion of rural people belongs to these holding categories. It is hardly disputed that the income of these farms cannot be raised up to desired level to suffice alleviation of poverty unless existing crop production systems are diversified through inclusion of high value crops under rainfed rice and wheat occupies about 60% and 28% of total cropped area and contribute about 55% and 25% of total production respectively. Out of this area, that area can be shifted to oilseed crops, which is lacking in water resources in meeting out the water requirement of rice and wheat crops (Singh, 2008*b*).

3. Poverty alleviation

Micro level studies suggest that diversification can be an important source of income augmenting. The evidences showed that a shift towards high value crops benefited the poor by directly generating employment and raising agricultural productivity. One of the main channels through which diversification towards high-value crops can reduce poverty is via the participation of small farmers. However, although smallholders have the benefits of proportionally larger labor pools; this may be offset by constraints such as lack of access to credit. Thus, there is continued debate as to whether smallholders can successfully diversify into the high-value sector. The main determinants of poverty are: i) lack of income and purchasing power contributable to lack of productive employment; ii) the continuous increase in price of food, specially food grains which account for 70-80 percent of the consumption; and iii) inadequacy of social infrastructure, affecting the quality of life of the people and their employability.

4. Environmental sustainability

Crop diversification proved to be paramount importance in mitigating the environmental problem arising on account of monoculture. Inclusion of certain cops in sequential and intercropping system has been found to reduce some weeds to a considerable extent, thereby reducing herbicide needs to a great extent in areas where such weeds have assumed alarming proportion. Legume intercropping in cereals grown with wider row spacing has been reported to reduce nitrate leaching. It is reported that with continuous rice-wheat cropping system, the bulk density of soil increased, especially in the 30-45 cm soil profile. Soil organic carbon and available N, P and K increased markedly, when the wheat in rice-wheat cropping system was substituted with a legume (Prakash, 2008).

5. Sustainable agricultural development

The diversification and intensification for sustained productivity of crops and cropping system have in practice especially with the development of new generation multi-crop bed planting. The FIRB planting technique provides an opportunity for crop diversity through inclusion of various crops as well as feasibility of inter or relay cropping thereby leading to resource efficient cropping system.

6. Conservation/ management of natural resources

Impact of diversification on natural resources and environment is debatable. It is often argued that intensive cultivation of market oriented high value crops demand more of water, chemical fertilizers and pesticides which may affect natural resources and environment adversely. Data in table 2 indicates differential water requirement of different crops. Water required to produce 1 kg of high value crops is much lower than cereals crops, however, livestock production is water intensive.

Diversification can also contribute positively towards sustainability of the natural resources. The leguminous crops may enrich the soil fertility. Livestock and poultry manure are rich in N, P and K. Diversification also helps in better pest management, crop rotation and intercropping break pest cycle especially in monoculture areas. Water saving technologies like sprinkler and drip irrigation may be used in fruits and vegetable production not only for high WUE but also for higher yield.

7. Weed management

Diversified intensive cropping systems viz., rice-vegetable pea-wheat-greengram and rice-potato-wheat result in reduction of *Phalaris minor* population. Similarly, interruptions of rice-wheat system through rice-mustard-greengram and rice-berseem once in three years were also found to be quite effective in reducing the *Phalaris minor* population by 48 & 55 % at Modipuram and 30 & 64 % at Faizabad (Fig. 1 & Table 1).

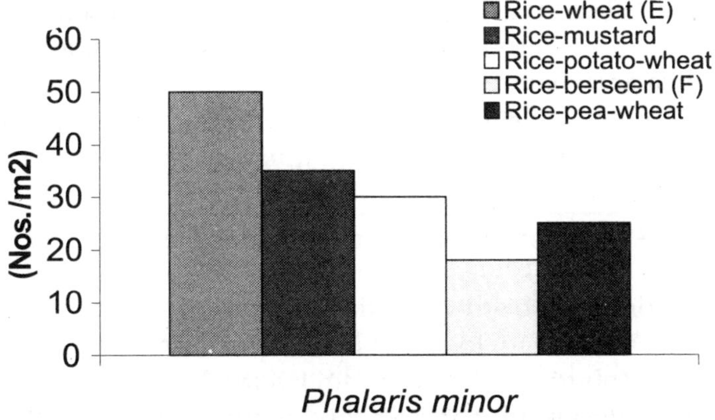

Fig. 1. Effect of crop interruption on *Phalaris minor* population at 30 DAS in wheat at Faizabad

Considering component aspects for improvement in rice-wheat production system strategy, raising of legumes, wheat and mustard on raised bed, irrigation in rice at hair line cracking stage or one day after disappearance of irrigation water instead of continuous submergence, summer ploughing and interruptive cropping of crops like berseem, mustard and pea are very much helpful in improving system productivity, achieving higher profitability and maintenance of soil fertility with effective natural management of important weeds like *Phalaris minor*.

Factors responsible for crop diversification

1. Shift from traditionally grown less remunerative crops to more remunerative crops

Within Agriculture Sector, the diversification is considered a shift from one crop to another or from one enterprise to another in terms of area, production, income, and transfer of resources. Based on these concepts, the nature of diversification can be classified as any one, or all of the following three situations.

- Shift from farm to non-farm activities,
- Shift from one crop to another or one enterprise to another, and
- Use of resources in diverse but complementary activities.

2. Market infrastructure development

Market infrastructure development and certain other price related supports also induce crop shift. It is one of the important factors responsible for diversifying traditionally grown crops to the crops having readily available market. It can be explained by illustrating the recent cultivation of Lemon grass in Bihar and Madhya Pradesh. "Lemon grass cultivation is a boon for farmers (in Bihar and Patna district in particular) because the crop grows well in less fertile and marginal lands, is drought resistant, and is not attacked by animals and insects or pests, as the crop is bitter in taste. Also, selling the harvested produce (lemon grass oil) is not a problem because of its ever increasing demand and the large number of buyers. Selling the aromatic oils, extracted from aromatic crops such as Mentha and lemon grass, is easier than selling some other crops, because of well developed markets in nearby states such as Uttar Pradesh and Madhya Pradesh. Most of the produce is being purchased by the traders located in these states, and so far no difficulty has been encountered by the farmers. Some traders have also set up shops in Bihar, as the production of aromatic oils in the State rises at par with the neighboring States such as Uttar Pradesh, which have had a 30 years' lead over Bihar in initiating their cultivation (Prabu, 2008).

3. Higher profitability

In this regards, higher profitability and the resilience/stability in production also induce crop diversification, for example sugar cane replacing rice and wheat. Besides adoption of proper input management technologies for traditional rice-wheat cultivation, diversification of the system through introduction of crops of diverse nature may be a good preposition to break the monotony of the system and help to sustain productivity over a period of time. One of the approach suggested to address these problems is the adoption of alternative efficient cropping systems, which ensures high productivity, profitability and resource use efficiency (Gangwar and Prasad, 2005).

4. The crop shift (diversification) also takes place due to governmental policies

The crop shift (diversification) also takes place due to governmental policies and thrust on some crops over a given time, for example creation of the Technology Mission on Oilseeds (TMO) to give thrust on oilseeds production as a national need for the country's requirement for less dependency on imports.

5. Crops are practiced in rainfed lands

Crop diversification and also the growing of large number of crops are practiced in rainfed lands to reduce the risk factor of crop failures due to drought or less rain.

6. Crop substitution and shift

Crop substitution and shift are also taking place in the areas with distinct soil problems. For example, the growing of rice in high water table areas replacing oilseeds, pulses and cotton; promotion of soybean in place of sorghum in vertisols (medium and deep black soils) etc.

Estimation of crop diversification

There are number of methods, which explain either concentration (specialization) or diversification of commodities or activities in a given time and space by a single indicator. Important ones include:

- Simpson Index
- Herfindahl Index
- Entropy Index

To examine the nature of crop diversification in agriculture, temporal changes in area, production and value of different crops, and quantity (and/

or value) or livestock can be assessed. To estimate the pace of diversification, annual compound growth rates of area, production and yield of different crops can be computed by evaluating the given table 1 & 2 (Tewari, 2008)

Table 1. Existing cropping pattern in India

States	Crop Area % to Gross Crop Area (TE 2004-05)		
	More than 30%	10% to 30 %	2 % to 10%
J & K	MAIZE	PDY	-
Punjab	PDY - k, WHT	OTH	CTN
Haryana	WHT	PDY - k, CCL-k	PLS-r, OLS-r, CTN, SCN, OTH
Uttar Pradesh	WHT	PDY –k	CCL-k, PLS-k, PLS-r, OLS-r, SCN, OTH
Rajasthan		WHT,CCL-k, OLS-r, OTH	PLS-k, PLS-r, OLS-k, CTN
Maharashtra		CCL-k, CCL-r, PLS-k, CTN, OTH	PDY, WHT, PLS-r, OLS, OLS-k, CCN
Andhra		PDY-k, OLS-k, OTH	PDY-r, CCL-k, CCL-r, PLS-k, PLS-r, OLS-r, CTN
Karnataka		CCL-k, CCL-r, PLS-k, OTH, OLS-k	PDY-k, PDY-r, WHT, PLS-r, CTN, SCN
Bihar	PDY-k	WHT	CCL-k, CCL-r, PLS-k, PLS-r, OTH
West Bengal	PDY-k	PDY-r, OTH	WHT, PLS-r, OLS-k, OLS-r, JUTE

Symbols; PDY= Paddy, WHT= Wheat, CCL= Coarse cereals, CTN= Cotton, OLS= Oil seeds, PLS= Pulse, OTH= Other Crops , SCN= Sugarcane; k = Kharif, r = Rabi (Mehta 2005)

Sources of agricultural growth in different regions of India during 1980s and 1990s (per cent)

Crop diversification was the dominant source of agricultural growth during both the decades of 1980s and 1990s followed by rise in prices and yield in this region (Table 3). However, the contribution of yield to agricultural growth was less during the 1990s than 1980s, while the trend was just the reverse for prices. Among commodities, pulses, oilseeds and fruits & vegetables made a significant contribution (76 percent) to agricultural growth during 1980s. During 1990s, production portfolio included wheat, pulses, oilseeds, cotton and fruits & vegetables, with contribution of about 80 percent growth in the crop sector. The contribution of fruits & vegetables, wheat and cotton increased during the 1990s, while that of pulses and oilseeds decreased. Interestingly, crop diversification and prices were mainly responsible for the growth in wheat and fruits & vegetables, while these were crop diversification and higher yields in the case of cotton. The region had witnessed a transformation as a result of investment in the watershed programs and incentives on adopting water-saving technologies. These initiatives in the region led to crop diversification and higher crop yields, which eventually contributed to agricultural growth. Southern region, this region witnessed crop diversification as the major source of agricultural growth and its pace was much faster during the 1990s than 1980s (Table 3). Rice, oilseeds, fruits & vegetables and spices substituted coarse cereals during the 1980s, with a contribution of roughly 80 percent to the agricultural growth. During the 1990s, their contribution improved to 88 percent, substituting oilseeds and cotton. Coarse cereals, particularly maize also boomed in the region for the poultry sector. Increase in the yield of major crops also emerged as an important source of agricultural growth in the region, as a major part of it was characterized as rainfed and technologically laggard. However, greater emphasis on the watershed development in the southern and western regions had augmented water availability, leading to adoption of improved technologies, and consequently higher yields of crops (Joshi, 2003).

Table 2. Changes in cropping pattern in India

Period/crops	TE 1964-65	TE 1971-72	TE 1981-82	TE 1991-92	TE 1999-00	TE 2004-05*
Rice and wheat	31.3	33.7	35.9	36.1	37.53	36.57
Coarse cereals	28.1	27.5	24.5	19.6	15.67	13.76
Pulses	15.4	13.8	13.7	12.8	11.79	11.90
Food grains	74.9	75.1	73.1	68.5	64.99	64.03
Oilseeds	8.9	8.9	9.2	14.1	13.42	12.94
Cotton	5.1	4.9	4.6	4.1	4.72	4.34
Sugarcane	1.5	1.6	1.7	2.0	2.15	2.17
Other crops	9.6	9.5	10.4	11.3	14.72	16.44
Non-food grains	25.1	24.9	25.9	31.5	35.01	35.97
GCA in mha	157.65	164.41	172.99	182.70	190.76	186.05

(Tiwari, 2008*a*)

Diversification within crop sector

The crop sector is depicted a steady diversification in India with replacement of food grain crops with nonfood grain crops. The share of non-food grain crops in terms of area went up from 30 percent in TE 1981-82 to 35 percent in TE 1998-99, and in value terms it had a significant rise, from about 52 to 60 percent. Several non-food grain crops such as oilseeds, fruits, vegetables, spices and sugarcane have substituted mainly coarse cereals in the farmers' pursuit for higher income. Regional patterns in crop diversification in India were quite stark (Table 4).

Table 3. Sources of agricultural growth in different regions of India during 1980s and 1990s (per cent)

Region	Period	Sources of agricultural growth			
		Area	Yield	Price	Diversification
Northern	1980s	1.04	75.4	-6.5	29.7
	1990s	10.01	16.3	44.0	28.2
Eastern	1980s	17.8	49.7	11.8	19.7
	1990s	-29.7	38.7	45.8	42.7
Western	1980s	11.6	36.5	7.3	39.0
	1990s	13.4	24.8	25.7	35.8
Southern	1980s	10.4	39.5	16.8	32.1
	1990s	-8.7	36.2	39.3	45.0
All India	1980s	10.1	54.0	7.7	26.6
	1990s	4.0	29.3	35.2	30.7

Table 4. Share of food grain and non-food grain crops in cropping pattern and value of output in India at constant prices (per cent)

Region	TE 1981-82				TE 1998-99			
	Food grains		Non food grains		Food grains		Non food grains	
	Area	Value	Area	Value	Area	Value	Area	Value
Northern	77.42	54.92	22.58	45.08	76.86	53.74	23.14	46.26
Eastern	81.63	51.73	18.37	48.27	73.83	43.04	26.17	56.96
Northeast	70.11	44.43	29.89	55.57	65.06	35.80	34.94	64.20
Western	71.92	44.44	28.28	55.56	61.85	36.10	38.15	63.90
Southern	62.86	41.82	37.14	58.18	53.08	28.20	46.92	71.80
All India	70.44	48.08	29.66	51.95	65.44	39.65	34.56	60.15

Northern region

The northern region in India has more specialized in rice and wheat crops. The favorable government pricing policies, assured procurement, high-yielding technologies, and irrigation development have encouraged farmers to allocate more area in favor of these crops. Rice and wheat have replaced coarse cereals and pulses in this region and it is diversifying only marginally towards non-cereal commodities. With the availability of short-duration black gram, green gram and pigeon pea, pulses had started regaining area in this region (Joshi *et al.* 2000). Other important crops of the region were sugarcane, vegetables and fruits. There are reports, however, that extensive cultivation of rice and sugarcane was causing negative externalities related to soil and water resources. The soil fertility with respect to macro- and micronutrients was declining, and the water resources were depleting. These negative externalities have adversely affected the total factor productivity of rice-wheat based cropping system in this region (Kumar *et al.* 1998). It has potential for cultivating a variety of fruits and vegetables, but its exploitation depends on developing appropriate infrastructure with suitable linkages between production and consumption.

Eastern region

The eastern region of India is the most backward in terms of per capita income, agricultural growth and infrastructure development. The yield levels are low because of the uncertain production environment and poor adoption of improved varieties and technologies. Overall, the region was food-based concentrating largely on rice, with little diversification. The humid atmosphere and high rainfall make cultivation of rice more favorable in this region. In the limited non-rice areas, however, there was high diversity. It is an important vegetable growing area with about 44 percent share in the total vegetable area in the country in TE 1998-99 (Government of India, 2001). The cultivation of fruits was also increasing gradually. It has emerged as an oilseeds producing region too with cultivation of crops like rapeseed-mustard, groundnut, sesame and soybean.

Western and southern regions

The southern region depicted maximum diversification followed by the western region. It may be mentioned that these were the regions, which had accomplished higher agricultural growth during the 1990s over the preceding decade. These regions moved swiftly towards more non-cereal crops, which perhaps contributed to the acceleration of agricultural growth. These regions were relatively less developed in irrigation and largely relied on rainfall. Since pulses and oilseeds require less water, they had found niches in these regions. Like pulses and oilseeds, these regions witnessed substantial increase in area under fruits and vegetables also. The government-supported programs had promoted the cultivation of fruits and vegetables. Among others, watershed program had facilitated conservation of rainwater and gave higher priority to the cultivation of fruits and vegetables. Among cereal crops, maize was picking up fast in the southern region and to some extent in the western region also, largely as poultry feed. Among the crops gaining in these regions, oilseeds were under serious threat in the wake of import liberalization of edible oils, as the cost of imported oils, especially palmolean, was much lower than their domestic price. To sustain oilseed production, technical efficiencies in their production and processing will have to be innovated with better management and technology options.

Analysis of crop diversification

Table 5. Annual compound growth rate of crop area (% per annum) of major crops in India

Crops	1984-85 to1994-95	1994-95 to 2003-04
Rice	0.53	-0.05
Wheat	0.91	0.2
Coarse Cereals	-2.14	-1.09
Maize	0.49	1.67
Pulses	-0.21	-0.59
Oilseeds	4.03	-1.9
Cotton	0.74	-0.97
Sugarcane	2.71	0.95

Source: Metha (2005)

Diversification of agriculture and resource endowments

The relative level of diversification of agriculture across regions within a country will vary, depending on agro-climatic conditions, resource endowments and infrastructure (Table 6).

In quadrant 1 diversification is high as a risk mitigating strategy against production risks due to harsh and unpredictable agro-climatic conditions. Regions in quadrant 2 are agro-climatically better endowed, but diversification is low due to lack of infrastructure, technology and institutions. Regions in the third quadrant have high levels of irrigation; uses of modern inputs are common and have access to infrastructure and institutions. Here, agriculture is more specialized and market oriented, and diversification as an option for risk reduction or income enhancing strategy is not perceived to be 3 important. However, there may be sustainability problems associated with over specialization (due to mono cropping) leading to diversification in the long run. Finally, in quadrant 4 we have a situation where commercial diversification is high as an income augmenting strategy to meet the growing demand for high value commodities. Here, diversification gets strengthened with availability of latest technology and required infrastructure. The pace and nature of diversification in a country would thus vary from region to region and over time requiring different strategies.

Table 6. Diversification of agriculture and resource endowments

Level of diversification		
High	SUBSISTENCE agriculture; low and erratic rainfall; poor infrastructure; low irrigation; low population density. Diversification is a risk mitigation strategy	MARKET ORIENTED diversification towards commercial crops /high value crops. Driven by demand for high value commodities. Supply side factors favorable to Diversification.
Low	AGRO-CLIMATICALLY better endowed region. Lack of adequate infrastructure. Access to input and output markets constrained. Low adoption of improved technologies	SPECIALISED AGRICULTURE: High levels of irrigation; inputs; mechanization; low income/ market risk. Low relative profitability of substitute enterprises
	Low	**High**
	Resource endowments, inputs, infrastructure	

The objective of this paper is not to look at the whole gamut of diversification (definitions, scope etc.) but instead, the focus will be on selected high value commodities that are contributing to diversification of agriculture with particular reference to urbanization as a driver of agricultural diversification towards high value commodities (Rao *et al.* 2004).

Diversification of rice-wheat cropping system

The rice (*Oryza sativa L.*)-wheat (*Triticum aestivum L.*) is the largest agricultural production system ,which occupies 13.5 million ha in South Asia, including 10.3 million ha in India, extending from indo-Gangetic plains to Himalayan foot hills. Rice-wheat contributes 77.7% of the total food grain production of the country. In India 23 and 40 % of total rice and wheat area respectively is represented by the rice-wheat system alone. However, due to continuous cropping of rice-wheat during last three decades certain new problems are posing serious threat to its sustainability. These problems are: (i) Disturbed soil structure and formation of sub-surface hard pan due to widely adapted practice of puddling in rice cultivation, (ii) Buildup of *Phalaris minor* population in wheat (Malik, 1996) and *Echinochloa colonum* in rice, (iii) Declining factor productivity specially for fertilizer use (Yadav *et al.*, 1998).

(iv) Decline in water table in certain areas (Singh, 1994). (v) Development of multiple nutrient deficiencies and (vi) Decline in profitability due to enhanced cost of inputs used. The adverse effects of continuous rice-wheat cropping on soil conditions and reduction in productivity have also been reported by Nambiar and Abrol (1989).

Under these threats, the major issue is that how to tackle these problems and what should be our strategy. In this regards, besides adoption of proper input management technologies for rice-wheat cultivation, diversification of the system through introduction of crops of diverse nature may be a good preposition to break the monotony of the system and help to sustain productivity over a period of time. One of the approach suggested to address these problems is the adoption of alternative efficient cropping systems, which ensures high productivity, profitability and resource use efficiency (Gangwar and Prasad, 2005).

Diversification will not only help in reducing ill effects of continuous rice-wheat cropping but also result in; (i) much needed increase in productivity of oilseeds and pulses, which had been pushed back to low productivity in marginal lands, (ii) improvement in intake of protein component in the diets of vegetarian population and fodder quality of animals, (iii) increase in economic returns of farmers and (iv) exploitation of legumes for their atmospheric N-fixing capacity and quality residue for recycling so as to improve organic matter in the soil and to reduce the fertilizer requirements with good productivity of the system.

Any change to diversify the existing system may only be possible or desirable if it can solve or minimize new emerging problems on one hand while equally competitive in terms of productivity and profitability on the other. An attempt was therefore made to identify the efficient cropping systems based on the comparative advantage using intensification and partial substitution approach of diversification. Accordingly, based on studies conducted through cropping systems research programmes at 17 locations, the most efficient alternative cropping systems were identified so as to address major problems including 2nd generation problems of rice-wheat system.

High productivity and remunerative alternatives

In central plain zone of Punjab, rice-potato-sunflower system was found to be more remunerative (Rs. 70,262/ha/year as net return) closely followed by rice-vegetable pea-late wheat (Rs. 67,540/ha) and rice-potato-wheat (Rs. 67,089/ha) when compared to existing rice-wheat Rs. 37,020/ha (Table 7). Similarly, replacement of rice through groundnut once in three years was also desirable to sustain the system with average system productivity of 70.7 kg grain/day/ha and economic efficiency of Rs. 184.5/ha /day.

Table 7. Resource use efficiency under diversified rice based cropping systems (Orissa)

Cropping system	Land use efficiency (%)	Production efficiency (Kg REY/ha/day)	WUE (turns Kg REY/ ha-cm	Man days/ ha	Net returns (x 10^3 Rs/ ha)	Returns per reinvestment (Rs)
Rice-toria-fallow	64.20	35.72	56.0	340	16.81	1.58
Rice-toria-greengram	83.84	40.21	71.5	432	29.28	1.76
Rice-groundnut-greengram	73.33	39.94	63.9	423	20.85	1.55
Rice-groundnut-cowpea	90.96	41.81	72.2	515	29.25	1.62
Rice-maize-greengram	84.57	49.60	81.4	542	38.21	1.83
Rice-maize-cowpea	86.30	50.73	84.3	550	40.42	1.85
Rice-sunflower-greengram	84.20	37.93	62.8	442	20.65	1.47
Rice-rajmash-sesame	85.02	40.41	72.3	430	26.62	1.63
Rice-fieldpea-sesame	85.02	46.83	84.1	430	38.62	1.94
SEm±		0.86			1.45	0.04
CD(P=0.05)		2.57			4.30	0.11

In eastern plain zone of Haryana, diversification of rice-wheat cropping system through inclusion of greengram in *summer*, vegetable pea as catch crop before wheat and mustard as interruptive crop once in three years instead of wheat on raised bed planting system was found to be most suited for the zone. The rice-wheat-greengram system gave average productivity of 41.4 kg/day/ha with profitability of Rs.152.5/day/ha while rice-vegetable pea-wheat-greengram sequence gave system productivity of 44.2 kg/ha/day with profitability of Rs. 156.0/ha/day (Table 8).

In western plain zone of Uttar Pradesh, rice-vegetable pea-wheat-greengram (grain+green leaf manuring as residue), rice-potato-wheat and rice-mustard-greengram once in three years followed by rice-wheat-greengram were identified to be potential systems for obtaining higher productivity and profitability instead of existing rice-wheat system. The average productivity of these systems was 44.2, 62.5 and 42.8 kg grain equivalent yield/day/ha with profitability of Rs.118.8, 192.1 and 136.6/ha/days respectively (Table 8).

In central plain zone of U.P., inclusion of cowpea and greengram during *summer* for grain and residue incorporation as green leaf manure in rice-wheat system was proved to be most beneficial for increasing system productivity and maintaining soil fertility. Rice-wheat-greengram system was however most productive (151.0 q/ha as rice grain equivalent yield and productivity of 41.4 kg/ha/day) and remunerative (Rs. 45,438/ha /year as net return and profitability of Rs. 124.5 /ha /day) (Table 8). Growing of field pea in alternate year instead of wheat during *rabi* was also found to be more productive over years giving about 130.8 q/ha as system yield in terms of rice grain equivalent which was higher to the tune of 22.4% and remunerative with mean net return of Rs. 40,728 /ha/year and profitability of Rs. 111.6/ha/day.

Table 8. High productivity and remunerative alternatives

State/ Zone	Promising cropping sequences	Net returns	Conventional cropping pattern
Central plain zone of Punjab	Rice-potato-sunflower rice-vegetable pea-late wheat	Rs.70,262/ha/year (Rs. 67,540/ha)	Rice-wheat (Rs. 37,020/ha)
Eastern plain zone of Haryana	Rice-wheat-greengram rice-vegetable pea-wheat-greengram	Rs.152.5/day/ha Rs.156.0/ha/day	Rice-wheat
Western plain zone of Uttar Pradesh	Rice-vegetable pea-wheat-greengram rice-potato-wheat Rice-mustard-greengram	Rs 118.8 /ha/day Rs 192.1 /ha/day Rs 136.6 /ha/day	Rice-wheat
Central plain zone of U.P	Cowpea and greengram during *summer*	Rs. 124.5 /ha /day	Rice-wheat
Alluvial zone of West Bengal	Rice-potato-greengram	Rs. 179.8/ha/day	Rs.21.6 /day/ha
Himachal Pradesh	Maize (baby corn)+ asparagus beans-pea-summer squash	Rs.380533/ ha/357 days	Maize-Wheat Rs.68625/ha/257 day

(Gangwar, 2008)

In Eastern plain zone of Uttar Pradesh, rice-vegetablepea-wheat system was distinctly better than existing rice-wheat with productivity of 52.3 kg/ ha/day as rice equivalent grain yield and profitability of Rs. 202.8/ha/day followed by rice-potato-wheat with productivity of 45.5 kg/ha/day and profitability of Rs. 136.0/ha/day.

In part of new alluvial zone of West Bengal, inclusion of legumes like greengram and groundnut (oilseed cum legume) during *summer* and potato during *rabi* were identified to be distinctly better choices to diversify the existing system rice-wheat. However, rising of these crops on broad bed and furrow system was better to improve system productivity and resource use efficiency. Rice-potato-greengram gave highest productivity of 72.4 kg/ha/day in terms of rice grain equivalent and profitability of Rs. 179.8/ha/day while it was lowest in existing rice-wheat system (23.6 kg/day/ha productivity and Rs.21.6 /day/ha as profitability). Therefore, system deserves to be substituted. Among six cropping systems evaluated in terms of energetic at Modipuram, the overall performance of rice-mustard-greengram system was adjudged to be most efficient and profitable.

Significance of *kharif* pulses in crop diversification

Mere inclusion/replacement of pulses in a system is not sufficient to provide stability in yield. Diversification in managemental practices is also equally important. In India, most of the pulse crops are grown on flat bed by broadcasting with flood irrigation or under continental monsoonal type climate resulting in low productivity. Recent study conducted by Connor *et al.* (2003) in a farmer's participatory trial revealed that pulses grown on raised bed gave more increase in yield viz. urdbean (33.6%), mung bean (21.8%), Pigeonpea (46.7%) and gram (37%) than their cereal counter part viz. rice (6.2%) and wheat (6.4%) (Table 9). Moreover, there was a saving of 26-42% water under raised bed as compared to flat planting.

The significance of various Kharif pulses in meeting the N-need of the succeeding wheat crop revealed that as much as 50-60 kg N/ha may be supplied by the pulses to the succeeding wheat. By economizing nitrogen (N), Kharif pulses may save 12-48x10^8 joules/ha of energy. Singh and Singh (2006) observed that 25% of recommended dose of fertilizer (120:60:60 kg N, P_2O_5 & K_2O/ha respectively) can be saved in wheat only by growing urdbean as preceding crop and applying 20 kg of P_2O_5/ha to it. Response of kharif pulses to application of phosphorus is outstanding. Experiments conducted throughout the country revealed that response to 20kg P_2O_5/ha application varied as 42-186, 67-163 and 32-234 kg grain/ha in mungbean, urdbean and pigeon pea respectively.

Table 9. Effect of *kharif* crop diversification on the productivity and economics of wheat

Crop sequence	Grain yield (q/ha)	Net return of (Rs./ha) Wheat	Sequence	Benefit Wheat	Sequence
Rice-wheat	42.9	19585	35433	0.92	0.85
Mungbean- wheat	42.4	19175	23879	0.90	0.61
Soybean- wheat	41.2	18002	20584	0.84	0.57
Pigonpea- wheat	48.1	24724	34988	1.17	0.92
Urdbean- wheat	47.4	24062	24331	1.14	0.66
Cowpea- wheat	50.0	26811	34154	1.31	0.84
C.D. at 5%	4.3	1800	4280	-	-

Table 10. N-Fixation by *kharif* pulses

Crop	N-fixed (kg/ha)
Urdbean	119-140
Cowpea	09-125
Mungbean	50-66
Pigeonpea	04-200
Soybean	32-97

Challenges, opportunities and prospects of crop diversification

1. Constraints in crop diversification

Crop diversification in the country is taking the form of increased areas under commercial crops including vegetables and fruits since independence. However, this has gained momentum in the last decade favoring increased area under vegetables and fruits and also to some extent on commercial crops like sugar cane, cotton and oilseeds crops specially soybean. The major problems and constraints in crop diversification are primarily due to the following reasons with varied degrees of influence:

- Over 117 m/ha (63 percent) of the cropped area in the country is completely dependent on rainfall.
- Sub-optimal and over-use of resources like land and water resources, causing a negative impact on the environment and sustainability of agriculture.
- Inadequate supply of seeds and plants of improved cultivars.
- Fragmentation of land holding less favoring modernization and mechanization of agriculture.
- Poor basic infrastructure like rural roads, power, transport, communications etc.
- Inadequate post-harvest technologies and inadequate infrastructure for post-harvest handling of perishable horticultural produce.
- Very weak agro-based industry.
- Weak research - extension - farmer linkages.
- Inadequately trained human resources together with persistent and large scale illiteracy amongst farmers.
- Host of diseases and pests affecting most crop plants.
- Poor database for horticultural crops.
- Decreased investments in the agricultural sector over the years.

2. Globalization and crop diversification

With the liberalization of trade and providing the market access of agricultural produce between the different countries, the country will be required to promote much more diversified agriculture. For crops on which we have substantial area and production, specially foodgrains, the import market has to be insulated through increased productivity which gives us a kind of comparative advantage and also a level playing field so that large scale importation is contained and farmers interests are protected. The crops which are traditionally exported like basmati rice and spices and condiments also need to be supported in terms of area expansion and quality improvement to look towards much more opportunity for export. Accelerated growth in fruits and vegetables production is also required for improved nutrition of the country's population. In future, with improved living standards along with increased purchasing power, more and more people will look for nutritional and quality foods which will also call for greater crop diversification. There are some production areas such as food crops, plantation crops, poultry, dairy, sugar, cotton and oilseeds in which India has made its mark. There are some in which its emerging strength is already evident - sericulture, marine and inland fisheries for example. There are also others which now attract less attention, but in which the competitive advantages that India possesses can put it on the top of the world. No country grows such a wide range of fruits, vegetables and flowers and in such abundance as India and yet it has no record worth mentioning in horticultural exports.

3. Emerging technology and crop diversification

It is being increasingly realized that agriculture is no longer a subsistence activity carried out by peasant cultivators, but rather an enterprise and manufacturer of biomass using land, water, genetic material and the latest in technology. The agriculture of the twenty first century will increasingly be farmers' entrepreneurship harnessing technologies to optimize returns from his land and investments he makes on it. Biotechnology and genetic engineering in crops with focus on primary productivity and also on many quality traits will go a long way to improving the yield and quality of many important crop plants. With the advent of such emerging technologies and consequent scope for increased economic returns, the diversification in favor of such crops will be the future focus. Many other related technologies and their adoption will also inject an added dimension in crop diversification. Decision support systems, governmental policies, geographic information system, application of information technology leading to market information etc., will also lead to crop diversification primarily on economic considerations.

4. Research and developmental support for crop diversification

Future agriculture will be much more knowledge and skill based rather than the traditional subsistence agriculture. In the wake of globalization and opening up of the global market, there will be much more opportunity for entrepreneurship development in agriculture. This also calls for paradigm shifts in research and technology development and also the transfer of technology for successful crop diversification. The research system not only needs to address the issues connected with continuance and indulgence and knowledge in the areas of emerging technologies but also create a cadre of scientists through the continuous upgrade of skills and human resource development. The researchers also need to popularize the technologies, impart knowledge and skills to the extension functionaries for the transfer of technologies to the farmers. This knowledge-based farming will call for much more interaction between the researchers, extension workers and farmers. The fruits of the innovative technologies should reach the farmers at the earliest and also spread in the quickest possible time.

5. Institutional and infrastructure developments towards crop diversification

To sustain and operationalize crop diversification, institutional support is required to the two thirds of the country's crop area which is rainfed. Crop diversification in terms of reducing the risk of rainfed farmers is also very vital to a country like India, where two thirds of the farmers are also resource poor. However, crop diversification in well endowed area is more of an economic consideration. The National Agricultural Research System with its Crop and Commodity based Institutions, Natural Research Management Based Institutions and State Agricultural Universities are jointly addressing the issues connected with the crop diversification. The government of India has also developed a counter support mechanism through the establishments of Crop Directorates for each of the major crops and groups of crops like Oilseeds and Pulses for developmental and technology transfer focus on each of these crops and commodities. These Directorates act as a coordinating agency between the research and development activities on the one hand and between the Federal Government at the Central and Regional Governments at the State level for technology transfer and other promotional activities.

6. Government policies and strategies for crop diversification

Considering the importance of crop diversification in the overall developmental strategy in Indian agriculture, the government of India has taken several initiatives for agricultural development in general and crop diversification in particular (Hazra, 2003b). These initiatives are as follows:

- *Launching a Technology Mission for the Integrated Development of Horticulture in the Northeastern Region*: The programme will establish effective linkages between research, production, extension, post-harvest management, processing, marketing and exports and bring about a rapid development of agriculture in the region.
- *Implementing National Agriculture Insurance Scheme*: The scheme will cover food crops and oilseeds and annual commercial and horticulture crops. Small and marginal farmers are eligible for 50 percent subsidy under the Scheme.
- *Operationalizing Technology Mission on Cotton*: The Technology Mission will have separate Mini-Missions on technology generation, product support and extension, market infrastructure and modernization of ginning and pressing units.
- Provision of Capital Subsidy of 25 per cent for construction/ modernization/ expansion of cold storages and storages for horticultural produce.
- *Creation of Watershed Development Fund*: At the National level for the development of Rain fed lands.
- Infrastructure Support for Horticultural Development with emphasis on post harvest management.
- *Strengthening Agricultural Marketing*: Greater attention to be paid for development of a comprehensive, efficient and responsive marketing system for domestic marketing as well as exports by ensuring proper quality control and standardization.
- *Seed Crop Insurance*: A pilot scheme on Seed Crop Insurance has been launched which will cover the risk factor involved in production of seeds.
- *Seed Bank Scheme*: About 7-8 percent of certified seeds produced in the country will be kept in buffer stock to meet any eventualities arising out of drought, floods or any other form of natural calamities.
- *Cooperative Sector Reforms*: Amendment to the National Cooperative Development Corporation (NCDC) Act, 1952, and Replacement of the Multi-State Cooperative Societies (MSCS) Act, 1984.

All these measures will lead to crop diversification and increase the production and productivity of crops.

Conclusion

Diversification needs are highly location specific depending upon the existing resources and driving forces. Therefore, location specific strategies need to be worked out for obtaining beneficial advantages of diversification. Agricultural diversification is our old age asset but it succumbed in light of green revolution for the sake of food security. Now we are food sufficient citizen and can think for diversification towards HVA in the new area of globalization. The demand side drivers like income and urbanization and supply side drives like irrigation, roads and processing infrastructure play significant role in promoting crop diversification. Though middle and large farm holdings favor crop diversification, small holders may also participate in diversification up to some extent. Diversification towards high value crops is helpful in poverty alleviation, food security, employment generation, efficient management of natural resources and enhancing the international trade. Rice is the most important food crop in Asia and will remain so for many years to come. However, in marginal and upland areas of Asia, rice-based cropping systems have low returns. Shifting marginal areas out of rice into more profitable crops is seen as a solution. However, commodity-specific approaches are risky and, given the weak infrastructure and marketing base, may not achieve much for resource poor farmers in rainfed areas. Alternatively, flexible cropping systems for upland farmers that feature production of more income-elastic goods like horticultural products are a means of diversifying their income sources. Indeed, crop diversification away from rice to high-value crops such as oil seeds, pulses, fruits and vegetables has been successful in many countries including India, although quality and timely production are usually crucial to economic success with such high-value crops. Diversification should need based, area specific and should be encouraged in marginal and waste lands, coastal areas and also in hills. India has a great potential for diversification and it needs to be encouraged by intervention in Govt. policies towards high value agriculture.

References

Connor, D.J., Gupta, R. K., Hobbs, P. R. and Sayre, K. D. 2003. A resource book. Rice- wheat consortium for I.G.P. p.103-108.

Gangwar,B. 2008. Diversification of Rice-Wheat Cropping System for Sustained Production; *Proceedings of training Course,* Crop Diversification for Sustainable crop Production, Nov.8th 2008, Centre of Advanced Studies in Agronomy College of Agriculture, G.B. Pant University of Agriculture and technology Pantnagar; p. 7-8

Gangwar, B. and Prasad, K. 2005. Cropping system management for mitigation of second Generation problems in agriculture. *Indian Journal Agricultural. Science.* 75 (2): 65-78.

Government of India. 2001. Report of the Special Group on *Targeting ten million employment opportunities per Year.* Planning Commission, New Delhi.

Gunasena, H.P.M. 2003. Intensification of Crop Diversification in the Asia-Pacific Region, *Crop Diversification in the Asia-Pacific Region,* Food and Agriculture Organization of the United Nations Regional Office for Asia and the Pacific Bangkok, Thailand.

Hazra, C.R., 2003*a*. Crop Diversification In India, *Crop Diversification In The Asia-Pacific Region;* p.32

Hazra, C.R., 2003*b*. Crop Diversification in India, *Crop Diversification In The Asia-Pacific Region;* p.44.

Joshi, P. K. 2003.Crop Diversification in India: Nature, Pattern and Drivers (Book); p.5-6

Joshi, P.K., Asokan, M., Datta, K.K. and Kumar, P. 2000. Socio-economic constraints to legumes production in rice-wheat cropping systems in India; In *Legumes in rice and wheat cropping systems of the Indo-Gangetic Plain* (eds. Johansen, C., J.M. Duxbury), 176-184

Kumar, A., Tripathi H. P., Yadav, R. A. and Singh, D.S. 2008. Diversification of rice(*Oryza sativa*) –wheat (*Triticum aestivum*) cropping system for sustainable production in eastern Uttar Pradesh. Indian Journal of Agronomy 53(1): 18-21.

Kumar, P., Joshi, P.K., Johansen, C. and Asokan, M. 1998. Sustainability of rice-wheat based cropping systems in India. *Economic and Political Weekly* Sep.26: A-152-157.

Malik, R. K. 1996. Herbicide resistance weed problem in developing world and method to overcome them. (*In*): Proc. 2nd International Weed Science Congress. Copenhagen, Denmark, FAO, Rome, Italy. p. 664-673.

Mehta, R. 2005. National Conference on Agriculture for Kharif Campaign-2005 22nd - 23rd March, 2005, Ministry of Agriculture Government of India New Delhi.

Nambiar, K. K. M. and Abrol, I. P. 1989. Long term fertilizer experiments in India: An Overview. *Fertilizer News*. 34: 11-20.

Prabu, M.J.2008. Crop diversification only increases, not crops, your income, *Farmer's Notebook;* THE HINDU, Online edition of India's National Newspaper Thursday, Jun 12, 2008.

Prakash, A. 2008. Crop Diversification in Wheat Based Cropping System; *Proceedings of training Course,* Crop Diversification for Sustainable crop Production, Nov.8th 2008, p-13.

Rao, P. Parthasarathy., Birthal, P.S., Joshi, P.K. and Kar, D., 2004. Agricultural Diversification in India and Role of Urbanization; Mtid Discussion Paper No. 77; International Food Policy Research Institute; p-3

Reddy B. N. and Suresh G., 2008. Crop diversification with oilseeds for higher profitability; *National Symposium on New Paradigms in agronomic Research,* Nov.19-21, 2008 Navasari, Gujarat.

Satyasai, K J. S. and Viswanathan, K.U. 1997.*Commercialisation and Diversification of Indian Agriculture,* Occasional Paper-5: 14

Singh, A. and Singh, N. P. 2006. *Indian Journal of Agronomy* 51 (2):97-99.

Singh, N.T. 1994. Land and water management in north and north-western India. (*In*): *Management of land and water resources for sustainable agriculture and environment, Bulletin*, New, Delhi, India: Indian Society of Soil Science. 95-120.

Singh, R.P. 2008*a*,b. Recent Advances in Crop Diversification for Oilseed Production; *Proceedings of training Course*, Crop Diversification for Sustainable crop Production, Nov.8th 2008. Centre of Advanced Studies in Agronomy College of Agriculture, G.B. Pant University of Agriculture and technology Pantnagar; p. 33-34.

Tewari, S.K. 2008.Land Use Dynamics, Crop Diversification and Food Security in India; *Proceedings of training Course*, Crop Diversification for Sustainable crop Production, Nov. 8th 2008. Department of Agricultural Economics, College of Agriculture, G.B. Pant University of Agriculture and technology Pantnagar. p.1-2.

Tewari, S.K. 2008*a*.Land Use Dynamics, Crop Diversification and Food Security in India; *Proceedings of training Course*, Crop Diversification for Sustainable crop Production, Nov. 8th 2008. Department of Agricultural Economics, College of Agriculture, G.B. Pant University of Agriculture and technology Pantnagar. p.5.

Yadav, R.L., Yadav, D.S., Singh, R.M. and Kumar, A. 1998. Long-term effects of inorganic fertilizer inputs and crop productivity in rice-wheat cropping system. *Nutrient Cycling in Agro-ecosystems*. 51: 193-200.

□□□

9

Farming System Approach : Its Principles and Role Towards Sustainability

M.S. GILL

Farming system research has grown to the point where there not only a single method, but rather a number of alternative paradigms are designed for different purposes in response to different institutional, economic or political influences. Indeed, it is one of the strengths of the farming movement which is broad enough to attract around a common set of principles, agricultural research specialists with varying interest; the international or national systems of different kinds; individual corps or entire farm improvement; emphasis on on-farm research or more general concern with research priorities and planning. These common principles include the farmer-centred research; the holistic nature of farming systems; the need for multidisciplinary approach; the importance of on-farm work; and the iterative, continuous nature of agricultural research.

Our country is fortunate to have different types of agro-climatic regions, varying from temperate to tropical and desertic to humid climates alongwith vast seasonal and soil variabilities. Each agro-climatic region in the country has suitability for the best productivity of a certain production system. However, sustaining household food security has been an issue of prime importance to majority of the farmers belonging to the category of small and marginal holdings. Improved technologies, even when considered technically sound, find limited acceptance by majority of resource poor farmers because such technologies are mainly targeted to the resource endowed production systems. To solve the problems of farmers who operate in complex-diverse-risk-prone-environment the need is being felt to develop more holistic, client oriented and interdisciplinary approach of farming system research. It envisages development and transfer of appropriate technologies by keeping in view the farm resources in farmers' participatory mode. The integrated farming system approach is the only way to conserve natural resources and save the farming community from the vicious cycle of malnutrition and poverty. The farming system which is an appropriate combination of farm enterprises viz., cropping system, livestock, fishery, forestry, poultry and the other means available with the farmers to integrate them for better profitability. It interacts adequately with the environment without dislocating the ecological and socio-economic balance on one hand and attempts to meet the national goals on the other (Ahlawat and Sharma, 1987).

With the globalization and economic integration of nations, the trade and agriculture would be governed by new rules of the game. The metamorphic changes taking place globally would require a paradigm shift in our approach towards farming system. India, with its vast genetic resource base, undoubtedly possesses enormous potential, which if exploited appropriately, would prosper the country at globally dominating position in agriculture and trades, provided we analyze our true strength with respect to huge genetic resources, diverse agro-climate, cheap labour and enterprising farmers.

At present, some weaknesses such as fragmented land holding, low level of mechanization, poor-economic condition of farmers, illiteracy, low use of inputs, largely un-irrigated cultivation and traditional farming systems warrant attention of planners, researchers and extension functionare to develop and promote integrated farming system models, which may prove as an effective tool for the upliftment of farming community by increasing their per unit productivity levels and also by increasing the employment avenues.

Farming system is designated as a set of agricultural activities organized into functional unit(s) to proficiently harness solar energy while preserving land productivity, environment quality, maintaining desirable level of biological diversity and ecological stability. It aims for higher productivity, profitability, sustainability, balanced food, clean environment, recycling of resources and income round the year together with capability for adoption of new technology, solving energy, fuel and fodder requirement. It further attempt to deal with the issues of deforestation, employment generation, literacy rates, input-output efficiency, enhanced opportunity for agriculture oriented industries and standard of living of the farmers.

A judicious mix of two or more of these farm enterprises with advanced agronomic management tools may compliment the farm income together with help in recycling the farm residues. The selection of enterprises must be based on the cardinal principles of minimizing the competition and maximizing the complementarity between the enterprises. In general, farming system approach is based on the following objectives :

- To develop farmhouse hold systems involving rural communities on a sustainable basis
- To improve efficiency of farm production system
- To raise the farm and family income
- To look after welfare of farm families through satisfying their basic needs

Embedded general principle is a essential five-step procedure for farming system research is explained below :

- *Classification* : Classification is concerned with the identification of homogenous group of farmers with similar natural and socio-economic characteristics. It forms the basis for the setting of priorities and for targeting of research and extension to particular farm types. Its importance is reflected in the economist' concern with "recommendation domains" and in the agronomists' concern with "representative trial sites".
- *Diagnosis* : Diagnosis deals with identifying the limiting factors and development opportunities of particular target farm types identified at the classification stage.
- *Generation of recommendations* : Recommendations involves drawing on the existing body of knowledge, but in most field situations it is likely to involve experimentation, either at the farm level or at the research station or at both, as prerequisite.
- *Commitment to implementation* : Commitment is usually found in farming systems programmes often directly in the form of seed multiplication for example, but also directly through support to the extension service.
- *Evaluation* : Evaluation is an important component of all programmes reflecting the importance attached to continuous contact with the farmers. Evaluation will normally lead to reappraisal and therefore to a reinitiation of the cycle.

These five principles amply evinced the implementation procedure of farming system research, which is open ended and iterative.

Farming system approach vs. Farm Community

- It is the only approach involved in developing farm household systems.
- It is built on the principles of productivity, profitability, stability and sustainability.
- It complements to component-oriented approach to development
- It involves active participation of rural communities.
- It emphasizes understanding of farmhourse hold community inter-linkage.
- Farming System Approach (FSA) reviews constraints and assesses potentials.
- FSA combines improvements desired from better technology
- FSA needs efficient support services and requires better plicies.
- It is continuous, dynamic and iterative learning process based on analysis, planning, testing and evaluation

Farming system approach strategy

In view of limitations on horizontal expansion of land and agriculture, the only alternative left is for vertical expansion through various farm enterprises requiring less space and time but giving high productivity and ensuring periodic income specially for the small and marginal farmers located in rainfed areas, dry land arid zone, hilly areas and tribal belts.

Land is the major resource in India, which provides livelihood to the rural population. However, about 11.3 per cent of the rural families are landless with over 69.4 per cent land holders own less than 1 ha and 21.3 per cent own between 1 and 2 ha. These small land holders are not able to generate remunerative employment throughout the year. Likewise, water is another critical resource required to enhance the agriculture production. Out of the total arable land of 169 million ha, only 28 per cent area is under irrigation and rest is dependent on rainfall cropping area is located in low rainfall regions where the employment opportunities exist hardly for 40-50 days in a year. Similarly, livestock is another important resource owned by the poor. Generally, the rural families maintain 1-2 cows or buffaloes for milk, a pair of bullocks for scarce, most of the poor families maintain large herds by grazing on common lands.

Trees play vital role in Indian agriculture. Besides their productive role in five Fs, i.e. forage, fuel, fruits, food and fertilizer, tree render numerous services like conservation of soil and moisture, provide shade and shelter, live fencing and barriers, protection of the environment etc.

Keeping in view all these components of factor productivity, the following farm enterprises can be combined easily with the existing resources.

Agri - Livestok

Agri - Livestock-Poultry
Agri - Horticulture Systems
Agri - Horticulture- Sericulture Systems
Agri – Forestry –Silvi,- Pastoral System
Rice - Fish
Rice – Fish + Mushroom
Floriculture + Apiary(bee keeping)
Fishery + Duckery + Poultry
Fishery + Piggery + Poultry

For meaningful execution of integrated farm enterprises, the following steps should be undertaken by a multidisciplinary team and extension professional together with farmers participation and involvement at all stages:

- Through understanding of existing farming systems and their components.
- Assessment of resource availability in the farm environment and identification of bio-physical, socio-economic, institutional, administrative and technological constraints.
- Development of economically viable and efficient integrated farming systems suitable for various domains.
- Diffusion of improved technology and receiving feed back for further improvement of the system as a whole.
- Continuous improvement in component technology to fit into a given farming system
- Improvement in quality of farming system
- Research-extension linkage through "On-farm Adaptive Research"
- Development of national and international linkages in accordance with WTO regime.

Present status of IFS in India

India's post independence progress in food grains production has rightly been recorded as from begging bowl to the bread basket. Development of various agronomic production techniques in Mexican wheat varieties for high yield realization such as shallow seeding at 3-5 cm depth, delay in sowing from mid October to mid November, first irrigation at CRI Stage and adequate and balanced NPK fertilization had resulted in a quantum jump (45% increase) in wheat production from 11.4 mt in 1966-67 to 16.54 mt in 1967-68 (Singh, 2001). Notwithstanding these achievements, the problems of food security, poverty and equity and sustainability are continuing. Thus, there is an urgent need to exploit the natural resources rationally and start a new era of integrated farming for increasing the factor productivity.

In areas, where rice is the predominant crop during rainy season, less area, except rice fields bunds (hugh elevation), are available for forage cultivation, which comprises about 5-10 per cent of the total rice area of the country (Thakuria and Gogai, 2000). Cultivation of high yielding ecologically suited forage species may increase the production of quality forage for animals without affecting the crop productivity. In Assam, Setaria (cv. Kazungula) grass grown on rice bunds gave 18.6, 44.5, 52.7 and 63.1 per cent more yield over setaria (cv. Nandi) hybrid bajra, para (local species grown by the farmers) and guinea, respectively.

At Banasthali (Rajasthan) growing of crops under inter row space of fruit plants during establishment years, it was observed that grain yield of wheat and pod yield of groundnut followed the order : anola> pomegranate>guava>jamun>ber (Singh, 2000), Similarly, drip irrigation helped to increase the grain yield of wheat by 10.8 per cent over furrow irrigation method in fruit plants. The corresponding increase in pod yield of groundnut was 15.8 per cent. Gupta *et.al* (1999) conducted a field study at Kalyanpur village in Rajasthan to green the sandy wasteland. They used an integrated farming system approach involving agri-horticulture, silvipasture, energy plantation, arable and other unconventional economic crops. They found all these new measures useful for meeting the desired goals. The adoption of an integrated farming system with cropping, rainfed fruit trees and goat rearing on drylands at the Regional Research Station, Aruppukottai (Tamilnadu) resulted in a considerable increase in income of small and marginal farmers. The crops grown were cotton, black gram, sorghum, cowpea and fodder, while the fruits include ber, custard, apple and an amla (Senthilvel *et al.*, 1998). Prasad (1993) gave guidelines to identify agricultural, horticultural, livestock and agri-horti-silvi pastural based farming system research approach for the NEH hill region of India. Growing crops with fruit trees are highly economical and liable to minimize the risk factor occurring with sole cropping under low rainfall. Trees help in nutrient pumping from lower soil strata to the crop root zone. Gill *et al.* (2000) reported that mango occupies largest area (43%) among fruit trees in the country. Except in arid and temperate condition, mango can be grown successfully in all agro-climatic conditions. During the initial phase of plantation, its inter-space can be successfully exploited by growing leguminous cowpeas fodder. Additionally, 200 plants/ha of *Subabul* (4 month old seedling) were grown in mango plant inter space (10 x 10 m). *Subabul* was cut for fodder and fuel every year after the harvest of cowpeas fodder crop. Growing of cowpeas in the inter row space of mango orchard proved to be an advantageous proposition.

Another farming system approach is agroforestry which is used for different land use systems and technologies where woody perennials (trees, shrubs, palms, bamboo etc.) are deliberately used on the same land management unit as agricultural crops and/or animals in some form of spatial arrangement or temporal sequence. Gill and Burman (2001) defined agroforestry system as dynamic, ecologically based natural resource management system that, through the integration of trees in farmland and rangeland, diversifies and sustains production of small land holdings for increased social, economic and environmental benefits.

The unique features of agroforestry which distinguished it from other farming systems is the deliberate introduction of trees into the landscape, these trees are not the conventional timber producing trees of forestry but rather the multipurpose tree species of agroforestry. The tree has the potential to maintain sustainability and improve productivity of farm lands in different agro-climatic regions. Dwivedi (1992) suggested combination of agroforestry components in different region of India.

Watershed management

Watershed management in itself is a farming system approach on individual as well as community basis. It is a scientific approach and now has become a medium of central and state, Governments for planning of rural development schemes. It includes conservation, development and optimum utilization of natural resources like soil, water, vegetation etc. to reduces soil erosion, moderate floods, alleviate drought, improve water availability for increasing food, fodder, fuel, timber and fibre production on sustainable basis. Besides, it also helps in employment generation, development of human and other economic resources of village to promote savings and income generation activities. The main aim of watershed is to manage the land and water resources for sustained production. In watershed management programme, an action plan is formulated based on information collected through participatory rural appraisal exercise on topography, slope, soil, climate, vegetation, land use, socio-economic status of farmers, conservation measures at places, need of fodder, fibre, fuel, food etc. (Shakir *et al.*, 2000).

Technologies for adoption in watershed

- Land development including *in-situ* soil and moisture conservation measures (contour bunds, graded bunds, bench terraçing).
- Drainage line treatments through mechanical, bio-engineering and bio-mechanical measures like spillways at gully heads.
- Development of small water harvesting structures like low cost farm pond, gully bund, stock dam etc.
- Construction of percolation tank.
- Plantation of fodder, fuel-wood, timber and horticultural trees on community land, degraded land, government land, private land and at the sides of road etc.
- Pasture development alone or in combination with plantation
- Development of agroforestry.
- Crop demonstration for popularizing new crops, varieties or innovative management practices.

Thus, farming system approach with a focus on precision farming should be aimed to meet the challenges of food requirements in the next millennium. Integrated nutrient, water and pest management is the key for managing the natural resources and evolving location specific technologies, involving different viable agri-based entrepreneurs (dairy, poultry, piggery, goatry, fishery, bee-keeping, mushroom cultivation, growing of fruits, vegetables, timber tree etc.) for small and marginal farm families. This approach needs to be appropriately strengthened for achieving ever green revolution.

References

Ahlawat I.P.S. and Sharma, R.P. 1987. Agronomic terminology, Indian Society of Agronomy, IARI, New Delhi pp, 66.

Dwivedi, A.P. 1992. agro-forestry Principles and Practices. Oxford and IBH Pub. Co. Pvtg. Ltd., New Delhi, India, pp 60-72.

Gill, A.S. and Burman, D. 2001. Nutrient management in Agro-forestry Systems. *Fer. News* **46** (3): 25-28 and 31-36.

Gill, A.S. Deb, R. Roy and Ajit 2000. Forage production potential of cowpea in mango orchard. *Indian Farming* **50** (1) : 17-18.

Gupta, J.P., Rathore, S.S. and Sharma, B.M. 1999. Greening sandy wastelands the Kalyanpur experience. Proc. Symp. held in India on Recent advances in management of arid ecosystem. Pp. 499-504.

Hagde G. and Narayan 2001. Food and environmental securities rural poor should be the key player. *Indian Farming* **50** (10) : 20-23.

Prasad R.N. 1993. Farming system approach a solution for sustainable agriculture for hill. Proc. of national seminar on resource management for hill agriculture held at Barapani, Meghalaya. *Indian Journal of Hill Farming* **6** (1) : 19-27.

Senthilvel, T., Latha K.R. and Gopalasamy N 1998. Farming system approach for sustainable yield and income under rainfed verti sols. *Journal Madras Agri.* **85** (1) : 65-67.

Shakir Ali, Singh, K.D. and Prasad, S.N. 2000. Watershed management a strategy for conservation of nutural resources and sustained bio-mass production. *Indian Farming* **50** (7) : 30-33.

Singh, Y.P. 2000. Production of groundnut and wheat under interspace of fruit plants grown with two irrigation methods. *Indian Farming* **50** (9) : 12-14.

Singh, Punjab 2001. Sustainable crop production-limitations and strategies. *Indian Farming* **50** (10) : 7-10.

Thakuria, K and Gogoi, P.K. 2000. Possibility of growing perennial grasses on rice field bunds. *Indian Farming* **50** (4) : 28-29.

□□□

10

Crop Diversification for Profitability in Jute and Allied Fibre Crops

M.K. SINHA, SABYASACHI MITRA, T. RAMASUBRAMANIAN AND B.S. MAHAPATRA

Jute (*Corchorus capsularis* L. & *C. olitorius* L.) is the main commercial crop of the Eastern and North Eastern India providing livelihood security to about 5.0 million people comprising of 4.0 million farmers, 0.25 million mill workers and 0.50 million people engaged in jute based ancillary sectors. It is grown in an area of little over 0.8 million hectare, producing nearly 10 million bales (1 bale @ 180 kg) of fibre, which is about 40 percent of the worlds' production. Mesta, (*Hibiscus cannabinus* L. & *H. sabdariffa* L.) is grown in an area of 0.15 million ha with a production of 1.0 million bales. The major jute growing states are West Bengal, Bihar, Assam, Orissa, Uttar Pradesh, Meghalaya and Tripura while mesta is mostly cultivated in Andhra Pradesh, Maharastra, Orissa and Bihar. Sunnhemp (*Crotalaria juncea* L.) is cultivated mostly in Bihar, Uttar Pradesh, Madhya Pradesh, Rajasthan, Maharastra and Tamil Nadu. Since independence, there is a two fold increase in productivity of jute and mesta which has been possible mainly due to the development of high yielding varieties along with improved production technologies and also due to proper policy support from the government. Despite tremendous increase in productivity, there is still a wide disparity in the yield level of the crop among the states as well as the agroclimatic zones of the same state. For example in jute, among the 87 jute growing districts, only 9 have the productivity above the national average (22 q/ha) and the rest are well below that. Besides, the area of jute, mesta and sunnhemp is confined in a few states and has not spread much, though they are of commercial importance. Similarly the acreage of other fibre crops like ramie and sisal has not increased substantially though ramie and sisal fibres are several times costlier than raw jute (jute and mesta), sunnhemp and cotton. Jute and allied fibre production under the present system of cultivation is labour intensive and costly, around 65-70% of the total cost of production being spent for weed management and retting. As a result, farmers are not getting better price which is definitely influencing their choice for the crop. So to enhance the profitability of jute and allied fibre farming, we have to introduce high value crops as components of jute based intercropping or multiple cropping. Besides, it is imperative to introduce the cultivation of jute and allied fibres to non-traditional areas. The ICAR funded Network Project on ramie, clearly showed that ramie crop can be grown successfully and profitably in the non-traditional areas like Nilgiri hills, Goa and Dapoli, both a sole crop and as an intercrop.

Crop diversification in jute, as in other crops, may be achieved through several ways such as:

- Single crop to multiple / mixed crop
- Introduction into non-traditional areas
- Effective utilization of less productive lands with remunerative crops
- Processing and value addition

Approaches for crop diversification in jute and allied fibres

Single crop to multiple crops

During the initiation of first five year plan, total acreage under jute was only 0.26 million hectares with a total production of o.7 million bales of raw jute. The jute varieties available during that period, i.e., D154 (*white* jute) and Chinsurah Green (*tossa* jute) were of 150-160 days duration with yield potential of 15-18 q ha^{-1} only. The increase in irrigation potential and subsequent advent of intensive agriculture and introduction of multiple cropping systems exposed jute crop to stiff competition from other major crops and it was necessary to fit jute crop in multiple cropping sequences and make it more remunerative through the development of short duration, high yielding varieties with judicious manipulation of the agrotechniques like advancing sowing time; optimising planting geometry, fertiliser dose, plant protection measures, retting techniques etc. Development of pre-mature flowering resistant varieties, viz., JRO-878, JRO-7835 and JRO-524 and reducing the crop duration from 150 to 120 days, not only advanced the sowing time to as early as middle of March, but also enabled jute to be fitted in multiple cropping sequence with transplanted Aman paddy. This led to a paradigm shift in jute agriculture as the entire jute growing area of the country was taken over by the *tossa* jute. (Das and Maiti, 1998).

On rainfed uplands, jute-lentil, jute-blackgram, jute- greengram, jute-toria, jute-blackgram-toria/wheat, jute-rice-toria, jute-mustard-cowpea sequences were found to be promising in West Bengal, Assam, Orissa, Bihar and Uttar Pradesh, whereas in rainfed lowlands, jute-paddy-lentil sequence was found promising in Bengal, Bihar, Orissa and Uttar Pradesh. Under assured irrigation facilities, jute-paddy-potato, jute-rice-potato/wheat/pea, jute-rice-mustard, jute - paddy - vegetables, etc. have shown promising results in upland condition (Table 1) (Mitra *et al.*, 2006).

Table 1. Jute based cropping sequences for different land situations

Land situation	Crop sequences	Region
Rainfed uplands	Jute-lentil jute-blackgram jute- greengram jute-toria	North and South Bengal
	Jute-blackgram-toria/wheat jute-rice-toria jute-toria	Assam
	Jute-mustard-cowpea	Orissa, Bihar, UP
	Mesta - horsegram	AP, TN
Irrigated uplands	Jute-rice-potato jute-rice-vegetables jute-rice-mustard Jute-rice-potato/wheat/pea	North and South Bengal,
	Jute-rice-wheat	Assam
	jute-radish jute-onion jute-berseem	Katihar, Bihar
Rainfed lowlands	Jute-rice-lentil	Bengal, Bihar, Orissa, UP

Source : Kundu and Mandal (1995), Roy and Choudhary (1989), Mitra et al. (2006)

Under irrigated condition, the net return and benefit : cost ratio were found maximum in jute-paddy-tomato (Rs. 38,898/- & 1.53) followed by jute -paddy-mustard (Rs. 25,191/- & 1.43), jute-paddy-pea (Rs. 24,453/- & 1.35) and jute-paddy-rajmas (Rs. 23,597/- & 1.35) in farmer's field at North 24 Pargana district of West Bengal. Thus the introduction of high value crops in the conventional jute-rice cropping sequence increased the profitability of the system by Rs. 4,825 in case of rajma to Rs. 19,436 in case of tomato (Table 2) which implies that choice of component crop is very important (Chapke and Jha, 2006). Growing sunnhemp as green manure crop in *kharif* rice in jute - rice-potato, jute - rice - lentil and jute - rice - mustard cropping sequences at

Coochbehar, West Bengal resulted in an additional return of Rs. 2,944, 1,555 and 2,662 per hectare, respectively (Table 3) (AINPJAF, 2005). Mixed cropping of jute, red amaranth, white amaranth and summer raddish produced 36 q jute fibre, 19-29 q red amaranth (at 21 DAE), 7-9 q white amaranth and 5.7-9 q summer raddish per hactare under irrigated condition and fetched additional net return of around Rs. 4,000/- per hactare (Ghorai, 2007). Intercropping of jute + groundnut or jute + blackgram increased the fibre yield of jute with an additional grain yield of 3-4 q ha^{-1} from legumes. Intercropping of groundnut, soybean, or blackgram with roselle (*Hibiscus sabdariffa* L.) in 1:2 or 1:3 row ratios was found more remunerative than sole roselle when the intercrop was sown 15 days after the main crop. Similarly at Amadalavalasa, (A.P.) intercropping of roselle with redgram, maize and cluster bean increased the economic yield of the system under rainfed condition with maximum system yield (roselle equivalent yield) being observed in roselle + maize (3:1) intercropping (33.16 q ha^{-1}) followed by roselle + cluster bean (3:1) (32.08 q ha^{-1}) intercropping system, respectively (Table 4) (AINPJAF, 2008).

Table 2. Economics of jute based crop sequences in south Bengal under irrigated condition

Crop Sequences	Cost of cultivation (Rs ha-1)	Gross return (Rs ha^{-1})	Net return (Rs ha^{-1})	Benefit : Cost
1. Jute-Paddy-Tomato				
Jute	25282	38568	13286	1.53
Paddy	18945	25121	6176	1.33
Tomato	29752	49189	19436	1.65
Total	73979	112878	38898	1.53
2. Jute –Paddy-Mustard				
Jute	25143	37315	12172	1.48
Paddy	18901	25176	6275	1.33
Mustard	13896	20640	6744	1.49
Total	57940	83131	25191	1.43
3. Jute-Paddy-Pea				
Jute	25118	37895	12776	1.51
Paddy	19645	25207	5562	1.28
Pea	25354	31469	6115	1.24
Total	70117	94571	24453	1.35
4. Jute-Paddy-Rajmas				
Jute	25642	37928	12286	1.48
Paddy	19209	25695	6486	1.34
Rajmas	22480	27305	4825	1.21
Total	67331	90928	23597	1.35

Source : Chapke and Jha (2006)

Intercropping of maize with mesta (1:4) and strip cropping of blackgram with mesta (4:4) produced 31 q equivalent fibre yield (24 q fibre + 7692 cobs ha^{-1}) and 29 q equivalent fibre yield ha^{-1} (14 q fibre + 7.52 q blackgram seed ha^{-1}), respectively under rainfed environment. The sole crop (mesta) fibre yield was 20 to 21 q ha^{-1} in the similar environment (Ghorai *et al.* 2008).

Jute, on an average, adds about 15 t ha^{-1} of green leaves into the field, which enriches the organic matter as well as available nutrient status of the soil for utilization by the succeeding crop. Results of the 'Long term fertilizer trial' in jute – rice – wheat cropping sequence in new gangetic alluvial soil at

Table 3. Economics of cultivation of jute based crop sequences at Coochbehar, W.B.

Cropping sequences	Cost of cultivation (Rs ha^{-1})	Gross return (Rs ha^{-1})	Additional return* (Rs ha^{-1})
Jute – paddy + sunnhemp – potato	25,455	44,129	2,944
Jute – paddy + sunnhemp – lentil	25,309	43,082	1,555
Jute – paddy + sunnhemp – mustard	24,842	40,286	2,662
Jute – paddy– potato2	5,023	41,185	-
Jute – paddy– lentil	25,071	41,527	-
Jute – paddy– mustard	24,518	37,624	-

Source : AINPJAF (2005)
NB: * - over corresponding crop sequences without sunnhemp

Table 4.Effect of row ratio on yield attributes and yield of mesta-redgram intercropping systems

Treatment No.	Mesta fibre Yield (q ha^{-1})	Intercrop yield (q ha^{-1})	Mesta eqivalent yield (q ha^{-1})
Mesta sole crop	28.51	-	28.51
Redgram sole crop	-	4.67	8.73
Mesta : redgram (2:1)	19.61	1.54	22.51
Mesta : redgram (3:1)	25.13	1.23	27.44
Mesta : redgram (1:2)	13.92	2.06	17.83
Mesta : maize (2:1)	21.20	19,500*	28.61
Mesta : maize (3:1)	28.41	12,500*	33.16
Mesta : Cluster bean (2:1)	20.57	5.15**	23.82
Mesta : Cluster bean (3:1)	30.23	2.95**	32.08
CD(P=0.05)	2.86	-	2.99

- NB: no. of cobs for maize; ** - green wt. for cluster bean
Source : AINPJAF (2008)

Barrackpore revealed that the soil organic carbon after initial depletion showed an increasing trend after 36 years of continuous cropping when FYM was applied along with NPK (Saha *et al.*, 2008). Replacement of rice - rice - fallow crop sequence by jute-rice-jute sequence maintained better organic Carbon status of the soil in Orissa (Sadanandan and Mahapatra, 1975), while total P in the soil increased after each cycle in jute-rice-rice and groundnut - jute - rice sequences (Sadanandan and Mahapatra, 1972). In soybean - jute - rice sequence, the net gain of total N was 27 kg ha^{-1}, while in fallow - jute - rice system total N showed a deficit balance in Bangladesh (Patwary *et al.*, 1989). Thus the fertility status of soil can definitely be improved through judicious selection of crops and crop sequences along with appropriate management practices.

Introduction of a new crop (ramie) in non-traditional areas

Introduction of new crops in non-traditional areas is an effective way of crop diversification. The adaptability of the crop in the new agroclimatic conditions and the profitability of the crop / cropping system over the existing ones are the key to their acceptance. It is essential to develop the complete package of practice and to demonstrate the performance of the system to the farmers for its successful introduction in that area.

Ramie fibre is the longest, strongest and costliest textile fibre of plant origin and has a huge demand in the international market, but the supply of this fibre is well below the demand. Globally the trade of this high value fibre is dominated by China, as it ranks first regarding the area (0.085 million ha), production (0.133 million Mt) and productivity (15.37 q ha^{-1}) followed by Brazil and Phillipines. In India, there is a huge potential for ramie cultivation as our ecological conditions are quite congenial for its growth and development. However, ramie cultivation in India, as on date, is confined in a small area of about 200 ha in the North Eastern Hill states like Assam, Meghalaya and Nagaland and has not spread as it should have though the economics of ramie cultivation has been found to be quite impressive in northeastern region. The cultivation of sole ramie in this region fetched a net return of Rs. 25,000 to Rs. 35,000 ha^{-1} year^{-1} during 2nd to 4th year while in the 5th year the net return was around Rs. 3,00,000 ha^{-1} year^{-1} of which Rs. 15,000 to Rs. 17,000 ha^{-1} year^{-1} came from sale of fibre and the rest from the sale of rhizome (around 200 q ha^{-1}).

The edapho-climatic conditions of Nilgiri hill areas, Goa, Dapoli and other areas of the region has been found suitable for successful cultivation of ramie provided suitable location specific agro-techniques are available. Ramie has been introduced successfully in these areas by CRIJAF and ramie based multi-tier cropping system was also established in these zones. Ramie + rubber and ramie + cinnamon intercropping in Tamil Nadu, ramie + coconut and ramie + coconut + black pepper in Goa, ramie + mango and ramie + mango+ black pepper at Ratnagiri, Maharashtra as well as ramie + arecanut at Assam were particularly found to be encouraging (Table 5).

Table 5. Remunerative ramie based cropping systems identified

Area	Cropping systems
1. Traditional area	
Assam	Ramie + areca nut
2. Non-traditional areas	
Tamil Nadu	Ramie + rubber, Ramie + cinnamon
Goa	Ramie + coconut, Ramie + coconut + black pepper
Maharastra	Ramie + Alphanso mango, Ramie + Alphanso mango + black pepper

Source : CRIJAF - Final Report, Ramie Network Project, 2008

The experimental results under ICAR Network Project of Ramie revealed that cultivation of ramie alone resulted into an additional profit of Rs. 8112 over sole areca nut cultivation (Rs. 25,138 ha^{-1} $year^{-1}$) while the net profit from ramie + areca nut cultivation was Rs. 38,807 ha^{-1} $year^{-1}$ more than that from sole areca nut in Assam (Table 6). Similarly at Goa, sole ramie plantation recorded an additional profit of Rs. 28,980 ha^{-1} $year^{-1}$ as compared to sole coconut plantation. However, in coconut + black pepper + ramie intercropping system recorded slightly lower return of Rs. 40,270 ha^{-1} $year^{-1}$ as compared to sole ramie but it was considerably higher than sole coconut. Thus introduction of ramie increased the profitability of the cropping system both in the traditional and the non-traditional areas (Table 7) (CRIJAF, 2008).

Table 6. Economics of ramie + areca nut multi-tier cropping at Sorbhog, Assam

Cropping systems	Net return* (Rs ha^{-1} yr^{-1})
Sole areca nut	25,138
Sole ramie	33,250
Ramie + areca nut	63,945

Source : CRIJAF - Final Report, Ramie Network Project, 2008. * - mean of 3 years

At Barrackpore which is situated in southern part of West Bengal, application of irrigation based on IW/CPE ratio of 0.6 to 0.9 during pre- and post monsoon periods produced 15-17 q fibre ha^{-1} in 5 cuts as compared to

10-12 q fibre ha^{-1} in 3-4 cuts under rainfed condition in North Eastern States like Assam and Nagaland (Mitra and Sarkar, 2006).

Table 7. Economics * of coconut + black pepper + ramie multi-tier cropping at Goa

Cropping systems	Net return* (Rs ha^{-1} yr^{-1})
Sole Coconut (no.)	28,140
Sole ramie	57,120
Coconut + Black pepper + Ramie	40,270

Source : CRIJAF - Final Report, Ramie Network Project, 2008. * - mean of 3 years

Effective utilization of less productive lands

Introduction of sisal in the marginal and drought prone areas

Agriculture in the dry land areas of the country is risk prone with poor return due to the climatic uncertainty, which often endangers the livelihood security of the resource poor farmers of the region. In these areas, particularly poor rainfall along with very high temperature range, make majority of the lands unproductive or less remunerative for arable crops. Sisal due to its xerophytic nature can adapt well in such situation and with certain management gives adequate fibre yield, the fibre being extracted from its leaves through decorticator machine. The sisal fibre can be successfully utilized for manufacturing of diversified products and introduction of sisal in these semi-arid drought prone areas of our country may have a huge positive impact on the socio-economic condition of the people. Western Orissa, is a typical drought prone area where agriculture is mostly non-remunerative. The case study of sisal based agroforestry models established at Sisal Research Station, Bamra, situated in the western part of Orissa, revealed that both *Agave sisalana* and Hybrid sisal recorded better growth and yield when planted in double row system in between two rows of teak (*Tectona grandis* L.) or gambhar (*Gmelina arborea* L.) at a planting density of 4742 plants ha^{-1} for sisal and 470 trees ha^{-1} for teak or gambhar, respectively. Similarly sisal could be introduced successfully at in exisiting and / or new teak plantations at Amadalavalasa, Andhra Pradesh at similar planting density and application of N: P_2O_5: K_2O @ 120: 30: 60 kg ha^{-1} recorded significantly higher number of sisal leaves with increased length and breadth (AINPJAF, 2008).

Sisal crop, on an average, produces 1.5 to 2.0 t ha^{-1} of dry fibre, which is extracted from the sisal leaves starting from 3rd year and continues upto 12th year. Besides, sisal crop also produces around 100 t ha^{-1} of green biomass which can be recycled after composting and the fertilizer cost can be significantly reduced. The long term experiment on sisal cultivation at Sisal Research Station, Bamra, Orissa revealed that the net income from a well managed sisal plantation of 4 hactare area was around Rs. 1,00,000 to 1,20,000 per year with an additional earning of Rs. 50,000 at the end of 11th or 12th year from the sale of bulbils.

Value addition to end products: Golden incentive for diversification in future

Jute, recently, is facing stiff competition from cheaper synthetic fibres, particularly in the sacking and packaging sector and it is indeed a matter of serious concern to us. But the silver lining is that the consumers today, particularly in the developed countries, are more concerned about the environment and prefer products from natural sources. So to reap the benefit, it is the high time to think for product diversification and value addition in jute instead of sticking to only hessian and sacks. Blending of jute fibres with ramie, cotton, or other natural and synthetic fibres for apparel making, jute or jute and sisal blended geo-textiles, jute based handicraft items, etc. should be the focal items as these products have huge market both within and out side the country. The export earning of jute based products has increased from 233 crores in early eighties to 1200 crores in 2005-06. The market share of value added jute diversified products (JDP) has increased from 2% in mid eighties to 18% in 2005-06. The present international market potential for jute geo-textiles (JGT) alone is 95 billion US $ which must be tapped. The paper and newsprint production is around 5.8 million MT and demand is likely to increase to 8.53 million MT in the year 2010. The whole jute / kenaf plants having similar strength of hard wood, have emerged as potential pulp and paper material and can meet the shortage of raw material. A farmer cultivating a kenaf variety MT-150 for paper pulp in place of fibre, can earn an additional profit of Rs. 40,000 ha^{-1} within a duration of 140 days.

The ramie fibre produced in traditional and non-traditional areas may go to hand spinning and weaving sectors in the rural areas and the rest to major industries. This will create a significant amount of employment generation as well as infrastructure build up. The huge amount of biomass wasted during decortication (fibre extraction) can be effectively utilized for paper pulp making, composting, bio-gas generation, etc., which will certainly lead to further value addition. Ramie produces significant quantum of leaves which has high nutrient content and can be a potential source for natural anti-oxidants, food supplements, etc. or can be successfully used for developing animal feed. The gum content in Indian ramie fibre ranges from 18 - 29 per cent. The spent liquor after microbial degumming of ramie fibre is a rich source of pectinolytic enzymes, and liquid or solid enzyme formulations can be prepared from it which has tremendous use in textile and food processing industries. The semi-solid waste after degumming can be utilized for preparing biofertilizers.

Similarly in sisal, the excellent characteristic features of the fibre enables it for manufacturing of diversified products like ropes, fancy fibre articles, carpet yarns, reinforcement materials, etc. The huge amount of sisal biomass can also be effectively utilized for generating different value added products, as in case of ramie.

So to make the cultivation of fibre crops profitable and to create a sound market for natural fibres as well as for natural fibre based diversified products we have to adopt a holistic approach adaptation of location specific good agricultural practices for enhanced productivity and improved quality. Further there is need to develop value chains for these natural fibres which, once established, will certainly assure the supply of quality ramie fibre and / or yarn to handlooms and powerloom sectors; create market for fibre, yarns, etc., at local or regional level as well as for the diversified products at national and international levels; and will also create downstream industries leading to entrepreneurship development in rural, semi-urban and urban areas. Besides, these chains will connect the growers of these fibre crops with the end users so that profit of value addition at the end point of the chain can effectively infiltrate to the farmers. Such successful value chains can be the best models for diversification for jute and allied fibre crops in both traditional and non-traditional areas.

Acknowledgement

The author is thankful to Director, CRIJAF for all his support.

References

AINPJAF. 2005. Annual Report, pp. 40.

AINPJAF. 2008 Annual Report (in Press)

Chapke, R. R. and Jha, S. K. 2006. Introduction of jute based multiple cropping sequences in farmer's field under irrigated conditions. In: *CRIJAF Annual Report*, pp. 105-106.

CRIJAF. 2008. Final Report of ICAR Network Project on Development of ramie (*Boehmeria nivea* L. Gaud.) for different agro-climatic conditions and technological improvement of the fibre for textile, 77 p., Barrackpore, Kolkata, India.

Das, B. B. and Maiti, S. N. 1998. Jute (*Corchorus species*) and allied fibres research in India. *Indian Journal of Agricultural Sciences* 68 (8, Special Issue): 484-493.

Ghorai, A. K. 2007. Weed management of jute (*Corchorus olitorius*) by soil solarisation. *Indian Journal of Agricultural Sciences* 77 (6): 390-2.

Ghorai, A. K., Thokle, J.G., More, S. R., Saha, S. and Nawale, P.A. 2008. Inter and strip cropping of maize and blackgram with mesta in Maharastra. In: *Progress Report, CRIJAF (Technology Mission on Jute)*, pp. 41.

Kundu, A. L. and Mandalm N. C. 1995. Intensive cropping system in tarai zone of West Bengal. In: *Agroecosystems Management Proceedings of National Seminar at Viswa Bharti*, Shantiniketan, West Bengal, pp - 161-165.

Mitra, Sabyasachi and Sarkar, Sitanshu. 2006. Improved water management in ramie. In: CRIJAF Annual Report, pp. 72-73.

Mitra, Sabyasachi, Maiti, S. N. and Sarkar, Sitanshu. 2006. Recommendations for jute and allied fibre crops - An endeavour of All India Network Project. p-28, CRIJAF, Barrackpore, Kolkata.

Patwary, S. U., Haque, Q. and Badruddin, M. 1989. Role of legume on nitrogen balance and A-value of soil under different sequential cropping systems. *Thai Journal of Agricultural Science* **22** (3): 213-221.

Roy, R. K. and Choudhary, R. C. 1989. Production potential and economics of jute based cropping system in mid land situation of Kosi division. *Jute Development Journal* **9** (4): 18-22.

Sadanandan, N. and Mahapatra, I. C. 1972. A study on the effect of multiple cropping on the organic carbon status of upland alluvial soils. *Agriculture Research Journal of Kerala* **13** (1): 33-38.

Sadanandan, N. and Mahapatra, I. C. 1975. A study on the effect of multiple cropping on the phosphorus content of upland alluvial soils. *Agriculture Research Journal of Kerala* **10** (2): 71-74.

Saha, M. N. and Ghorai, D. 2004. Long term fertilizer experiments proves the importance of jute cultivation to enrich soil nutrient status. *JAF News* **2** (2): 13-14.

Saha, M. N., Saha, A. R., Saha, S., Jana, A. K., Maji, B. and Maitra, D. N. 2008. Nutrient management, soil health and sustainable crop production. In: *Jute and Allied Fibre Updates*, pp. 140-161. Karmakar, P.G., Hazra, S.K., Ramasubramanian, T., Mandal, R. K., Sinha, M.K. and Sen, H. S. (Eds). CRIJAF, Barrackpore, Kolkata, India.

□□□

Balancing Nutrition

11

Integrated Nutrient Management — A Need for Sustainable Commercial Production

A.S. CHANDEL

Integrated nutrient management refers to combined use of chemical fertilizers, organic manures and bio-fertilizers with an express objective of obtaining sustainable and environmentally safe crop production. The concept relies on the sound crop husbandry, harnessing the natural nutrient capacity through bio-engineering and mining of soil reserves. It also stresses on search for more efficient nutrient sources. It is the complimentary use of chemical fertilizers, organic manures and bio-fertilizers.

Impact of fertilizers

Fertilizers are still the 'King Pin' in the present system of commercial agriculture. Scientific and balanced use of fertilizers assume vital importance in commercial and sustainable crop production. Efficient use of fertilizers pay back to the farmers more profit per unit investment. A survey report of FAO (2005) indicated that between 1965 and 1976, fertilizers were responsible for over 50 per cent increase in the crop production in developing countries. This is truly so in India where contribution of fertilizers into total grain production can be placed at 1 per cent in 1950, 32 per cent in 1980 and estimated 53 per cent in 2010. This indicates that fertilizers is the key input in the strategy of achieving target of commercial production.

Changing scenario of nutrient management

Introduction of high yielding varieties with increased use of commercial fertilizers and improved irrigation facilities have resulted in spectacular increase in yield of crops. The recent energy crisis and the consequent price hike of fertilizers coupled with low purchasing power of farmers in mass have again renewed the interest of organic recycling throughout the world. The nutrient management has gone through three phases of development.

Phase I : The period 1950-1965 witnessed the increased awareness of chemical fertilizers along with due consideration for the practice of application of organic sources of nutrients and green manuring.

Phase II : The beginning of 'Green Revolution' in mid sixtees initiated a sudden spurt in the use of chemical fertilizers and gradual decline in organics and green manuring. This is evident from the fertilizer use growth from 7.84 lakh tones in 1965-66 to about 15 million tones in 1995-96 and 20.34 million tones in 2005-06.

Phase III : It can be traced to mid seventies, mid eightees ownward and period since 1992, when oil crisis witnessed globally with shortage of fertilizers, renewed the interest in India in propagation of integrated nutrient management involving greater recycling of organics into agriculture. Dependence on INM got greater interest with the concern of environmental pollution due to heavy use of inorganics for commercial production.

Table 1. Nutrients consumption in India

Nutrient	Consumption (kg/ha)
N	60.86
P_2O_5	22.04
K_2O	7.14
Total	90.04

Table 2. Fertilizer (in terms of nutrient) use in India (2005-06)

Nutrient	Fertilizer use (000t)
N	12723
P_2O_5	5204
K_2O	2413
Total	20340

Source: India Year Book, 2007

Table 3. Fertilizer (in terms of nutrient) consumption zonewise

Zone	Fertilizer consumption (t)
South	4677781
West	4923159
North	5745045
East	2723741

Nutrient need of intensified commercial cropping

Crop production is the most important source of renewable wealth and development. An integrated plant nutrient supply system involving an appropriate mixture of organics, biological N fixation, phosphate solubilizing microbes and need based chemical fertilizers would be crucial for the sustainability of commercial production.

Intensive cropping systems can annually remove 500-900 kg N + P_2O_5 + K_2O/ha per year alongwith substantial quanties of secondary and micronutrients. A paddy-potato-wheat rotation has been shown to remove Fe (4640 g), Mn (1234 g), Zn (615 g), Cu (325 g), B (305 g) and Mo (17.5 g) per ha. Thus, intensive cropping has widened the scope and need of integrated nutrient management.

Intensification of cropping system with greater use of high analysis chemical fertilizers has resulted in the mining of soils leading to deficiencies of secondary (sulfur) and micronutrients (Zn, Mn, Cu, B and Mo).

Facts about integrated nutrients management

- *Plant needs several nutrients for growth and production*: These nutrients can be supplied to the plants through different sources. Fertilizers constitute the main source. Organic manures also supply nutrients but their content is very low. This supply can also be augmented with certain auxillary nutrient generating bio-fertilizers.
- *Integrated nutrient supply emphasis on use of all resources for supply of nutrients* : Fertilizers are essentially chemical in nature and organic manure are largely of bio-origin (animal and human waste, plant residue and by-products). City waste (urban compost and sewage and sludge) represent house-hold wastes and are not strictly of bio-origin. Green manuring of bio-origin is also included in integrated nutrient supply.
- *Fertilizers are more concentrated source for supply of nutrients than organic manures* : To supply any required level of nutrient to a crop, fertilizers can meet the requirement with far less quantity whereas more bulk will be needed to supply that much of the nutrient from the organic manures.

- Organic manures contain several nutrient elements : However, thus far it is their N value which has received the major emphasis: Research results have established that organic manures have the ability to nip several budding nutrients deficiencies especially secondary and micro-nutrients.

Table 4. Average nutrient composition of common organic manures

Organic manure	Primary nutrients (%)			Micronutrients (Mg/kg)					
	N	P	K	Fe	Zn	Mn	Cu	B	Mo
FYM	1.0	0.5	0.90	2600	57	250	2.5	2.1	0.7
Pig manure	1.88	2.13	0.67	1200	50	70	8.9	-	-
Poultry manure	1.89	1.90	1.60	1400	90	210	7.1	5.0	-
Coat/sheep manure 0.50	0.65	0.50	0.03	-	2570	150	61	4600	-
City compost	1.50	0.50	1.0	-	400	560	150	15	9
Sewage sludge	2.40	1.20	0.002	-	2459	262	643	9	0.6
Press mud	1.20	1.96	2.20	1140	94	450	-	-	-
Green manure (*Sesbania*)	2.25	0.37	-	140	17	80	3.2	20	0.2
Azolla	4.03	0.27	1.70	-	-	-	-	-	-

Source: J.C. Katyal (1992)

- *The greatest challenge before India will always remain to produce more food to meet growing demand created by ever increasing population* : In modern agriculture, fertilizers have contributed upto 50 per cent rise in food grain out put. Therefore, axing on fertilizer use will never be a viable strategy. However, there is always a need to maximize the efficiency of fertilizer use which is presently quite low (<50 per cent for N and 10-30 per cent for P).
- *The nutrients needs for higher yields can not be met exclusively through organics and bio-fertilizers* : The nutrients supply from organic manure is not only low but also slow. In integrated nutrient supply, these materials should be considered as help to top up supply of nutrients from chemical fertilizers.
- Organic manure's decomposition products mediates the improvement in soil's physical, chemical and biological environment and thus enhance further the merit of harmonizing the use of chemical fertilizer and natural manure.
- *Integrated nutrient supply approach is economically viable and environment friendly* : However, fertilizers will continue to be the dominant contributor for meeting nutrients requirement of crops in India.

Conclusion

For commercial agriculture production, considering environmental conservation, the integrated nutrient management seems to be the most sustainable system of nutrient management. However, its package may vary from region to region and from farming system to farming system.

□□□

12

Organic Farming for Sustainable Agriculture : Global and Indian Perspective

B.S. MAHAPATRA, T. RAMASUBRAMANIAN AND H. CHOWDHURY

Organic farming is developing rapidly and is now practiced in more than 120 countries of the world. Globally 30.4 million hectares, which constitutes 0.65 per cent of the agricultural land, are currently managed organically. India ranks 33rd in terms of total area under organic cultivation and 88th in terms of agricultural land under organic crops to the total farming area. India has now become a leading supplier of organic herbs, spices, basmati rice, etc to the developed nations. The export realization during 2007- 08 was around 78 million US $ registering a 200% growth over the previous year. Though organic farming is an inevitable phenomenon, it alone cannot feed the billion population of India. The combination of both organic and inorganic is undoubtedly the best option as on today for developing country like India. In order to ensure the food and nutritional security, rather than promoting organic farming *en masse* and universally, it would be desirable to delineate areas carefully for organic farming. In India, there is ample scope for pure organic farming in the rainfed areas, where there is a little or no use of fertilisers and other agrochemicals. The Government of India has taken considerable efforts to promote the organic farming without compromising the food grain production to feed the ever growing population in India. Launching of "National Programme for Organic Production" (NPOP) and the dedication of "National Institute of Organic Farming" (NIOF) were the landmark achievements in the arena of organic farming in India.

Green revolution of India has undoubtedly changed the scenario of food grain production from 'ship-to-mouth' status to self sufficient and surplus status. The food grain production in India has been doubled during the post green revolution period with virtually no increase in net cultivated area; it increased from 95 million tonnes in 1967-1968 to 209 million tonnes in 1999-2000 from the same net area under cultivation (140$\pm$1 m ha) (Fertiliser Association of India, 2004). This marvellous achievement in agricultural production was mainly due to the increased use of inputs like fertilisers, pesticides and farm machinery. As time went by, extensive dependence on chemical farming has shown its darker side. The land is losing its fertility and is demanding larger quantities of fertilizers. Pests are becoming resistant and warrant frequent sprays, which resulted in residues much above the safety levels. Further, stringent standards for non-organic food in European and US markets have led to the rejection of many Indian food consignments in the recent past which brought to the attention of ill-effects of modern agriculture, and paved the way for organic farming.

Organic farming has entirely different perspective between developed and developing countries. Developed countries are endowed with surplus resources, but limited population with high purchasing power. Hence, a consumer in a developed country expects enough food of highest possible quality that has been produced without any serious ecological impact and harmful residues. Thus organic farming may be the best answer to fulfil such expectations in rich countries (Marwaha and Jat, 2004). However, on the other hand, for a resource poor, highly populous country like India, with ever growing demand for food, the primary concern is the enough availability of food at affordable prices with reasonable quality. In this backdrop, considerable efforts have been taken to promote the organic farming in India without compromising the food grain production to feed the ever growing population. This chapter deals with the issues and opportunities in implementing the organic farming in India.

Area under organic farming: global statistics

Organic agriculture is developing rapidly and is now practiced in more than 120 countries of the world. Its share of agricultural land continues to grow in many countries. Globally 30.4 million hectares, which constitutes 0.65 percent of the agricultural land, are currently managed organically. The continent with most organic land is Australia with almost 11.9 million hectares, followed by Europe (7 m ha), Latin America (5.8 m ha), Asia (almost 2.9 m ha), North America (2.2 m ha) and Africa (almost 0.9 m ha) (Source:

SOEL-FiBL Survey 2007). As of 2006, Australia ranked first in terms of area under organic agriculture (12 m ha), followed by China (2.3 m ha) and Argentina (2.2 m ha) (Table 1). Liechtenstein (29.1%) ranked first in terms of percentage share of organic land to the total agricultural land. This was followed by Austria (13.0%), Switzerland (11.8%) and Italy (9.0%) (Table 2). The percentage share of organic land to the total agricultural land in India was 0.3% in 2006 (Willer, 2008).

Table 1. Top ten countries in terms of area under organic farming (as of 2006)

Rank	Country	Area under organic agriculture (ha)	% of organic land to total agricultural land
1.	Australia	12,294,290	2.8
2.	China	2,300,000	0.4
3.	Argentina	2,220,489	1.7
4.	USA (2005)	1,620,351	0.5
5.	Italy	1,148,162	9.0
6.	Uruguay	930,965	6.1
7.	Spain	926,390	3.7
8.	Brazil	880,000	0.3
9.	Germany	825,539	4.8
10.	UK	604,571	3.8

(*Source:* FiBL Survey, 2008)

Table 2: Top ten countries in terms of percentage of organic area to the total agricultural land (as of 2006)

Rank	Country	% share of agricultural land
1.	Liechtenstein	29.1
2.	Austria	13.0
3.	Switzerland	11.8
4.	Italy	9.0
5.	Estonia	8.8
6.	Greece	7.6
7.	Portugal	7.3
8.	Swedon	7.1
9.	Latvia	7.0
10.	Timor Leste	6.9

(*Source:* FiBL Survey, 2008)

Area under organic farming: Indian scenario

Currently, India ranks 33rd in terms of total land under organic cultivation and 88th for agricultural land under organic crops to total farming area. As per 2006-07 data, the cultivated area under certification was 3,39,113 ha. including wild collection area the total area under certification was around 2.8 million ha. (www.apeda.com). Madhya Pradesh has 1.63 lakh ha of area under organic certification followed by Maharashtra (1.15 lakh ha), Orissa (74,585 ha), Jammu and Kashmir (32,541 ha), Rajasthan (24,868 ha) and Kerala (14,744 ha) (Table 3). Among the state governments, Madhya Pradesh took early lead in organic farming and Uttarakhand followed the suit and the state has been declared as organic state.

Table 3:. Area under organic certification in different states of India

S.No.	State	Area (hectares)		
		Certified area	Under conversion	Total
1	Andhra Pradesh	5561.17	4925.9	10487.07
2	Arunachal Pradesh	65.87	632.77	698.64
3	Assam	2526.61	540.24	3066.85
4	Bihar	0	0	0
5	Chhattisgarh	279.16	28.72	307.88
6	Delhi	3632.63	1830.35	5462.98
7	Goa	4100.5	2849.8	6950.3
8	Gujarat	7102.31	658.51	7760.82
9	Haryana	3382.54	15.78	3398.32
10	Himachal Pradesh	69.03	9507.7	9576.73
11	J & K	32541.79	0	32541.79
12	Jharkhand	10.5	2253.35	2263.85
13	Karnataka	8735.06	2976.78	11711.84
14	Kerala	11631.93	3112.73	14744.66
15	Manipur	913.68	5105.87	6019.55
16	Maharashtra	41390.48	72238.44	113628.92
17	Madhya Pradesh	87536.03	59875.81	147411.84
18	Mizoram	0	16802.5	16802.5
19	Meghalaya	0	304.4	304.4
20	Nagaland	0	878.89	878.89
21	Orissa	66625.42	7959.69	74585.11
22	Punjab	347.6	698.36	1045.96
23	Rajasthan	15034.26	9697.53	24731.79

Contd...

24	Sikkim	274.82	1531.91	1806.73
25	Tripura	0	0	0
26	Tamil Nadu	3414.09	1652.39	5066.48
27	Uttar Pradesh	5589.56	1700.57	7290.13
28	Uttarakhand	3178.63	5250.88	8429.51
29	West Bengal	7332.75	3147.18	10479.93
30	Other	510.52	966.32	1476.84
	Total	**311786.94**	**217143.4**	**528930.31**

Global organic movement

At global level, the roots of organic farming can be traced back to the first quarter of the early 20th century in Europe (Stockdale *et al.*, 2001). In 1924, the Austrian philosopher Dr. Rudolf Steiner conceptualized and advocated organic agriculture and in 1927 a trademark 'Demeter' was introduced for organically produced food. The formation of the International Federation of Organic Agriculture Movements (IFOAM) in 1972 gave framework for the discussion and codification of internationally recognized principles of organic farming. The main principles of organic farming are :

- to produce food of high quality in sufficient quantity,
- to interact in a constructive and life-enhancing way with natural systems and cycles,
- to consider the wider social and ecological impact of the organic production and processing systems,
- to encourage and enhance biological cycles within the farming system, involving micro-organisms, soil flora and fauna, plants and animals,
- to maintain and increase the long-term fertility of soils,
- to maintain the genetic diversity of the production system and its surroundings, including the protection of wildlife habitats,
- to promote the healthy use and proper care of water, water resources and all life therein,
- to use, as far as possible, renewable resources in locally organized production systems,
- to give all livestock conditions of life with due consideration for the basic aspects of their innate behaviour,
- to minimize all forms of pollution and
- to progress towards an entire production, processing, and distribution chain which is both socially justifiable and ecologically responsible.

Organic movement in India

Organic farming was practiced in India since thousands of years. The great Indian civilization thrived on organic farming and was one of the most prosperous countries in the world. In traditional India, the entire agriculture was practiced using organic techniques, where the fertilizers, pesticides, etc., were obtained from plant and animal products. At that time organic farming was the backbone of the Indian economy. The country had witnessed a tragic Bengal famine in 1942-43 that claimed millions of lives due to starvation. India's annual food grain production in 1950-51 was only about 50 million tonnes. In this backdrop, modern agriculture played a vital role in alleviating the hunger through green revolution. India's achievement in agricultural production after the green revolution has been exemplary and mainly due to the increased use of inputs like fertilisers, pesticides and farm machinery. The food grain production in India itself has been more than doubled during the post green revolution period with virtually no increase in net cultivated area. The intensive modern agriculture for the past five decades has led to several environmental and health issues. The land is losing its fertility and is demanding larger quantities of fertilizers. Pests are becoming resistant requiring the farmers to use stronger and costlier pesticides. The over-use of pesticides resulted in residues much above the safety levels and this brought to the attention of ill-effects of modern agriculture, and paved the way for organic farming. According to the Agricultural and Processed Food Products Export Development Authority (APEDA), around 2.8 million hectares of land was under organic farming in India in 2007. Further, there are over 15,000 certified organic farms in India. India therefore is one of the most important suppliers of organic food to the developed nations. No doubt, the organic movement has again started in India.

Status and scope of organic farming in India

According to Marwaha and Jat (2004), the organic farming probably is seldom practised and there is dearth of data on comparative yield and economics of crops with pure organic farming versus intensive farming. Whatever little organic farming that is being practised in India seems to be taken up mainly as a part of contract farming. In India, the area under organic farming is estimated to be only 0.3% of total cultivated area. Even in Europe and other developed countries of the world, despite serious efforts and incentives, the area under organic farming is generally less than 5%. Organic farming may be far from reality for three reasons. Firstly, the limited availability of organic sources; secondly, for getting optimum yields, crop plants require nutrients in sufficient quantities and in readily available form. This could be best possible through fertilisers, and hardly through organic sources. Thirdly, for plant protection, the organic sources and biological methods that have been tried so far have not met with considerable success in the country. Marwaha and Jat (2004) were of the opinion that, organic farming may be promoted in selected rainfed areas and for export oriented crops in India.

In India, only about 38% of the cultivated land is irrigated and in rest of the rainfed areas (62%) there is little or no use of fertilisers and other agrochemicals on account of associated risk in farming and poor resources. So, almost 62% of such areas of the country may be regarded as akin to organic farming. Therefore, there could be ample scope for pure organic farming in those areas. Similarly, some areas under tea, coffee, spices and cashew may be easily brought under organic farming with a thrust on export of organic produce. Thus, rather than promoting organic farming *en masse* and universally, it would be desirable to carefully delineate areas for organic agriculture, which provides better profitability and boost to the export (Marwaha and Jat, 2004).

Several states in India had already initiated steps to encourage the organic farming among the farming communities. Among the state governments, Madhya Pradesh took early lead in organic farming followed by Uttarakhand (declared as organic state) and other states. The 'Organic Farming Council of Punjab' established in 2006 is taking lead in promoting organic farming in Punjab. Green manuring with *Sesbania/Crotolaria*–Pusa Basmati– Chickpea cropping system under organic mode has been identified as a complete rotation to be followed in *Tarai* region of Uttarakhand. Wheat, lentil and mustard were some of the important crops grown successfully under organic mode with yield realization of 25-27 q, 12-14 q and 10 q, respectively at Pantnagar. An overview of organic agriculture in India is presented in Table 4.

Table 4. An overview of organic agriculture in India (as of 2004-05)

Area under certification	:	2.5 m ha.
Total certified product	:	115,238 MT.
Number of processing units	:	158
Accredited inspection and certifying agencies	:	11
Number of products exported	:	35
States involved in organic export		
Kerala	:	1232 MT.
West Bengal	:	937 MT.
Karnataka	:	476 MT.
Tamil Nadu	:	471 MT.
Punjab	:	541 MT.
Himachal Pradesh	:	521 MT.
Maharashtra	:	375 MT.
All India total organic export	:	6472 MT.
Premium collected against organic export	:	Rs. 80-90 crores (tentative)

(*Source:* Bhattacharyya and Chakraborty, 2005)

Regulation of organic production: global perspective

Organic agriculture is based on the commitment of farmers and processors to work according to set standards and regulations, which define the organic production system. Further, organic agriculture is based on transparency to make the production system comprehensive and reliable. Finally, organic agriculture needs the consumer confidence for market development. Certification and accreditation systems primarily serve as tools to enhance trade, market development and foster confidence among the consumers (Herrmann and Rundgren, 2007). The first international standards for organic products were published by IFOAM in 1980. The first legislative initiatives were developed by some European countries (e.g. Austria, France) in the 1980s. In 1991, the EU passed the organic regulation 2092/91 and set standards with major implications for international trade, and included not only production standards, but also standards for labelling and inspection. Various countries in Europe, Latin America and Asia introduced legislation in the 1990s. In 1999, Codex Alimentarius approved the first guidelines for organic plant production. In the new millennium, most major economies have implemented legislation on organic production. Currently more than 60 countries have a regulation. In India the regulations for organic agriculture is implemented only for export commodities (Huber et al., 2007).

As of 2006, around 395 organizations worldwide offer organic certification services. Most certification bodies are in Europe (160), followed by Asia (93) and North America (80). Many of the certification organizations also operate outside of their home country. 40 per cent of the certification bodies are approved by the European Union, 32 percent have ISO 65 accreditation, and 28 percent are accredited under the US National Organic Program (Herrmann and Rundgren, 2007).

Lack of acceptance and recognition among the different certification and accreditation systems can contradict the objective of enhancing trade, market development and fostering confidence. An important initiative for international harmonization is the IFOAM Accreditation Program, which assesses certification bodies against the IFOAM norms. The International Organic Accreditation Service (IOAS) is a non-profit independent organization registered in Delaware, USA that offers international oversight of organic certification, through a voluntary accreditation process for certification bodies active in the field of organic agriculture. The IOAS implements the IFOAM Accreditation Program. Under this program, applicant certification bodies are assessed against the IFOAM Norms - the Criteria for Certification Bodies and the IFOAM Basic Standards (Herrmann and Rundgren, 2007).

Current status of regulatory mechanism in India

National programme for organic production

There is tremendous scope for cultivation and export of wide range of organic produce in India. Looking into its importance, Ministry of Commerce, Government of India launched National Programme for Organic Production (NPOP) in 2000. Declaring standards for organic production, policy on accreditation system, inspection and certification process and institution of logo were the major objectives of this programme at the time of initiation. A national organic logo "India Organic" has also been developed that has been in use for certified organic products. During July, 2002, Ministry of Agriculture, Government of India has also circulated guidelines for promotion of organic farming (Bhattacharyya and Gehlot, 2004).

Accreditation agencies

As of 2004, Agricultural and Processed Food Product Export Development Authority (APEDA), Coffee Board, Spices Board, Tea Board, Coconut Development Board, and Cocoa and Cashewnut Board were authorised by the Ministry of Commerce, Government of India as accreditation agencies for organic agricuture (Bhattacharyya and Gehlot, 2004).

Certifying agencies

At present there are 16 accredited certification bodies in India (www.apeda.com) (Table 5). In 2001, a group of organizations and corporate bodies took the initiative to set up the Indian Organic Certification Agency, INDOCERT. It has become an important element of the organic movement in India and mobilizes new forces and partnerships. In 2003, together with other partners, INDOCERT created the International Competence Center for Organic Agriculture (ICCOA), a service provider for networking, capacity building and market development in the organic sector in India. Among other activities, ICCOA implements the Indian Organic Market Development Project (2005-2007), the activities of which were focussed on awareness raising, market intelligence, developing organic market initiatives, and the India Organic Trade Fair (Kilcher, 2007).

National institute of organic farming

The Indian Central Government set up a National Institute of Organic Farming in October 2003 in Ghaziabad, Uttar Pradesh. The National Biofertiliser Development Centre was renamed as the National Institute of Organic Farming (NIOF). The purpose of this institute is to formulate rules, regulations and certification of organic farm products in conformity with international standards.

Further details on current status of regulatory mechanism in organic farming in India may be obtained from Bhattacharyya and Gehlot (2004).

Table 5. List of accredited certification bodies under NPOP

Name of the certification agency	Validity of current accreditation	Scope of accreditation
Bureau Veritas Certification India Pvt. Ltd. (Formerly known as BVQI (India) Pvt. Ltd.)	14-09-2009	NPOP; USDA; NOP
ECOCERT India Pvt. Ltd	22-08-2008 (Extended till next NAB meeting)	NPOP; USDA; NOP
IMO Control Pvt. Ltd.	27-09-2010	NPOP; USDA; NOP
Indian Organic Certification Agency (INDOCERT)	24-10-2008 (Extended till next NAB meeting)	NPOP; USDA; NOP
Lacon Quality Certification Pvt. Ltd.	31-09-2008 (Extended till next NAB meeting)	NPOP; USDA; NOP
Natural Organic Certification Agency (NOCA)	23-05-2009	NPOP; USDA; NOP
OneCert Asia Agri Certification Pvt. Ltd.	26-10-2009	NPOP; USDA; NOP
SGS India Pvt. Ltd.	01-05-2011	NPOP; USDA; NOP
Control Union Certifications (Formerly known as SKAL International (India))	28-05-2008 (Extended till next NAB meeting)	NPOP; USDA; NOP
Uttarakhand State Organic Certification Agency (USOCA)	13-11-2009	NPOP; USDA; NOP
APOF Organic Certification Agency (AOCA)	09-01-2010	NPOP
Rajasthan Organic Certification Agency (ROCA)	09-10-2010	NPOP
Vedic Organic Certification Agency	30-09-2011	NPOP
ISCOP (Indian Society for Certification of Organic Products)	30-11-2011	NPOP
Food Cert India Pvt. Ltd	30-09-2011	NPOP
Aditi Organic Certifications Pvt. Ltd	30-09-2011	NPOP

(*Source:* www.apeda.com)

Characteristics and components of organic farming

"Organic agriculture is a production system that sustains the health of soils, ecosystems and people. It relies on ecological processes, biodiversity and cycles adapted to local conditions, rather than the use of inputs with adverse effects. Organic agriculture combines tradition, innovation and science to benefit the shared environment and promote fair relationships and a good quality of life for all involved" (IFOAM). Unlike chemical farming, organic farming aims at "feeding the soil" rather than "feeding the plant". It is a means of giving back to the nature what has been taken from it. Since organic farming aims at maintaining soil health, the highest yields can be obtained in a sustainable and eco-friendly manner in the long run.

A large number of terms are used as an alternative to organic farming. These are: biological agriculture, ecological agriculture, bio-dynamic and natural agriculture. Organic farming is characterized by the maximal, but sustainable use of local resources and minimal use of purchased inputs. It facilitates the basic biological functions of soil-water-nutrient-human continuum. In organic farming the ecological balance is maintained as the bio diversity is being sustained. Organic farming is a form of agriculture that relies on crop rotation, green manure, compost, biological pest control, and mechanical cultivation to maintain soil productivity and control pests, excluding or strictly limiting the use of synthetic fertilizers, chemical pesticides, plant growth regulators, livestock feed additives, and genetically modified organisms. Enhancing soil health is the cornerstone of organic farming.

Though the organic farms produce lower yields, they require no synthetic fertilizers and pesticides. The decreased costs on those inputs, along with the premiums, which consumers pay for organic produce, create higher profits to organic farmers. Moreover, it has already been realized in certain areas that the yields in organic farming are far greater than those obtained by chemical farming in the long run. There is often a loss in yield while converting a chemical farm in to organic farm, but after few years the yield will certainly increase and get stabilised at a level often higher than that achieved under a chemical regime. It is therefore, recommended to convert gradually over a period of three to four years, if income from the farm is a key issue (www.satavic.org).

Indigenous seeds, crop rotation, intercropping, multicropping, biological control of pests and diseases, botanical pesticides, reduced tillage, mulching, green manuring and green leaf manuring, composting, vermicomposting, biofertilizers, soil and water conservation are some of the techniques and practices integral to organic farming. This list is not complete in itself, and farmers can innovate and adapt to suit their surroundings. There are certain researchable issues need to be addressed for popularization and subsequent successful implementation of organic farming in India. The issues are:

- Suitable land races / local cultivars of high value crops need to be identified for organic cultivation.
- Improved methods need to be standardized for enriching the nutrient status of the composts.
- Different cropping sequences need to be evaluated under various agro-ecological situations.
- The economics of compost preparation, cropping systems and seed production need to be worked out under various farming systems.
- Soil health, weed, pests and disease incidence need to be monitored regularly.
- The possibilities of incorporating location-specific livestock components in organic farming need to be studied.
- The indigenous technical knowledge (ITKs) need to be documented for further exploitation in organic farming.

Benefits of organic farming

The various benefits of organic farming all over the world include high premium, low capital investment and the ability to use traditional knowledge. Given below are some of the advantages of organic farming for small farmers in India (http://www.organicfacts.net):

- *High premium*: Organic food is normally priced 20-30% higher than conventional food. This premium is very important for a small farmer whose income is just sufficient to feed his/her family with one meal.
- *Low investment*: Organic farming normally does not involve capital investment as high as that required in chemical farming. Further, since organic fertilizers and pesticides can be produced locally, the yearly costs incurred by the farmer are also low. It should also be noted that while shifting from chemical farming to organic farming, the transition might be costlier.
- *Traditional knowledge*: Small farmers have abundance of traditional knowledge with them and within their community. Most of this traditional knowledge cannot be used for chemical farming. However, when it comes to organic farming, the farmers can make use of the traditional knowledge. Further, in case of organic farming, small farmers are not dependent on those who provide chemical know-how.

In India more than 80% of the farmers are small and marginal farmers and many small farmers are practicing organic farming; however, since they are unaware of the market opportunities they are not able to reap the benefits of organic farming mentioned above. Hence necessary steps need to be taken to address this issue.

Limitations of organic farming

Though there are several benefits in adapting the organic farming, certain barriers hinder the adoption at farmers' level. Although the land resources can move freely from organic farming to conventional farming, they do not move freely in the reverse direction. Initial crop loss is one of the major constraints while transforming the land from chemical to organic agriculture. The biocontrol agents may take three to four years to build up in a crop ecosystem. Moreover, their rate of success is in general, more in horticultural eco system but, in annual crop ecosystem their establishment is always questionable. Above all, farmers may afraid to adopt the organic agriculture without adequate support from the government and premium price for their produce. Thus, organic agriculture has its own limitations. These limitations were studied by the Office of Evaluation and Studies (OE), International Fund for Agriculture Development (IFAD). They include the following:

- Organic farming is labour intensive. Hence it is beneficial for a small farmer who has abundant labour in his/her family. A small farmer who does not have sufficient labour at home may find it difficult to hire external labour and therefore organic farming may not be feasible for him/her.
- Organic farming may also turn out to be expensive given the situation when a small farmer has to carry out the transition, modify the soil structure drastically or get a certification.
- Though a farmer can use a good deal of traditional knowledge in organic farming, since the practice is not as common as chemical farming, it is difficult for good scientific organic farming practices to propagate to the small farmers.
- Since organic produce is not traded in many markets, marketing the organic produce may be difficult for a small farmer.
- In organic farming, soil fertility needs to be replenished with the application of organic fertilizers regularly. If the organic fertilizers are not produced locally, it might be difficult for the small farmer, who would purchase the fertilizers in small quantities to obtain them from elsewhere. Further due to low economies of scale, there might be a huge transportation cost associated with the organic fertilizers obtained from long distances.

- It has also been observed that organic food prices are not stable and keep fluctuating from time to time. In such a scenario a low price might affect a small farmer drastically.
- Since the volume of organic food production is low (about 1-2% of total food production), it is very difficult to implement social security measures such as minimum support price.
- Certification is also an important aspect of organic farming. In many cases, providing education related to necessary certifications and the guidelines that need to be followed for obtaining the certifications may be difficult. Since certification cost is exorbitantly high and inorganic input use is prohibited in the nearby areas, organic farming may not be so easy to practise by the small and marginal farmers who constitute about 80% of the Indian farming community (Fertiliser Association of India, 2004).

Organic farming in India: controversies

Organic farming and its implications on India's food security were well discussed by Chhonkar and Dwivedi (2004). Some of the salient issues are presented hereunder:

- The widely publicised advantage of high income through export of organic produce to the western markets appears lucrative, but not so easy in real sense. Indigenous organic food market is in infancy and may remain so in the future. There is a remote possibility of a resource-poor small farmer of a remote area getting premium price for his organic produce after successfully overcoming all the steps and formalities beginning from farm certification to the final export. The family demands of such farmers may compel them to ultimately dispose-off their organic produce in local markets without any premium.
- Chhonkar and Dwivedi (2004) warned that if a section of pro-change farmers of agriculturally important areas currently engaged in food grain production began to convert their farms to organic with the intention of getting premium price, there will be no option left for the country other than importing food grains as it has practiced a few decades back.
- Nitrate is the main end product of organic manure decomposition; moreover it is continuously released from the organic matter under decomposition. Since nitrate release is not synchronised with either crop demand or its uptake, it tends to accumulate in excess in soil and pose environmental risk. The ions irrespective of their origin whether organic or inorganic will behave similarly. There is no evidence that nitrate ions

from organic sources are less mobile and have lower denitrification potential than from inorganic fertilisers. In addition, the trace elements and heavy metal concentrations present in the animal wastes and sewage sludges vary widely and at times very high and often exceed concentrations normally found in the inorganic fertilisers. Field application of these manures in large quantities may pollute the soil and there will be every chance to enter in to food chain and create health hazard.

- The biggest myth about the organic farming as indicated by Chhonkar and Dwivedi (2004) is that the country has enough organics available to replace chemical fertilisers to sustain present level of crop production. All tappable nutrients from organic sources will be barely able to meet the deficit of nutrients in soil after crop removal at present level of production.
- To maintain the present level of food grain production in India without input of inorganic fertilisers, the net additional area to be brought under cultivation will be more than the total geographical area of the country.
- There are contentions that organic farming is unsustainable. Lower yields and dependence of manure from low-yield cattle has prompted criticism that organic farming is environmentally unsound and incapable of feeding the world population. Among these critics are Norman Borlaug, father of the "green revolution," and winner of the Nobel Peace Prize, who asserts that organic farming practices can at most feed 4 billion people, after expanding cropland dramatically and destroying ecosystems in the process.
- A few case studies pertaining to organic production of wheat, cotton, tea and coffee in India were reported by Marwaha and Jat (2004). The overall economics of these initiatives was not in favour of organic farming. It was also reported that in India, organic farming may be a myth, because of limited availability and restricted use of organic resources and ever growing need for food and feed.

Production and demand for organic products

Consumer demand for organic products is increasing across the globe. Global sales of organic food and drink have increased by 43% from 23 billion US-Dollars in 2002 to 33 billion US-Dollars in 2005. Although all regions reported healthy increase, the two major engines driving growth are the European and North American markets. Global production of organic crops has not kept pace with demand, with many sectors reporting undersupply. The North American market has been experiencing supply shortages for a number of years; however, Europe and Asia are now also affected. Thus, large volumes of imports are coming in from other regions (Sahota, 2007).

The increasing demand for organic food products in the developed countries and the extensive support by the Indian government coupled with its focus on agri-exports are the drivers for the Indian organic food industry. The Indian government is committed towards encouraging organic food production. It allocated Rs. 100 crore during the Tenth Five Year Plan for promoting sustainable agriculture in India. With the domestic consumption being low, the prime market for Indian organic food industry lies in the US and Europe. APEDA (Agricultural and Processed Food Export Development Authority) coordinates the export of organic food in India. The NPOP standards for production and accreditation system have been recognized by European Commission and Switzerland as equivalent to their country standards. Similarly, USDA has recognized NPOP conformity assessment procedures of accreditation as equivalent to that of US. With these recognitions, Indian organic products duly certified by the accredited certification bodies of India are accepted by the importing countries like US and European markets (www.apeda.com).

India has now become a leading supplier of organic herbs, spices, basmati rice, etc. India produced around 58,597 MT of certified organic products which include all varieties of food products namely cereals, pulses, honey, tea, spices, coffee, oil seeds, fruits, vegetables and their value added products. The production is not limited to the edible sector but also produces organic cotton fibre, garments, cosmetics, functional food products, body care products, etc. India exported 35 items last year (2007-08) with the total volume of 19,456 MT (Table 6). The export realization was around 78 million US $ registering a 200 % growth over the previous year. As many as 98 agencies are involving in the export of organic products in India. Organic products are mainly exported to EU, US, Australia, Japan, Switzerland, and Middle East. Basmati rice leads among the products exported (5139 MT) (www.apeda.com). In February 2009, the export duty on export of Basmati rice has been withdrawn by the Government of India (Notification No. 10/2009-Customs) with a view to promote the organic products.

Table 6. Present status of organic products in India*

Total production	585970 M.T.
Total quantity exported	19456 M.T.
Value of total export	Rs. 30124 Lakhs
Total area under certified organic cultivation	339113 Ha.
Number of Farmers	141904

*Status as per 2006-2007 records *(Source: www.apeda.com)*

Conclusion

In India, only about 38% of the cultivated land is irrigated and in rest of the rainfed areas (62%) there is little or no use of fertilisers and other agrochemicals on account of associated risk in farming and poor resources. Therefore, there could be ample scope for pure organic farming in those areas. Thus, rather than promoting organic farming *en masse* and universally, it would be desirable to delineate areas carefully for organic agriculture, which provides better profitability and boost to the export (Marwaha and Jat, 2004). Moreover, there is a lacuna to show the comparative crop performance under pure organic farming and intensive agriculture in terms of crop productivity and food quality. Hence efforts need to be taken to address these issues with a view to promote the organic farming with great confidence in India. During the last four decades of the 20th century, the global population doubled from 3 to 6 billion and is estimated that by the year 2020, it will reach the 8 billion mark. Food and nutritional security is therefore a global concern. Neither conventional farming with inorganic alone nor the organic farming only with the use of organic inputs can face this challenge. The combination of both organic and inorganic is undoubtedly the best option as on today for developing country like India. Above all organic agriculture must be discussed with an open mind, with advantages and disadvantages being clearly considered in order to create positive mindset among the decision makers, because the most important factor that will enable organic agriculture to usefully contribute to food security is the attitude of decision-makers (Wynen, 1998).

References

Bhattacharya, P. and Chakraborty, G. 2005. Current status of organic farming in India and other countries. *Indian Journal of Fertilisers* **1**(9): 111-123.

Bhattacharya, P. and Gehlot, D., 2004. Current status of regulatory mechanism in organic farming. *Fertiliser News* **49**(11): 33-38.

Chhonkar, P.K. and Dwivedi, B.S. 2004. Organic farming and its implications on India's food security. *Fertiliser News* **49** (11): 15-31 & 38.

Herrmann, G.A. and Rundgren, G. 2007. Certification and Accreditation. *In*: The world of organic agriculture: statistics and emerging trends 2007 (Willer, H. and Yussefi-Menzler, M., Eds.). International Federation of Organic Agriculture Movements (IFOAM), Bonn, Germany and Research Institute of Organic Agriculture (FiBL), Frick, Switzerland, pp. 67-74.

Huber, B., Kilcher, L. and Schmid, O. 2007. Standards and Regulations. *In*: The world of organic agriculture: statistics and emerging trends 2007 (Willer, H. and Yussefi-Menzler, M., Eds.). International Federation of Organic Agriculture Movements (IFOAM), Bonn, Germany and Research Institute of Organic Agriculture (FiBL), Frick, Switzerland, pp. 56-66.

Kilcher , L. 2007. How organic agriculture contributes to sustainable development. *In*: The world of organic agriculture: statistics and emerging trends 2007 (Willer, H. and Yussefi-Menzler, M., Eds.). International Federation of Organic Agriculture Movements (IFOAM), Bonn, Germany and Research Institute of Organic Agriculture (FiBL), Frick, Switzerland, pp. 82-91.

Marwaha, B.C. 2005. Is India in a position to switch over to pure organic farming in totality? *Indian Journal of Fertilisers* 1(7): 47-52.

Marwaha, B.C. and Jat, S.L. 2004. Status and scope of organic farming in India. *Fertiliser News* 49(11): 41-48.

Sahota, A. 2007. Overview of the global market for organic food and drink. *In*: The world of organic agriculture: statistics and emerging trends 2007 (Willer, H. and Yussefi-Menzler, M., Eds.). International Federation of Organic Agriculture Movements (IFOAM), Bonn, Germany and Research Institute of Organic Agriculture (FiBL), Frick, Switzerland, pp. 52-55.

Stockdale, E.A., Lampkin, N.H., Hovi, M., Keatinge, R., Lennartsson, E.K.M., Macdonald, D.W., Padel, S., Tattersall, F.H., Wolfe, M.S. and Watson, C.A. 2001. Agronomic and environmental implications of organic farming system. *Advances in Agronomy* 70:261-327.

Tiwari, K.N., Sulewski, G. and Portch, S. 2005. Challenges in meeting nutrient needs in organic farming. *Indian Journal of Fertilisers* 1 (4): 41-59.

Willer, H. 2008. Organic Agriculture Worldwide: Current Statistics. *In*: The world of organic agriculture: statistics and emerging trends 2008 (Willer, H., Yussefi-Menzler, M. and Sorensen, N. Eds.). International Federation of Organic Agriculture Movements (IFOAM), Bonn, Germany and Research Institute of Organic Agriculture (FiBL), Frick, Switzerland, pp. 23-46.

Wynen, E. 1998. Evaluating the potential contribution of organic agriculture to sustainability goals. Environment and Natural Resources Service, Research, Extension and Training Division, Sustainable Development Department, Food and Agriculture Organization of United Nations (FAO), Rome.

□□□

13

Commercial Application of Organic Farming

D.K. SINGH

The roots of organic farming can be traced to the Europe back to the first quarter of the early 20th century (Stockdale *et.al.*, 2001). In 1924, the Austrian philosopher Dr. Rudolf Steiner conceptualized and advocated organic agriculture and in 1927 a trademark 'Demeter', was introduced for organic food product. Soil fertility was maintained through crop rotation and careful management and use of animal manure.

The formation of the International Federation of Organic Agriculture Movements (IFOAM) in 1972 gave an international framework for the discussion and codification of internationally recognized principles of organic farming. Some of these are:

- To produce food of high quality in sufficient quantity.
- To interact in a constructive and life-enhancing way with natural systems and cycles.
- To consider the wider social and ecological impact of the organic production and processing systems.
- To encourage and enhance biological cycles within the farming system, involving micro-organisms, soil flora and fauna, plants and animals.

- To maintain and increase the long-term fertility of soils.
- To maintain the genetic diversity of the production system and its surroundings, including the protection of wildlife habitats.
- To promote the healthy use and proper care of water, water resources and all life therein.
- To use, as far as possible, renewable resources in locally organized production systems.
- To give all livestock conditions of life due consideration for the basic aspects of their innate behaviour.
- To minimize all forms of pollution.
- To allow every one involved in organic production and processing a quality of life which meets their basic needs and allows an adequate return and satisfaction from their work, including a safe working environment.
- To progress towards an entire production, processing and distribution chain which is both socially just and ecologically responsible.

Definition

In philosophical term, organic farming means "farming in spirits of organic relationship". In this system every thing is connected with every thing else. Since organic farming means placing farming on integral relationship, we should be well aware about the relationship between the soil, water and plants, between soil-soil microbes and waste products, between the vegetable kingdom and the animal kingdom of which the apex animal is the human being, between agriculture and forestry, between soil, water and atmosphere etc. It is the totality of these relationships that is the bed rock of organic farming.

A large number of terms are used as an alternative to organic farming. These are: biological agriculture, ecological agriculture, bio-dynamic, organic-biological agriculture and natural agriculture. According to the National Organic Standards Board of the US Department of Agriculture (USDA) the word 'Organic' has the following official definition (Lieberhardt, 2003): *"An ecological production management system that promotes and enhances biodiversity, biological cycles and soil biological activity. It is based on the minimal use of off-farm inputs and on management practices that restore, maintain and enhance ecological harmony"*.

According to Codex Alimenarius (FAO, 2001) organic agriculture is aholistic production management system which promotes and enhances agro-ecosystem health, including biodiversity, biological cycles and soil biological activity. The primary goal of organic agriculture is to optimize the health and productivity of interdependent communities of soil life, plants, animals and people (Scialabba and Hattam, FAO, 2002). The management practices of organic farming are given in Block 1 (Scialabba, 1998).

In another definition FAO suggested that "Organic agriculture is a unique production management system which promotes and enhances agro-ecosystem health, including biodiversity, biological cycles and soil biological activity and this is accomplished by using on-farm agronomic, biological and mechanical methods in exclusion of all synthetic off-farm inputs".

Principles of organic agriculture

- *Principle of health*: Organic agriculture should sustain and enhance the health of soil, plant, animal and human as one and indivisible.
- *Principle of ecology*: Organic agriculture should be based on living ecological systems and cycles, work with them, emulate them and help sustain them.
- *Principle of fairness* : Organic agriculture should build on relationship that ensures fairness with regard to the common environment and life opportunities (trade included)
- *Principle of care*: Organic agriculture should be managed in a precautionary and responsible manner to protect the health and well-being of current and future generations and the environment.

The world of organic agriculture (As per 2008 survey conducted by Soel and IFOAM)

Area: World over, almost 30.4 m ha land is managed organically by more than 700,000 farms. This constitutes 0.65% of the agricultural land of the countries covered under study. Oceania holds 42%, Europe 24%, Latin America 16%, Asia 11% and North America 7.2%. Amang countries Australia is largest with 12.3 million ha, followed by China 2.3 m ha, Argentina 2.2 m ha and the US 1.6 m ha.

Land use: Out of total 30 m ha for which land use details are available, more than 20 million ha areas is under permanent pastures. More than half of this grassland is in Australia followed by Latin America and Europe. Only one sixth (4.5 m ha) is arable land. Permanent crops account for 5% of the organic agricultural land. The main crop categories for arable land are Cereals followed by field fodder growing, other arable crops, set-aside/green manuring, protein crops, vegetables, oilseeds, industrial crops, medicinal and aromatic plants, root crops, seed production etc.

Wild harvest collection: Besides above, there is about 62 m ha of organic wild collection area with 979 organic wild collection projects, world over. The largest collection areas are in Europe and Africa (almost 27 million ha each). In terms of quantities the important wild collected products are: bamboo shoot (36%), fruits and berries (21%) and nuts (19%). The countries with the largest wild collection areas are Finland (mainly berries), Zambia (bee pastures) and Brazil.

Standards and regulation: Global view

Currently 69 countries have implemented regulations on organic farming and 21 countries are in the process of drafting a regulation. Today 468 organizations worldwide offer organic certification services. Most certification bodies are in Europe (37%) and North America (18%). The countries with most certification bodies are US, Japan, South Korea, China and Germany. Out of these, 36 bodies have received IFOAM accreditation.

National Programme on Organic Production (NPOP) notified under FTDR act regulates the certification of organic products in India. As on September 2008, there are 16 accredited certification agencies.

Global Market

The global market for organically produced foods in 2005 was estimated at US $ 31 billion (Rs. 14,300 crores) and is likely to increase to US $ 102 billion by 2020 (Bhattacharya and Gehlot, 2003). According to a global survey conducted by Ockologic and Landbau (SOUL) (Formation of Ecology and Agriculture) of Germany, organic food in 2003 was produced on only 5% of the world cultivated area (Table 4). Austria had the highest percentage of its cultivated area under organic farming followed by Switzerland, Italy, Finland, Denmark, Sweden and Czech Republic. India had the least percentage of cultivated area under organic farming. There is thus considerable scope to increase the area under organic farming in India. Further, since India's share in global market for organic food is currently less than 1% there is tremendous opportunity to increase it.

The global market of organic was 38.6 billion $ in 2006 and have crossed to 40.0 b $ in 2007. Consumer demand is concentrated in North America and Europe. These two regions comprise of 97% of global revenues in organic food products trade.

Table 1. Area under organic farming in relation to GDP and share of agriculture to GDP

Country	Arable land (10^6 ha)[1]	Area under organic farming		GDP in 2003 (10^0 Dollars)	Share of agriculture in GDP[2]
High income e conomics					
Austria	1.39	11.30	15.7	251	2 NA
Switzerland	0.41	9.70	4.0	309	3
Italy	8.29	7.94	65.8	1466	3
Finland	2.20	6.60	14.5	161	3
Denmark	2.28	6.51	14.8	212	2
Sweden	2.68	6.30	16.9	301	1
UK	5.75	3.96	22.8	1795	1
Germany	11.79	3.70	43.6	2401	2
Norway	0.87	2.62	2.3	221	4
Australia	48.30	1.94	111.6	518	3
Netherlands	0.92	1.66	1.8	511	3
Spain	13.74	1.61	22.8	836	1
Belgium	0.82	1.40	1.3	302	3
France	18.45	0.23	25.8	1748	2
USA	176.02	0.10	40.5	10882	3
Japan	4.42		0.4	4326	
Upper middle income economies					
Argentina	33.70	1.89	63.7	13.0	11
Hungary	4.61	1.80	8.3	8.3	4
Chile	1.98	1.50	3.0	72	9
Middle income economies					
Chaina		0.06	8.5	1410	15
Low income economies					
India	161.7	0.03	4.8	599	23

[1]Yr. 2002, [2]Yr. 2003

Source: Fertilizer News (2004) and Fertilizer Statistics 2003-04 (2004)

Indian scenario

Organic farming has received considerable attention in India and Ministry of agriculture and Cooperation, Govt. of India constituted a Task force on Organic Farming under the chairmanship of Dr. Kunwarji Bhai Yadav, Ex-Director of Agriculture, Gujarat in 2000. The Committee in its report emphasized on the need for consolidating the information on organic farming and its benefits. One of the steering committees constituted by this Task Force under the Chairmanship of Dr. M.S. Swaminathan, Chairman, Farmer's Commission has suggested taking up organic farming as a challenging national task and to take up this as a thrust area of the 10th Five Year Plan. The steering committee advocated giving boost to organic farming in the rainfed areas and in the north-eastern states where there is limited use of agricultural chemicals. Madhya Pradesh took early lead in this regard and Uttarakhand and Sikkim followed the suit and these states have declared themselves as organic states (Marwaha and Jat, 2004). The Ministry of Commerce launched the National Organic Programme in April 2000 and Agricultural and Processed Food Products Exports (APEDA) is implementing the National Programme of Organic Production (NPOP) (Gauri, 2004). Under the NPOP, documents like National Standards, accreditation criteria for accrediting inspection and certification agencies, have been prepared and approved by the National Steering Committee.

It is estimated that there is around 76,000 ha of certified organic food production at the farm level and 2.4 million ha of certified forest area for collection of wild herbs in India (Bhattacharya and Chakraborty, 2005), but the actual area under organics is much more. In Maharashtra alone about 0.5 million ha area is under organic farming since 2003, out of this only 10,000 ha is the certified area. In Nagaland, 3,000 ha is under organic farming with crops like maize, soybean, ginger, large cardamom, passion fruit and chilli. The state of Rajasthan has 5,631 ha under organic farming with crops like pearlmillet, wheat, mungbean, guar, mustard and cotton.

Growing certified area

Before the implementation of NPOP during 2001 and introduction of accreditation process for certification agencies, there was no institutional arrangement for assessment of organically certified area. Initial estimates during 2003-04 suggested that approximately 42,000 ha of cultivated land were certified organic. By 2005 India had brought more than 2.5 million ha of land under certification. Out of this while cultivable land was approximately 76,326 ha, remaining area was forest land for wild collection. Growing awareness, increasing market demand, increasing inclination of farmers to go organic and growing institutional support has resulted into phenomenal growth in total certified area during the last two years. In March 2008, the total area under organic certification process stood at 8.65 laka ha.

Important crops under organic cultivation

As in organic management multiple cropping systems have been given preference over mono-cropping, it is difficult to assess the area under certification for individual crops. Depending upon the market, climate and requirement of the region different inter or companion crops. Cotton is the single largest crop under organic management with an output of approximately 2.25 lakh MT seed cotton during the year 2007-08. Pulses, soybean (mainly as intercrops among cotton), rice, wheat, oilseeds and coarse cereals are other important crops. Cotton is being grown mainly in Madhya Pradesh, Maharashtra, Orissa and Andhra Pradesh. Spices are predominant in Kerala while Tea and coffee are important crops are important crops under organic management in West Bengal and Tamilnadu. Honey, the single largest organic commodity in export is being collected from wild from the forests of Madhya Pradesh and Uttar Pradesh.

Growing organic food market

Although no systematic information is available on size of organic food market by as per the survey conducted by the International Competence Centre for Organic Agriculture (ICCOA) in top 8 metro citites of India (which comprise about 5.3% of the households) the market potential for organic foods in 2006 in top 8 metros of the country is at Rs. 562 crore taking into account current purchase patterns of consumer in modern retail format. The overall market potential is estimated to be around Rs. 1452 crore.

Market network

India is at a very nascent stage, when it comes to export of organic produce. During 2004-05 the total export was 6,472 tonnes worth Rs. 80-90 crores (Tables 2 and 3).

Table 2. India organic: An overview (2004-05)

1.	Area under certified		2.5 million ha
2.	Total certified product		115,238 tonne
3.	Total project certified		332
4.	Number of processing units		158
5.	Accredited inspection and certifying agencies		11
6.	Number of products exported		35
7.	States involved in organic export :		
	i)	Kerala	1232 tonne
	ii)	West Bengal	937 tonne
	iii)	Karnataka	476 tonne
	iv)	Tamil Nadu	471 tonne
	v)	Punjab	541 tonne
	vi)	Himachal Pradesh	521 tonne
	vii)	Maharashtra	375 tonne
8.	All India total organic export		6472 tonne
9.	Premium collected against organic export		Rs. 80-90 crores (tentative)

Source: Indian Journal of Fertilizers (2005)

Table 3. Export of organic produce from India to the EU (2003-04)

Product	Quantity (tonnes)
Tea	1,997
Pineapple	1,320
Spices	625
Honey	526
Rice	432
Seasome	354
Cashew nut	126
Walnut	89
Fruit	46
Cotton	26
Pulses	12
Sugarcane	8

Source: Indian Journal of Fertilizers (2005)

Limitations of organic farming

- Technical know-how
- Non availability of organic inputs
- Initial decline in crop productivity
- Unaware of negativeness of in-organics
- Positive aspect of organic farming are unknown
- Unaware of market trade
- Local conditions

Future of organic agriculture in India

Trends

- More recognition of NPOP by other countries
- Growth in production
- More grower groups under certification

Challenges

Inmediate challenge

- Maintaining yield during conversion
- Crop/area specific package practices
- Suitable varieties

- Pest and disease management
- Machines and tools
- Post harvest

Intermediate challenges

- Maintaining yield during conversion
- Crop/area specific package practices
- Suitable varieties
- Pest disease management
- Machines and tools
- Post harvest.

Possible thrust areas of research in organic agriculture

A. *Soil health dimension of the value addition*: Agronomical, Ecological and Economic value additions made through farming organically. Scientific investigations into organic technologies/practices, validating the ways soil health the soil organic carbon lead to restoring the soil micro flora and fauna (biodiversity). Here the type and quantum of these soil borne organisms would determine the sustainability of production systems.

B. *Safe food dimension of the value addition*: Safe food dimension; it relates to adding value by way of producing safe food. It has two aspects, that the food produced does not contain any harmful chemicals, which pass into the plant either from soil with minerals (e.g. heavy metal contaminated water or compost) or though plant protection means i.e. pesticides and insecticides (analyses as residues in the products). Organic value chain here should enable us differentiate what bad additions organic farming helped avoid and what good additions it helped make as against the conventional produce.

C. *Economic benefits dimension of the value addition*: Advocating the adoption of organic agriculture in the present scenario of market economy means that it will become popular with farmers only if there exists an environment of economic profitability. That will happen only when a wider range of organic products, both niche products and staple food items, are physically and economically accessible to consumers at various places (regions in the country), in good volumes (scale economies for distribution and trade) and at all times. It is a practical issue, which is determined by the state of supply chain mechanisms, both type and vibrancy that need to be developed and put in place.

Value systems research is important in the development of organic foods and farming. To fulfill this role a broad range of methods and approaches is needed. It is great challenge to include very different methods, applied within a wide range of perspectives, in an interdisciplinary or trans-disciplinary research effort of this kind.

References

Bhattacharya, P. and Gehlot, D. 2003. Current status of regulatory mechanism in organic farming. *Fertilizer News* **49** (11) : 33-38

Laegreid, M., Beckman, O.C. and Kaarstad, O. 1999. *Agriculture, Fertilizers and the environment,* CABI Pub. Norsk Hydro ASA, Oxon, U.K. pp. 294.

Scialabba, N. 1998. Evaluating the potential contribution of organic agriculture to sustainability goals. FAO's technical contribution of IFOAM's Scientifc Conference, Mar del Plata. Argentina, Novenber 16-19, 1998.

Stockdale, E.A., Lampkin, N.H., Hovi, M., Keatinge, R., Lennartsson, E.K.M., Macdonald, D.W., Padel, S., Tattersall, F.H., Wolfe, M.S. and Watson, C.A. 2001. Agronomic and environmental implications of organic farming system. *Advances in Agronomy* **70**: 261-327.

Scialabba, E.N. and Hattam, C. (Eds.) 2002. *Organic Agriculture Environment and Food Security*. FAO, Rome pp. 252.

□□□

Managing Water
The Vital Input

14

On-farm Water Management — Issues, Problems and Strategies

SUBHASH CHANDRA, ANCHAL DASS AND GURVINDER SINGH

Sufficient water must be present in the root zone for seed germination, evapotranspiration (ET), nutrient absorption by the roots, root growth, and soil microbiological and chemical processes that aid in the decomposition of organic matter and the mineralization of nutrients. These factors are necessary for sustaining optimum crop growth on a particular field. At the same time, the root zone must be sufficiently dry to ensure adequate aeration and root extension. The root zone must also be dry enough to allow field access for performing cultural activities such as planting, intercultural operations, fertilization, pesticide and herbicide applications, and harvesting operations. In order to realize potential yields, water movement through the soil must exist; water movement leaches excess salts from the root zone. To satisfy the objectives of water control, adequate soil water content must be maintained in a field. This must be done in a timely manner and within constraints imposed by characteristics of the area. This invokes for proper management of water at the farm level.

Water management can be defined as the planned development, distribution and use of water resources in accordance with predetermined objectives while respecting both the quantity and quality of the water resources. It is the specific control of all human interventions concerning surface and sub-terranean water. Every planning activity relating to water can be considered as water management in broadest sense of the term [International Commission on Irrigation and Drainage, ICID, 2000].

On-farm water management (OFWM) can be defined as the manipulation of water within the boundaries of a farm, a farming plot or field. For example, in canal irrigation system, OFWM starts at the farm gate and ends at the disposal point of the drainage water to a public water course, open drain or sink. OFWM, generally, seeks to optimize soil-water-plant relationship in order to achieve yield of desired level. The farm managers or the farmers usually try to achieve this desired yield by minimizing inputs and maximizing outputs, so as to optimize profits. To realize such objectives, water has to be managed skillfully through certain practices covering areas of: soil and water conservation, water application, drainage, soil amelioration, and crop management. All this has to be done within the context of the socio-economic environment of the community and the framer's personal situation.

Irrigation status in India

Presently, roughly 40% of the cultivated area is irrigated. The major sources of irrigation include canals, tube wells etc. (Fig. 1). The area irrigated by tube wells has increased, while canal irrigated area over the period has decreased. The low efficiency of the canal system and farmers' attitude of having own irrigation source may be credited for this reduction. As the water availability from a source is governed by many factors, therefore different type of irrigation situations can be encountered by the growers.

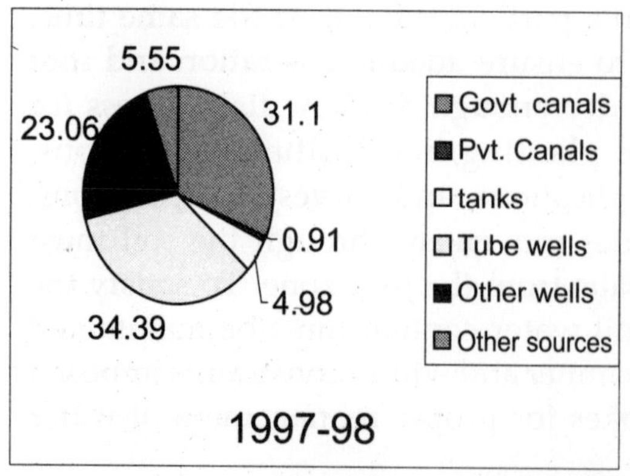

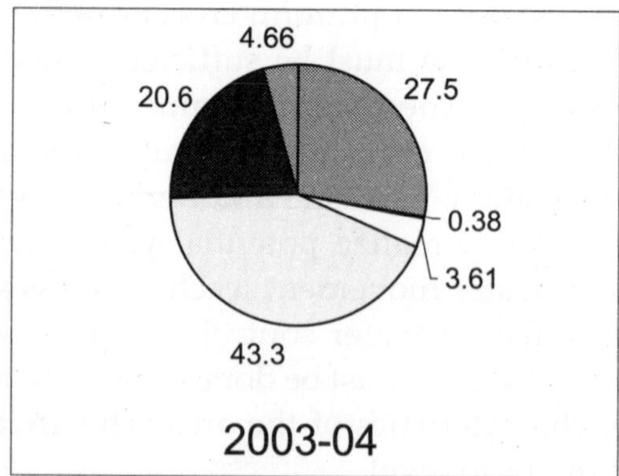

Fig. 1 Source wise irrigated area in India

Farming situations and water management

Water management strategies differ for different farming situations. In India, we have basically three types of framing situations from water availability point of view, namely, fully irrigated farms, under irrigated farms and rainfed farms. These farming situations may have different issues and priorities of water management. A larger chunk of agricultural land (farms) in India (approx. 60%) is rainfed and requires strategies for water conservation, water harvesting and recycling, and cultivation of less water requiring crops. The same is true for areas where availability of irrigation water is limited.

I. Irrigation source

1. Canal irrigation (Water distribution pattern)

Different water distribution practices are being practiced in India and other countries to ensure equity and to meet the crop demands.

(a) Warabandi method

This system has been successfully practiced in the Indo-gangetic plains and in Pakistan for past several decades. The basic principle under-lying the *warabandi* system of managing irrigation water is that the available water, whatever be its quantum, is allocated to the cultivators in equal proportions to their land holdings and not only to some of them to meet their total demand. This principle imposes water scarcity conditions, which necessarily leads to the adoption of more efficient water management practices by the farmer. Distribution of water is carried out by keeping the stream constant and regulating the time of its flow. The entire time in a week is allocated to the farmers. Time once lost can not be regained. So, if for any reason, a farmer is unable to receive his share of water there is no way of compensating him. Similarly, if the flow of water is less than authorized quantity due to some technical defects, the loss suffered on this account can not be compensated. It, therefore, becomes necessary that distributaries and water courses must flow as programmed.

Limitations

- Water may not be available at will of the farmers (as per availability of his time) and requirement of crop
- Tail-enders suffer severe water shortages
- Seepage loss in a water course is not reflected is this system
- No consideration for crops to be grown

(b) Shejpali and other systems of water distribution

The *shejpali* and block system of irrigations is practiced in Maharashtra, parts of Gujarat and Karnataka. In this system, estimates of expected water availability are made. Applications are invited from the farmers seeking information on the crop to be grown and the area to be irrigated under each crop. Water is then sanctioned taking into account the total demand and the water availability. A schedule called *"Shejpali"*, giving terms to different irrigations for the sanctioned crop area on outlets is prepared for each rotation of the canal system.

In the block system, a long term agreement for the supply of water for 6 to 12 years is done especially in case of perennial crops.

A system called *"Rigid shejpali"*, in which definite dates and durations for the supply of water to a particular field area are recorded on the passes issued to the farmers for sanctioned area has been introduced during 2nd half of 20th century. In the *shejpali* system, the farmers are supplied water in turns from tail to head, as per the schedule prepared and communicated in advance. This system gives consideration to the crops to be grown.

In southern and north-eastern states, West Bengal, Orissa, Bihar and Jharkhand, where paddy is the main crop, the irrigation water-flow below the canal outlet is allowed from one field to another through surface flooding. This system of water distribution is referred to as the "localized system" of irrigation for paddy areas. This results in insufficient water distribution and low water use efficiency. None of the systems gives consideration to conveyance losses in the water distribution systems.

Night irrigation in canal systems (Mainly in south and south east Asia)

Night flows are inefficiently used, unless efficient procedures are adopted for their effective utilization. The importance of efficient utilization of night supplies of canal systems has not received the right attentions in many of the systems. Darkness, cold, fear, departure from normal working hours, and desire for sleep deter irrigation staff, farmers and labourers from activities at night. Night irrigation usually requires more labour and costs. It requires smaller stream flows and well graded fields. Paddy, trees and other widely spaced crops are more adaptable to night irrigations, as compared to other crops. Inefficient water applications, breeches in channels and wasted water flowing into drains are common during night.

Remedy : Storage of water in night using nights as flow/ travel time.

2. Open wells/ tube wells irrigation

Water well is a hydraulic hole in the earth down to supply of water excavated for the purpose of bringing groundwater to the surface. Irrigation wells differ from those used to supply water for domestic use because of the large volumes of water that have to be pumped from them for irrigating even small farms. Efficient and economical utilization of ground water through wells depends on design of wells to best suit the characteristics of the wall bearing formations.

While the farmer who has a dug well or tube well is in position to irrigate his crop exactly when and to the extent he desires, the farmer who depends on canal irrigation is on the mercy of a system over which be has no control. The utility of irrigation is judged by the cropping intensity. In most part of the country, cropping intensity is 200% in the tube well or dug well irrigated land as against 100% or less in canal irrigated land. However, a large number of wells/ tube wells dug by individual farmer for irrigating crops like sugar cane, rice etc., has resulted in the depletion of ground water.

Table 1. Effect of canal and well irrigation on yield of wheat and rice

State	Crop	Grain yield (kg/ha) average of 5 years		
		Well	Canal	Tank
Punjab	Wheat	3069	2078	-
Haryana	Wheat	2201	1797	-
Andra Pradesh	Rice	4840	2700	1880
Tamil Nadu	Rice	4770	2140	2090

Source: Reddi and Reddy (1997)

Two general principles for the design of open well or tube well

(i) Water requirement

When the yield potential of a well is more than the field water requirement of the farm under the control of the well-owner, it is always economical to sell the additional yield to the neighbouring farmer rather than limiting the capacity of the well to the need of a small or medium size holding. Individual well for each small holding is often not economical. Thus high discharge irrigation wells should be installed and owned jointly by a group of farmers or by a suitable agency like state tube-well agency.

ii) Location of well

It should be based on:

- Topography of the farm
- Type of water conveyance system
- Method of irrigation
- Recharge possibility
- Safety of pumping set and other installation.

II. Selections of crops/cropping system

A major opportunity of increasing crop yields and WUE in particular a environment is to select plant species adapted to the available amount and distribution of water. High water requiring crops like rice, sugarcane, must be restricted when water supply is low. But it has been seen that farmers immediately shift to rice/ sugarcane from traditional crops once the irrigation facility is created due to associated gains in terns of higher yield easy market and high profits, in addition to preferred food (rice). This results in over exploitation, and thus, depletion of water resource. In such areas, crops like vegetables, cotton etc. could be promising. Whereas in areas like eastern and north eastern regions where ground water is available in plenty and where there is prevalence of water logging, cultivation of water loving crops including rice should be promoted.

III. Selection of land

Development of irrigation facility with in a farm or a large area should be done keeping in view the irrigability of land to be irrigated. If the land is barrren with poor soil conditions like rocky terrain, infertile soil, and other physiographic constraints, investing on irrigation for agricultural crops, will not yield intended benefits of crop yields and income. In such areas hardy crops like oilseeds, pulses, grasses or trees capable of growing and yielding satisfactorily should be given preference.

IV. Method of irrigation

Though the improved and efficient methods of irrigation like drip, sprinkler and subsurface methods are available now, but farmers still prefer to use flood method of irrigation in cereals, oilseeds and at the most, furrow irrigation in vegetable and other widely spaced crops. This is because the crop geometry of most close- growing crops like rice and wheat does not permit economically and technically the use of drip irrigation and its likes. Flood method, moreover, is very convenient and require less labour though there is much wastage of precious water.

The pressurized system of irrigation like drip, sprinkler methods are not suitable everywhere with all crops technically and environmentally, in addition to constraints of high initial and maintenance cost. Many a times, it has been noticed that development agencies/ research agencies are not competent enough to install and maintain such systems what to talk of our illiterate farmers. A complete training of end users and the trainers is essential for their successful use and adoption.

A phasic shift from traditional to micro irrigation must be enforced on. To begin with, the most sensitive states of north and north west, south west and central India bank on very high water use efficiency (about 90%) vis-à-vis tremendous water saving (40-50%) and hight yields should be considred.

Table 2. Efficiency of different irrigation methods

Method	Efficiency (%)
Surface irrigation	30-40
Sub-surface irrigation	40-50
Sprinkler irrigation	60-70
Trickle irrigation	80-90

Table 3. Water use efficiency and crop yield under different leveling indexes

Leveling index (cm)	Water use efficiency (kg/ha-cm)		Yield (q/ha)	
	Wheat	Bajra	Wheat	Bajra
1.2	166	120	46.6	37.3
2.0	138	106	42.2	34.2
2.5	128	99	39.3	32.9
3.0	116	92	36.5	31.4
3.7	110	84	34.9	27.9

$$\text{Leveling Index (LI)}\ \Sigma = \frac{\text{DLi- ALi}}{N}$$

Where, DL is desired level at i[th] grid point,

Al is achieved level at i[th] grid point

N is the total number of grid points

Field to field method of irrigation often results in over irrigation of plots near the delivery point and under irrigation at the tail end. Soil animals like rats etc. make under ground routes for loss of water to fields not desired to be irrigated.

V. Proper leveling and layouts

Land, especially in northern states like Jharkand, Himanchal, Uttarakhand, western ghats, eastern ghats region, north eastern region and many other parts of the country is mostly undulating with variable slopes which leads to non uniform spread and root zone storage of rain or irrigation water which results in poor water productivity. This necessitates proper leveling and layout of farm land. The WUE and yield of wheat and bajra was favoured by the better land leveling (Table 3). Lack of machinery and high cost involved in these operations distract the farmers to get their land leveled. Small and fragmented holdings or fields pose another serious problem for this operation.

VI. Natural hazards

Droughts, cyclones and floods are the most detrimental natural hazards damaging the crops on a large scale. Because of the greed of the farmers for more area or due to compulsion of growing size of family, the one time adequate sized bunds, embankments or channels have been encroached up on for cultivation or for constructions. This has increased the danger of floods at farm or regional level. Cyclones in coastal areas bring salty sea water to the crop lands and also spoil the surface water resources.

VII. Scheduling of irrigation

In a situation where adequate water is available on demand, farmers often irrigate the crops earlier than the time of actual need in their eagerness to obtain good growth and high yield of crops. Their attention is to produce more yields per unit land irrespective of the amount of water used. This leads to wastage of water and it may cause damage to crops and land. On the other hand, delayed irrigation due to lack of proper knowledge may enforce water stress to crop causing decline in yield. Optimum irrigation scheduling should be based on crop needs to avoid over and under irrigation, and to ensure high WUE. There are basically, three approaches for scheduling of irrigation.

1. Plant behavior approach

- Plant appearance
- Plant water potential
- Plant growth
- Indicator plants
- Leaf diffusion pressure

Out of these, plant appearance character is widely used. But it has been seen that considerable damage to the crop due to stress has already been occurred before the plant starts showing convincing water stress symptoms.

2. Based on soil water content

It is the most accurate and dependable method. However, determination of soil water from time to time is necessary to know when the soil water has reached the level at which the irrigation is best applied.

3. Climatological approach (IW/CPE ratio)

The approach is based on the close and direct relationship of corp ET with pan evaporation. It is assumed that when irrigation is applied water is lost from the soil through ET in the same way as the evaporation occurs from an open pan evaporimeter. Thus irrigation can conveniently be applied to a crop using this approach. Howevr, it may not be true completely and could be misleading in scheduling irrigation especially under shallow ground water situations.

4. Depth of irrigation

In general, depth of irrigation is kept as 5-7 cm, it is, however, possible that some crops like pulses vegetables, oil seeds etc. may actually require or may give satisfactory yield with only 3-4 cm irrigation depth. But farmers use flood method of irrigation in which giving 3-4 cm irrigation is not practically feasible. Thus, there is need to devise mechanism of applying irrigation of depths as low as 2-3 cm to improve crop and water productivity especially in case of sensitive crops.

VIII. Land quality degradation: water logging and salinization

Irrigation sometimes instead of being a boon becomes curse for the farmers as it leads to water logging and salinization of the area. Main points of concern here include:

- Excess and non-uniform irrigation
- Seepage from unlined and ill-maintained canal system cause long duration wet conditions in the adjoining agricultural lands resulting into rise in the water table and development of salinity and sodicity
- Unplanned construction of farm buildings and roads
- Under capacity drainage outlets
- Adoption of unscientific cropping practices such as growing more water requiring crops against the designed cropping pattern in a canal command result in application of more than the anticipated quantity of water over relatively smaller areas
- Use of poor quality waters in Gujrat, Rajasthan, parts of U.P. (Agra, Mathura) Punjab and Haryana

Salt problems in some parts of Punjab, Haryana, U.P. and Delhi are due to the combined effect of poor quality irrigation water, shallow water table and inadequate drainage. Ramganga canal catchments and western Yamuna canal catchments salimization are the typical examples of land degradation problems associated with irrigation.

IX. Threats to carrying capacity of irrigation channels

Over a long period of time, the water carrying capacity of field channels gets reduced due to deposition of silt, growth of weeds algae and other biological wastes (plant litter). Seepage and percolation losses also increase due to clogging and breaching of channels. This results in the decrease in area actually irrigated than designed initially.

X. Wastage of farm area and difficulty in farm operations

Open channels take 2-4% of land area out of cultivation and also the net work of channels interferes with farming operations.

XI. Pollution

In many parts of the country, NO^-_3 and $H_2PO^-_4$ or HPO^-_4 leaching with irrigation water has been reported to pollute ground water resources making it unfit for drinking purpose.

Farming Situation II (Farms with limited irrigation water supply)

Most of the issues and problem are alike to that of farming situation I. However, due to shortage of water, several modifications need to be made to improve upon the water productivity. The land modifications which result in the efficient utilization and management of water are discussed as follows.

Use of furrow irrigated raised beds

The constraints associated with this technology are:

- Farm machinery needed is not available with farmers
- Inconvenience for movement of labour in the farm due to V- shaped furrows
- Water many a times fails to seep up to the centre of bed hence germination is hampered
- Repair, maintenance or reformation of beds involve high cost
- Difficulty in ploughing permanent beds; machine adjustments are needed

Ridge and furrow method of irrigation

- Time consuming
- Accumulation of salts at top of ridge
- Poor germination
- Do not suit to closely planted crops
- Difficulty when large-size stream available for irrigation

Farming Situations –III

(Generally rainfed with provision of supplementary irrigation)

Water harvesting and recycling

Rain water harvesting and the collection, storage and recycling of rain water for irrigation and other uses is essential in such aresa. India is characterized with wide variations in physiographic, climatic, edaphic and socio-economic conditions. Therefore, rain water harvesting technology is highly location specific. The various rain water harvesting practices in vogue may be grouped as:

- In-situ retention of rainfall on the land itself by agronomic/ mechanical measures
- Harvesting surface runoff from land surface and its storage for later use
- Diversion of perennial surface and sub-surface water sources (springs and streams) in storage structures, and
- Roof top rain-water harvesting

a) The in-situ water harvesting practices include bunding and terracing, vegetative barriers, contour bunding, leveling, contour farming cover crop and mulching, conservation tillage and deep ploughing.

   Issues: Over designing of size of bunds, wastage of area and high costs, less durability of stone or earthen bunds, problem in establishment of vegetative barriers, fragmented and small holdings create problems in laying contour bunds etc.

b) Surface runoff harvesting can be achieved through dugout ponds, tanks, *Khadins, Ahars/Havelis*, diversion bunds and roof-top rain water harvesting system

   Major issues here include availability of land for such structures, conflict about site, improper design shape and size, no use if no rainfall occurs, losses through evaporation and deep percolation and seepage, and breach of dyke.

c) Stream flow/run off harvesting is practiced through nala -bunding, check dams, stop dams, weirs etc. Issues and problems as in (b)

Suggested means to improve on-farm water management

The thinking of people that "not the water, only land is a limiting resource" needs to be changed because the value of water has hardly been realized by the users irrespective of purpose. From agricultural point of view, the following points may help greatly in improving the farm and factor productivity

- Proper layout of irrigation channels taking into account the irrigability of land, area, slope and physical characteristics of soil
- Consolidation of land for convenient laying of irrigation network and water harvesting
- Formation of water users' association for proper and equitable distribution of water and for maintenance of the system
- The selection of the crops must match with water availability
- Micro irrigation system must be modified suiting to the local situations and be popularized by all possible means
- Installation of underground pipe line system, it will prevent loss of land and avoid difficulties in farm operations
- Flooding needs to be discouraged, however, if necessary, the size of borders or check basin must be optimum for achieving maximum water productivity
- Antecedent soil moistures and contribution of ground water may be taken into account before deciding volume of irrigation
- Land leveling using laser leveler is essential to off set the over irrigation
- In ridge and furrow planting alternate row may be irrigated in water scare areas
- The modern devices (tensiometer, moisture meter) for scheduling irrigation needs to be popularized at commercial level.

□□□

15

Hydroponics Agriculture : Status, Scope and Limitations

RENU PANDEY, VANITA JAIN AND K.P. SINGH

Hydroponics (Greek words *'hydro'* meaning water and *'ponos'* meaning labour) is a method of growing plants using mineral nutrient solutions without soil. It is also called as controlled environment agriculture (CEA) since raising plants hydroponically requires control of environmental factors such as light intensity and duration, temperature, humidity, pH of the solution/medium and mineral nutrients.

Historical background of hydroponics

The study of crop nutrition began thousands of years ago. According to the ancient history, various experiments were undertaken by Theophrastus (372-287 B.C.) while several writings of Dioscorides on botany dated from the first century A.D. are still in existence (Douglas & James, 1975). The classic work on growing terrestrial plants without soil was published by Sir Francis Bacon in 1627, the book named *'Sylva Sylvarum'*. After Bacon's work, water culture became a popular research technique. In 1699, John Woodward published his water culture experiments with spearmint. He observed that plants grew better in less pure water sources than plants in distilled water. Experiments of German botanists, Julius von Sachs and Wilhelm Knop (1859-65) resulted in a development of the technique of soilless cultivation. It was

Professor William Frederick Gericke (1937), who finally introduced the term *hydroponics* and wrote the book named *'Complete Guide to Soilless Gardening'*. Two other plant nutritionists, Dennis R. Hoagland and Daniel I. Arnon, at the University of California wrote a classic in 1938 in agricultural bulletin, *The Water Culture Method for Growing Plants Without Soil*. These two researchers developed several formulas for mineral nutrient solutions, known as Hoagland solutions. Modified Hoagland solutions are still used today.

One of the early successes of hydroponics occurred in 1930s on Wake Island where it was used to grow vegetables for the passengers of Pan American Airlines. Since the inception of hydroponics, research to refine the methodology has continued. In the late 1960s researchers at the Glasshouse Crops Research Institute (GCRI), Littlehampton, England developed the nutrient film technique along with a number of subsequent refinements (Graves, 1983). This research gave rise to the hydroponic systems used today. Jensen and Collins (1985) published a complete review of hydroponics highlighting many new cultural systems developed in Europe and the United States. In recent decades, NASA has done extensive hydroponic research for their 'Controlled Ecological Life Support System' (CELSS). Hydroponics intended to take place on Mars are using LED lighting to grow in different color spectrum with much less heat.

Techniques of hydroponic or controlled environment agriculture

There are primarily two types of hydroponics *viz.*, i) solution culture, and ii) medium culture. In solution culture, there is no solid medium for the roots to support, just the nutrient solution while the medium culture has a solid medium for the roots and is named for the type of medium being used, e.g. sand culture, gravel culture or rockwool culture.

A. Solution culture

There are three main methods of growing plants in solution culture:

(i) Static solution culture

In static solution culture, plants are grown in containers of nutrient solution, such as glass mason jars (typically in-home applications), plastic buckets, tubs or tanks. The solution may or may not be aerated but usually gentle aeration is required. In case of unaerated solution, the level of solution is kept low so that enough roots are above the solution to get adequate oxygen. A hole is made in the lid of the reservoir for each plant. The lid can be of styrofoam (thermocol) or any other material sufficient to support the weight of shoots. There can be one to many plants per reservoir. Reservoir size can be increased as plant size increases. Aeration can be provided by an aquarium pump, aquarium tubing and aquarium valves. Clear containers are covered with aluminium foil, carbon paper, black plastic or other material to exclude light in order to eliminate the formation of algae. The nutrient solution is either changed on a schedule or when the concentration drops below a certain level as determined with an electrical conductivity meter. Whenever the solution is depleted below a certain level, fresh nutrient solution is added. In raft solution culture, plants are placed in a sheet of buoyant plastic that is floated on the surface of the nutrient solution. That way, the solution level never drops below the roots.

(ii) Continuous flow solution culture

The nutrient solution constantly flows past the roots in continuous flow culture. The main advantage of this method is that it is much easier to automate than the static solution culture since sampling and adjustments to the temperature and nutrient concentrations can be made in a large storage tank that serves potentially thousands of plants. One of the popular methods is the nutrient film technique (NFT) in which a very shallow stream of water containing all the dissolved nutrients required for plant growth, is re-circulated past the bare roots of plants in a watertight gully, known as channels. The depths of re-circulating stream remains very shallow, little more than a film of water, and therefore called 'nutrient film'. This ensures that the thick root mat in the bottom of the channel, although moist, is in the air and that there is an abundant supply of oxygen to the roots of the plants. For proper design of NFT system, it is required that it should have right channel slope, flow rate and channel length. The main advantage of the NFT system over other forms of hydroponics is that the plant roots are exposed to adequate supplies of water, oxygen and nutrients. The result of these advantages is that higher yields of high quality produce are obtained over an extended period of cropping. A downside of NFT is that it has a very little buffering against interruptions in the flow e.g. power outages, but overall, it is probably one of the more productive techniques.

The same design characteristics apply to all conventional NFT systems. It is recommended that slopes of 1:30 to 1:40 are used along the channels. This allows for minor irregularities in the surface but, even with these slopes, ponding and waterlogging may occur. The flow rates for each gully should be 1L/minute. The flow rate may be 0.5L/min at planting, while upper limit of 2L/min appears to be the maximum. Flow rates beyond these extremes are often associated with nutritional problems. The channel length should not exceed 10-15 meters otherwise depressed growth rates of plants will occur.

(iii) Aeroponics

In this system, roots are suspended in air in a closed chamber and are saturated with fine drops, in the form of a mist or aerosol of nutrient solution intermittently. In this method no substrate is required. Plants are grown with their roots suspended in a deep air or growth chamber with the roots periodically wetted with fine mist of atomized nutrients. Main advantage of aeroponics is the excellent aeration. Aeroponic techniques have proved very successful for propagation, but have yet to prove themselves on a commercial scale. Aeroponics is also widely used in laboratory studies of plant physiology. Aeroponic techniques have been given special attention from NASA since a mist is easier to handle in a zero gravity environment than a liquid.

B. Medium culture

In this method, there are two main variations for each media, i.e., sub-irrigation and top irrigation. For all techniques, most hydroponic reservoirs are now built of plastic but other materials have been used including concrete, glass, metal and wood. The containers should be such that it excludes light to prevent algae growth in the nutrient solution.

(i) Passive sub-irrigation

This method of culture is also known as passive hydroponics or semi-hydroponics. In this method, plants are grown in an inert porous medium that transports water and fertilizer to the roots by capillary action from a separate reservoir as necessary, reducing labor and providing a constant supply of water to the roots. It is a very simple method in which a pot is kept in a shallow solution of fertilizer and water or on a capillary mat saturated with nutrient solution. The media (expanded clay and coconut husk) that contain more air space than traditional potting mixes deliver increased oxygen to the roots. Epiphytic plants such as orchids and bromeliads, whose roots are exposed to the air in nature can be grown using this method of hydroponics. Additional advantages of passive hydroponics are the reduction of root rot and the additional ambient humidity provided through evaporation.

(ii) Flood and drain sub-irrigation

In this method, a tray is kept on the reservoir of nutrient solution and is either filled with growing medium, i.e., clay granules or any other media and planted directly, or pots filled with media. A timer is used to cause the pump fill the upper tray with nutrient solution at regular intervals, after which the solution drains back down into the reservoir. This keeps the medium regularly flushed with nutrients and air.

(iii) Top irrigation

In Top irrigation method, nutrient solution is periodically applied to the surface of the media. This operation may be done manually once per day in large containers of some media, such as sand or gravel. Usually it is automated with a pump, timer and drip irrigation tubing to deliver nutrient solution as frequently as 5 to 10 minutes every hour.

(iv) Deep water culture

In deep water culture, the plant roots are suspended in a solution of nutrient rich, oxygenated water. In traditional methods, large plastic buckets and large containers are used with the plant contained in a net-pot or thick thermocol sheet suspended from the centre of the lid and the roots suspended in the nutrient solution.

Media

One of the most important decisions a hydroponic farmer has to make is the type of medium he should use. Different media are appropriate for different growing techniques.

Diahydro

Diahydro is a natural sedimentary rock medium that consists of the fossilized remains of diatoms. Diahydro is extremely high in Silica (87-94%), an essential component for the growth of plants and strengthening of cell walls.

Expanded clay

It is made by baking the clay pellets and known under the trade name of 'Hydroton' or LECA (light expanded clay aggregate). Hydroton or expanded clay pellets are suitable for hydroponic systems in which all nutrients are carefully controlled in water solution. The clay pellets are inert, pH neutral and do not contain any nutrient value. The clay is formed into round pellets and fired at high temperatures (1200°C) in rotary klins. This makes the clay to pop-up and become porous. The main advantage of hydroton is it is light in weight and does not compact over time. This is an ecologically sustainable and re-usable growing medium because of its ability to be cleaned and sterilized by washing in solutions of white vinegar, chlorine bleach or hydrogen peroxide and rinsing completely. But there is an opinion that clay pebbles are best not re-used even when they are cleaned due to root growth which may enter the medium. Breaking open a clay pebble after a crop has been grown will reveal this growth.

Rock wool

Rock wool, also called mineral wool is the most widely used media in hydroponics. It is an inert substrate for both free drainage and re-circulating systems. It is produced by aerosolization of molten mineral compounds which results in a fibrous medium accessible to capillary action that is not degraded by microbiological activity.

Coir

Coco peat, also known as coir or coco, is the leftover material after the fibres have been removed from the outermost shell of the coconut. Coir is a 100% natural growing medium.

Perlite

Perlite is made from volcanic rock after being superheated into very lightweight expanded glass pebbles. It is used either loose or in plastic sleeves immersed in water. It is also used in potting soil mixes to decrease soil density and facilitates drainage. Perlite generally holds more air and less water. If not contained, it can float if flood and drain feeding is used. It is a fusion of granite, obsidian, pumice and basalt. This volcanic rock is naturally fused at high temperatures undergoing what is called "Fusionic Metamorphosis".

Vermiculite

Like perlite, vermiculite is another mineral that has been superheated until it has expanded into light pebbles. Vermiculite holds more water than perlite and has a natural "wicking" property that can draw water and nutrients in a passive hydroponic system. If too much water and not enough air surround the plant roots, it is possible to gradually lower the medium's water-retention capability by mixing in increasing quantities of perlite.

Sand

Sand is the cheapest and easily available medium. However, the main disadvantages of using sand are that it is heavy, it does not always drain well and it must be sterilized between use.

Gravel/Quartz

Quartz or gravel of size <2mm can be used as a medium after washing it with dilute acid and properly rinsing with water. Indeed, plants growing in a typical traditional gravel filter bed, with water circulated using electric power head pumps, are in effect being grown using gravel hydroponics. Although it is heavy but it has advantages such as it is inexpensive, easy to keep clean, drains well and won't become waterlogged.

Brick shards

Brick shards have similar properties as that of gravel. But they have the added disadvantages of possibly altering the pH and require extra cleaning before re-use.

Nutrient solution

Plant nutrients used in hydroponics are dissolved in water and are mostly in inorganic and ionic forms. All the 17 elements *viz.* C, H, O, N, P, K, S, Mg, Ca, Fe, Mn, Cu, Zn, B, Cl, Mo and Ni essential for plant growth are supplied using different chemical combinations. Primary dissolved cations are Ca^{2+}, Mg^{2+} and K^{+} and the major nutrient anions in the solution are NO_3^-, SO_4^{2-} and $H_2PO_4^-$. Though numerous 'recipes' for hydroponic solutions are available but all combinations of chemicals reach to a similar total final compositions. Commonly used composition of nutrient solution, also called Hoagland solution, is given in Table 1. Chelating agents are used to keep Fe soluble. Many variations of the nutrient solutions used by Arnon and Hoagland have been named as 'modified Hoagland solutions' and are widely used. Variation of different mixes throughout the plant life cycle further optimizes its nutritional value. pH of the nutrient solution should range between 5.6 to 5.8. Once the plants are grown, it will change the composition of nutrient solution upon contact by depleting specific nutrients more rapidly than others, removing water from the solution, and altering the pH by excretion of either acidity or alkalinity. Care is required not to allow salt concentrations to become too high, nutrients to become too depleted, or pH to change far from the desired value.

Table 1. Composition of nutrient solution for hydroponic culture (maize)

Chemical	Final concentration (mM)	Molecular weight (g)	g/L (200x stock; use 5 ml/L)
Macronutrients KNO_3	1.25	101.11	25.28
$Ca(NO_3)_2$	1.5	236.16	70.85
$MgSO_4$	0.75	246.47	36.97
KH_2PO_4	0.5	136.09	13.61
NH_4Cl	1	53.50	53.50
Micronutrients			g/L (1000x stock; use 5 ml/L)
H_3BO_3	0.00005	61.83	3.0915
MnCl	0.00001	198	1.9800
$ZnSO_4$	0.000002	288	0.5760
$CuSO_4$	0.0000015	250	0.3750
$NH_4Mo_7O_{24}$	0.000000075	1236	0.0927
Fe-EDTA	0.000074	367	27.1580

Note: pH of nutrient solution should be adjusted between 5.6 to 5.8

Scope of hydroponics/ controlled environment agriculture

Growing crops in hydroponics under protected cultivation can be considered the most complex production system available today. In terms of farming systems, Ruthenberg (1980) classified hydroponic cultivation as a "high input - high output - high risk" system. In fact, the available techniques to date require considerable specialisation with sophisticated management and know-how as well as high financial inputs to realise expected production potential, otherwise the crop failures can be disastrous. Before going for a large scale hydroponic system, the growers should be much more critical in regard to site selection, structures, the growing system, pest control and markets.

Site selection

One of the important environmental factors depending on which site selection for CEA should be done is light intensity and duration rather than locating close to a population centre. Highest light levels are especially important if the greenhouse vegetables have to be grown during the winter, when tomato and cucumber prices are at their highest. Generally, a 1% decrease in light reduces yield by 1%. A greenhouse in a high light region can produce more than 500 tons of winter tomatoes per ha per year. Producing such yields in northern latitudes is only possible if the crops are grown through the summer period, when market prices are at their lowest.

Another important factor is temperature. For example, if tomatoes are selected as the crop to be grown year-round, low elevations must be avoided. Because it would be difficult to maintain desirable temperatures in the greenhouse during summers even with fan and pad cooling since the evaporative cooling is usually ineffective due to high ambient humidity. Given the high cost of fan and pad equipment, future hydroponic growers should select sites at specific elevations that have summer temperatures which do not require evaporative cooling, therefore sparing the costs of such cooling equipment. At the same time, an elevation should be selected that is not too high in order to avoid high heating costs in winter. As an alternative to fan and pad cooling a high-pressure fog system can be used. Recent experiences have proven this method of cooling desirable if the feed water is absolutely free of any un-dissolved or dissolved solids. If evaporative cooling systems are used, locating the greenhouse in a region of low outdoor humidity is important.

Important consideration for site selection is that the site should be free of insects that might be vectors for severe viral diseases. Growing in regions where there are mild winters normally increases the incidence of insects and diseases due to the continued life cycle of the pest. Selecting a site that is not already a major producer of vegetable crops is also advisable for economic returns purpose.

Energy and water

There are many choices available for energy sources such as natural gas, propane, fuel oil and electricity. Earlier hydroponic growers did not consider cost differences between the types of energy used in CEA. Many used natural gas and fuel oil. Coal was also used but air pollution standards and regulations made the use of this fuel prohibitively expensive.

Water quality is one of the major concerns for hydroponics. Plant growth is affected by the interaction of the dissolved chemical elements in the water supply, the chemical properties of the growing medium to which the water is applied, and the fertility program employed. Therefore, in selecting a greenhouse site, a grower must be aware of several chemical properties of water that might cause problems for greenhouse crops: pH, alkalinity, soluble salts, calcium, magnesium, boron, fluoride, chloride, sulphates, sodium, carbonate, and iron. The cleaner the water, the greater the opportunity to achieve maximum yields.

Structures and environmental control

Today the European glass structures are commonly being built for vegetable production in the south-western part of the United States and are very different from the polyethylene /fibreglass houses used in hydroponic production between 1965 and 1990. To achieve a more uniform growing environment without rapid temperature fluctuations, more total volume of space is being allotted within a greenhouse. The types of polyethylene sheet films retard the loss of infrared heat. These films are reported to reduce 20% of the heat loss from a greenhouse and are used commonly. Other glazing materials, such as fibreglass, polyvinyl chloride, have proven either inconvenient or much more expensive than polyethylene. Newer materials, such as polycarbonates and acrylics have become much more common, but their popularity has been offset by high costs.

Greenhouses are expensive and controlling the environment within a greenhouse requires considerable energy. Whatever be the source of energy, it should be conserved once it is in the greenhouse. In regions of cold winter weather, thermal curtains of porous polyester can be installed to reduce night heat loss by as much as 57%.

Within a greenhouse, computers can operate hundreds of devices such as vents, heaters, fans, hot water mixing valves, irrigation valves, curtains, lights, etc. by utilizing dozens of input parameters, such as outside and inside temperatures, humidity, outside wind direction and velocity, carbon dioxide levels and even the time of day or night. A computer can keep track of all relevant information such as temperature, humidity, CO_2, and light levels.

Such a data acquisition system enables the grower to gain a comprehensive understanding of all factors affecting the quality and timeliness of the product.

Selection of growing systems

While there are many types of growing systems, the two most popular growing media today are rockwool and perlite but in India, coco-peat along with perlite is most commonly used. The irrigation system can be activated more than 30 times per day depending upon the temperature. At University of Arizona, excellent tomato crops were grown in a container no bigger than 956 cm^3 and maximum yields were 12.8 kg of tomatoes per plant over a period of 6-months.

In order to reduce media cost and maximize control over mineral nutrition, pH, aeration and root diseases, little root volume should be provided. High salt levels are maintained in the root systems where the E.C. of feed solution approach 3.5 and the drain water at an E.C. of 4.5 to 5.0. In future, systems designed should be closed, with no drainage which will prevent any loss of mineral elements and the contamination of groundwater.

Pest control

White flies, leaf miners, pin worms, nematodes, *Cladosporium* leaf mold and viruses, as well as root diseases such as *Pythium* root rot and bacterial wilt are commonly reported in hydroponics/CEA. To prevent many of these problems, the drain solution is often sterilized (Runia, 1995) using heat treatment, ozone and ultraviolet radiation. To control pests, integrated pest management (IPM) is of particular interest. Crop production requires both the identification of possible crop disease and insect problems, and the ability to properly integrate disease and insect prevention and control practices into a total management plan.

Status of hydroponics

International status

The area under hydroponics/CEA began to expand significantly in Europe and Asia during 1950s and 1960s, and large hydroponic systems were developed in the deserts of California, Arizona, Abu Dhabi, and Iran (1970) (Fontes, 1973; Jensen and Teran, 1971). In these desert locations, the advantages of the technology were augmented by the duration and interest of solar radiation, which maximized photosynthetic production.

At present, the largest commercial hydroponics facility in the world is "Eurofresh" Farms in Wilcox, Arizona, which has sold 125 million pounds of tomatoes in 2005. Eurofresh has 318 acres (1.29 km^2) under glass and represents about a third of the commercial hydroponic greenhouse area in the United States. Eurofresh does not consider its tomatoes organic, but they are pesticide-free. They are grown in rockwool with top irrigation.

Almost 20 years have passed since the last real commercial interest in hydroponics, but today there is renewed interest among growers to establish CEA/hydroponics systems. In the last four years, nearly 40 ha of greenhouses have been built in Colorado, Nevada, and Arizona. Many more hectares are planned, not only in the Southwest, but in Mexico too. The future for hydroponics appears more positive today than any time over the last 50 years.

National status

Hydroponics did not reach India until 1946. The first research study was initiated at the Government of Bengal's Experimental Farm at Kalimpong in the Darjeeling district. Initially, a number of problems were faced but later on after a careful appraisal of salient problems, during 1946-47, the Bengal System of hydroponics was developed representing the efforts to meet Indian requirements. Last year (2008), about seven farmers in south Gujarat have adopted this technology for growing different varieties of exotic hybrid tea roses by Mr. Kumar Patel of 'Best Roses'. He currently produces 10 million rose stems worth about Rs 5 crore, all of which is flown to overseas flower markets in Japan, New Zealand and Europe. It is now planned to double the production to 20 million rose stems by installing the hydroponics system on 12 hectares in Kuched village of Navsari district. Localised experiments have also shown that other exotic crops like strawberry, green garlic and tomatoes can also be grown using this imported technology. A Pune-based turnkey consultants, Flora Consult, which had helped 'Best Roses' set up the system three years ago, is now helping a vitrified-tile industry to set up a cluster of seven projects of exotic hybrid tea roses in Varawav village near Idar, Sabarkantha district of North Gujarat. Another giant 'Landmark Agrotech' project is the second big hydroponics project in Gujarat and is currently under implementation.

In India, several tracts of wastelands having poor quality soil but plenty of water can be brought under hydroponics. All that will be needed is to create an impervious surface at the bottom and bunds to hold water. The technology used for polythene lining of canals can come in handy for creating large hydroponic farms to grow food crops, vegetables and other plants.

Advantages of hydroponics over conventional agriculture

Today, hydroponics is an established branch of agronomical science. Progress has been rapid and results obtained in various countries have proved it to be thoroughly practical and to have very definite advantages over conventional methods of agriculture/horticulture. The two chief merits of the soilless cultivation of plants are, first, much higher crop yields, and secondly, the fact that hydroponics can be used in places where ordinary agriculture or gardening is impossible. Besides these, there are other advantages listed below:

- Less space is required to produce same amount of crop as compared to that grown in the field.
- Less growing time is required. Growth of plant is faster as there is no mechanical impediment to the roots and all the nutrient elements are available to the plant in plenty. Further, to increase plant growth, lighting systems such as metal halide or high pressure sodium lamps are used to lengthen the day or to supplement natural sunshine.
- Labour and garden maintenance is reduced as the intercultural operation is often absent or is very less, fertilization and irrigation is automated and no hard manual work is required.
- Water conservation is the biggest advantage. Hydroponics saves an incredible amount of water since it uses as little as 1/20th the amount as a regular farm would to produce the same amount of food. Waterlogging never occurs.
- Saves money by recycling nutrients and water. In case of closed system of hydroponics such NFT, the nutrient is recycled thereby preventing loss of nutrient elements and preventing soil pollution. Large amounts of water can be recycled not used by the plant, after being aerated and eliminating anoxic conditions.
- Pest and disease problems can be controlled easily while weed is practically non-existent.
- Plants grown hydroponically avoids soil borne pests.
- More control over the plants rooting environment as the root zone's temperature, humidity, darkness, etc. can be easily manipulated.
- Higher yields can be obtained since the number of plants per unit is higher compared to conventional agriculture and produce can be obtained over an extended period of cropping (Table 2)
- Higher returns. Some plants can be raised out of season which can fetch higher income to the farmers.
- Excellent quality of produce with no dirt or smell.
- A hydroponic farmer in Virginia developed a Ca and K enriched head of lettuce claimed that their hydroponic lettuce uses 90% less water than traditional soil farming.

- Besides being a commercially useful technique, hydroponics is also a standard technique to be used in biological research and teaching.

Table 2. Hydroponic averages compared with ordinary soil yields

Crop	Bengal		United States	
	Hydroponic equivalent per acre	Agricultural average per acre	Hydroponic equivalent per acre	Agricultural average per acre
Wheat	8,000 lb.	5,600 lb.		
Oats	3,000 lb.	1,850 lb.		
Rice	12,000 lb.	750-900 lb.		
Maize	8,000 lb.	1,500 lb.		
Soya beans	1,500 lb.	600 lb.		
Potatoes	70 tons	8 tons	62.5 tons	6-10 tons
Beetroot	20,000 lb.	9,500 lb.		
Cabbage	18,000 lb.	13,000 lb.		
Peas	140,000 lb.	25,000 lb.		
Tomatoes	180 tons	5-10 tons	300 tons in Florida	5-10 tons for unstaked field plants; 50 tons for greenhouse cultured crops
Cauliflower	30,000 lb.	15-35,000 lb.		
French beans	42,000 lb. of pods for eating			
Lettuce	21,000 lb.	9,000 lb.		
Lady's finger	19,000 lb.	5-10,000 lb.		
Cucumber	28,000 lb.	7,000 lb.		

Source: Douglas and James (1975)

Limitations of hydroponics/CEA

Though there are many merits of hydroponics over conventional agriculture, there are some limitations too:

- Higher set up cost
- Growers require skill and knowledge to maintain optimum production in commercial applications
- Because each plant in a hydroponics system is sharing the exact same nutrient, diseases and pests can easily affect each plant.

- Plants react quicker to changes in the environment, however, if this change is for the worst, plants will quickly react to it; showing signs of deficiency or trouble.
- Hot weather and limited oxygenation may limit production and can result in loss of crops

As with most things, it is important to understand the advantages and disadvantages of hydroponics. This will allow one to make informed decisions on what application is "right" for his individual requirements. This technology allows for growing where no one has grown before, be it underground, or above, in space or under the oceans this technology will allow humanity to live where humanity chooses. If used for our own survival, hydroponics is and will be a major part of our collective future (Winterborne 2005).

References

Douglas, James S. Hydroponics. 5th ed. Bombay: Oxford Uni. Press, 1975. 1-3.

Fontes, M.R. 1973. Controlled-environment horticulture in the Arabian Desert at Abu Dhabi. HortScience 8:13-16.

Gericke, W.F. 1940. The complete guide to soilless gardening. Prentice-Hall. Englewood Cliffs. N.J.

Graves, C.G. The nutrient film technique. *Hort. Rev.* 5: 1-44.

Hoagland, D.R. and D.I. Arnon. 1938. The water culture method for growing plants without soil. *California Agr. Expt. Sta.* Circ. 347.

Jensen, M.H. and W.L. Collins. 1985. Hydroponic vegetable production. *Hort. Rev.* 7: 483-558.

Jensen, M.H. and M.A. Teran. 1971. Use of controlled environment for vegetable production in desert regions of the world. Hortscience 6:33-36.

Runia, W.T. 1995. A review of possibilities for disinfection of recirculation water from soilless cultures. *Glasshouse Crops Res. Sta.*, Naaldwijk, Holland. p. 9.

Winterborne J, 2005. Hydroponics - Indoor Horticulture.

□□□

Combating Pests Weeds, Insects

16

Strategies for Cost Effective Weed Management

S.S.L. TRIPATHI

In addition to significant crop losses, weeds make crop harvesting more difficult. If the harvested crop is contaminated with weed seeds, growers who save harvested seed grains to plant the next season will sow weed seeds along with crop seeds. The quality of the food and feed derived from the crop is lower if weed seeds contaminate harvested grains.

Most of the farmers in the country have relied largely on hand weeding and hoeing. High labour cost makes manual weed control uneconomical in large farmers. Manual weed control is being used less often in developing countries, because laborers are leaving rural areas for better paying jobs in urban areas. Other weed control techniques that can be used on a regular basis are cultivation, crop rotation, cover crops, mowing, mulching, planting geometry and planting time to give crops the competitive edge over weeds. In a few instances, bio-control agents, such as plant pathogens and insect herbivores are also used to manage weeds.

Integrated weed management (IWM) is modeled after the more familiar integrated pest management (IPM) used to control insects and plant pathogens. Both IWM and IPM are based on ecological and evolutionary principles. The ultimate objective is to achieve cost effective control while reducing inputs of synthetic pesticides. A fundamental concept in IWM is implementation of control measures based on economic thresholds. Crop plants can withstand some competitive pressure from weeds without a significant decrease in crop yield. Weed scientists have established the maximum number of various weed species that can be tolerated by a crop before significant decrease in yields–the economic threshold. Growers are encouraged to wait until the weed reaches the economic threshold before implementing control measures. In addition, IWM relies on using a variety of control measures to slow the evolution of resistance to a single control measure. Therefore, IWM is maximized when growers have access to the widest possible array of weed control tools.

Basic keys to effective weed management

Four keys to an effective weed management programme are knowing the types of weeds present, the problems they cause, estimating their economic thresholds and knowing their available management options.

3- dimension approach to weed management

An integrated approach which takes into account the growth characteristics of both desirable species and weeds is the key to success. The three main steps of a successful weed management programme are:

- Removal of weeds or reduce weed seed set
- Restrict weed germination
- Encourage competition from desirable species.

Deliberation

Determining the level of weed infestation will indicate whether control is warranted; which weeds are having the greatest impact; priority paddocks for weed control and their locations, and potential control and management tactics. For successful control, understanding of the seed production and germination periods of the major weeds and pasture plants is essential.

Diversity

A wide range of cost-effective 'restoration' tools are available to achieve these objectives, and may be broadly categorised as either 'weed removers' (which remove weeds from pastures and/or reduce weed seed set) or 'pasture improvers' (which encourage desirable and competitive species). Plan to use as many tools as possible in conjunction with each other, so that all the tools, such as tactical use of fertilizer, tactical herbicide use, biological control agents, can assist in weed management.

Diligence

Estimating the losses a weed may cause and its economic treatment threshold, are closely related. The economic threshold is the infestation level that causes an economic loss greater than the cost of management. The treatment threshold is the level where management must be applied to avoid reaching the economic threshold. If the weed population or level of control obtained does not result in enough additional production to cover the cost of the herbicide, then the treatment threshold has not been met and the spray should not be done.

In recent years, the focus of weed control has increasingly been placed on using a combination of methods—a concept known as Integrated Weed Management (IWM) or Best Practice Management. IWM is a multi-disciplinary approach; it includes chemical, physical and biological methods of weed management together with an understanding of ecology and effective education and extension. Each weed control method has advantages and disadvantages and, even though individual techniques have become more sophisticated, none has emerged as the ultimate tool for effective weed management at a reasonable cost.

Effective weed control can be achieved using various options at appropriate times. The physical control techniques include hand pulling (annuals and tree seedlings), cutting (vines, shrubs, tree saplings and trees), chaining (trees), mowing (annuals, tree saplings, shrubs and vines), tilling (annuals, shrubs and tree seedlings) and hoeing (plants regenerating underground).

Cost effective weed management strategies

1. Preventive measures

(A) *Use of weed-free crop seeds*: Many weed seeds having similar life cycle to that of crop may contaminate it. The weed-free crop seeds may be achieved by-

- Cleaning and testing of all the farm produce i.e. crop seeds
- Using clean agricultural implements
- Using well decomposed weed-free FYM and compost
- Avoid weed seeds carrying with irrigation water and feeding screenings, grain or hey containing weed seeds to live stock
- Avoid use of gravels, sand and soil from weed infested areas
- Use vigilence

(B) *Crop management practices*:

- Use of vigorous and fast growing crop varieties to compete with weeds
- Proper fertilizer placement to ensure more availability of nutrients to crop
- Proper crop rotation for checking weeds
- Optimum plant population per unit area
- Effective prevention of weed seed production in both cropped and non-cropped areas

(C) *Seed certification*: Always use certified seeds because these are pure material

(D) *Strict follow up of weed laws*: Prevent the spread of noxious weed seeds to enter in the country

(E) *Quarantine Laws*: Restrict the entry of interstate and international movement of imported plants capable of spreading diseases, insects and weed seeds.

Mechanical methods

(i) Hand Pulling

- Good for small areas like kitchen garden, lawns etc.
- Within rows and hills of certain crops in which the weeds are difficult to control with implements

(ii) Hoeing

- Practiced in row crops & gardens

(iii) Mowing

- It reduces food reserve in underground parts of plants

(iv) Flooding

- It kills plants by excluding air. High temperature in summers may kill protoplasm

(v) Burning

- Used mostly in non-cultivated lands

(vi) Mulching with non-living material

- Mulching prevents light and air to check the top growth of weeds

(vii) Summer Ploughing

- Summar ploughing dessicates the rhizomes, tubers and roots by exposure to solar radiation

2. Cultural methods

(I) Selection of competitive crops and varieties

(a) *Crop competition*: Crops like sorghum, corn, soybean, cowpea etc compete better with weeds for nutrients, soil moisture, sunlight and CO_2

(b) *Crop rotation*: Continuous growing of one crop in rotation may increase the population of weeds. Rotation must be changed

(II) Adjusting date of planting

Adjust the planting time in such a away that the crops become established before the growth of weeds

(III) Spacing

Narrow spacing may cause less weeds

(IV) Planting method

Planting should be done after pre-sowing irrigation so that weed seeds present in the field may grow and ploughed

(V) Fertilizer placement

In fast growing crops like maize and sorghum etc if the 'N' is applied in the higher amount then weed growth will be suppressed due to less response of 'N' to the weeds

(VI) Irrigation and drainage

- Frequent irrigation interval at initial crop growth stage induce several flushes of weeds
- Initial standing water in low land rice reduces weeds.
- Tubers of *Cyperus* and rhizomes of *Cynodon* are killed due to lack of O_2 in submerged conditions in rice

(VII) Soil Solarization

This is a technique of heating the soil using solar energy to kill soil born pathogens, nematodes and weed seeds. A duration of 4 or 6 weeks of solariztion is necessary for satisfactory control of most of the weeds.

- Gives excellent control of wide range of weeds
- Many annuals and some perennials can be controlled
- Soil solarization has been found superior over herbicides and manual weed control
- Crop raised on solarized soil always tend to have healthy and vigorous seedlings

(VIII) Incorporation of crop residues

- Incorporation of residues of crops like, wheat, rice and sorghum in the soil reduce biomass of many weeds like *Phalaris minor*

3. Biological methods

Natural enemies of plants (bioagents) are used which are harmless to desired plants. Bioagents may be insects, disease organisms, parasitic plants, selective graizing by live-stock in forest gardens and orchards.

This is a long time programme and needs extensive testing before the release of parasite for control of weeds. The bioagent must be host specific.

4. Some examples of biological weed control

Weed	Bioagent
Opuntia spp.	Moth borer (*Cactoblastis cactorum*)
Lantana camara	Lantana bug (*Teleonimia scrupulosa*)
Eichornia crassipes	Rhizoctonia blight

5. Use of bio-herbicides

Weed	Bio-herbicide
Morrenia odorata in citurs	De-vine
Aeschynomone spp.	Collego
Sorghum halepense	Bipolaris
Non-specific	Biolophos

Integrated weed management

Herbicides continue to be a powerful tool for controlling weeds. However, there is a growing desire to be less dependent on "herbicidal" control of weeds. Herbicides are viewed as expensive; some weeds have developed resistance to various herbicides, and there is growing public concern regarding the environmental/health effects of pesticides.

Integrated weed management (IWM) is an approach of weed control that attempts to address these concerns. The approach is driven by the need to maximize crop yields and profits while simultaneously protecting natural resources and minimizing environmental effects. IWM strives to be less dependent on herbicides by relying on the combination of many agronomic practices to control weeds. These practices include the following:

- Practices that favor crop competition
- Diversified production systems that make it difficult for weeds to adapt
- Practices that limit the spread of weed species
- Optimal herbicide use

Agronomic practices that favor crop competition

Seeding shortly after seed bed preparation

Seeding should occur as soon as possible after seed bed preparations are complete to prevent weeds from getting a "head start" on the crop. The earlier weeds emerge relative to the crop, the greater the crop yield loss.

Shallow seeding and good "on row" covering

Seeding as shallow as conditions permit is key to rapid emergence. But seeding deep, particularly into cold soils, will likely delay emergence, increasing the crop's susceptibility to seedling diseases and the effects of herbicide residues.

Drilling in a crop with good "on row" covering results in better crop establishment compared to broadcast seeding, particularly in a dry spring. On row covering squeezes the soil like a sponge, providing better moisture conditions for crop germination while leaving the soil between the rows loose and less than ideal for germination of weed seeds.

Row spacing

Reported effects of row spacing on crop yield and weed management have been variable.

Optimum seeding rate

The optimum seeding rate is influenced by a multitude of environmental and economic factors. During drought period, lower seed rate than normal may increase crop yield by reducing inter-plant competition for soil moisture. When conditions are ideal, higher than normal seeding rates can increase crop competitiveness and yield.

Tillage

The use of tillage to control weeds can be quite controversial. The adoption of reduced tillage practices has led to reduced costs of production and the conservation and improvement of soil resources. The majority of reduced till farmers would be quite reluctant to use tillage as a means of weed control for fear of losing these benefits.

The weed problem has intensified with the introduction of high yielding and input intensive technology. Due to development of herbicide resistance, weed shift, soil, water and environmental pollution and health problems it is imperative to use environmentally safe, sustainable and cost-effective weed management technology.

References

Kulshrestha, G. and B.S. Parmar, 1992. Resource management for sustainable crop production. Souvenir 1992, Indian Soc. Agron., pp. 339-343.

Kumar, B., N.T. Yaduraju, K.N. Ahuja and U. Prasad. 1992. Effect of soil solarization on weeds and nematodes. Ann. Weed Sci. Cof. Indian Soc. Weed Sci. Proc. 137.

Kumar, B.N.; N.T. Yaduraju; K.N. Ahuja and D. Prasad, 1993. Effect of soil solarization on weeds and nematodes under tropical Indian conditions. *Weed Res.* 33: 423-429.

□□□

17

Strategies for Cost Effective Insect Management

S.N. TIWARI

Management of insect-pests has become a costly affair in most part of India where different IPM modules are being copied blindly without proper analysis of insect-pest problems and evaluation of management tools/ techniques which are highly specific to agro-ecosystem. Moreover, even after spending a huge amount in crop protection, farmers are unable to tilt cost benefit ratio in their favor. According to most acceptable definition of IPM, pest management is an ecological approach to pest control in which all available techniques are evaluated and consolidated in a unified program to manage the pest population so that economic damage is avoided and adverse side effects on environment are minimized. Most unfortunately, in most of the agro-ecosystem available techniques are being consolidated without proper evaluation due to which neither economic damage nor adverse side effect on environment is being minimized even after putting a lot of resources.

Insecticides are the most important tool for the management of insect pests and presently it is not possible to control any major insect pest without its use. However, its application proves to be very costly if not used judiciously. In most of the agro-ecosystem the cost of protection increases mainly due to injudicious use of insecticides as insect pests quickly develop resistance against it which is also inherited to future generations. On the other hand, manipulation of cultural practices and natural control significantly check the insects to become pest at no extra cost. Under such conditions the insect management may be made cost effective, by strengthening the cultural and natural control and checking or delaying the development of resistance by adopting insecticide selectivity. However, such attempts will succeed only when it is based on ecosystem specific information.

A. Why the management of insect-pests is so costly at farmer's level?

The cost of protection is increasing day by day due to various reasons the most important of which are:

1. *Use of ineffective methods*: In most of the agro-ecosystem in India the farmers are spending a lot of money in implementation of ineffective control measure. Generally such control measures are implemented on the basis of their performance in other ecosystem. Since most of such measures are ecosystem specific, they fail to deliver desired result even after huge expenditure. For example, generally it is reported that parasitoids such as *Trichogramma chilonis* or *japonicum* are useful in management of Yellow Stem Borer of Rice. Although, such parasitoids parasitize the eggs of YSB in some agro-ecosystem, they fail completely in others. Similarly, so many bio-pesticidal formulations of neem, bacteria, viruses and fungi are being used at large scale in so many agro-ecosystem, although, their performance is not appreciable everywhere. The farmers also use so many insecticides which are not effective due to various reasons.
2. *Non-implementation of cultural control methods*: Purposeful manipulation of agronomic practices such as tilling and cultivation of soil, use of resistant varieties, time of sowing and harvesting, use of clean seed, regulating irrigation, controlling crop growth with fertilizers, clean cultivation, cropping scheme and trap crop, pruning and thinning and destruction of crop residue etc. have been found to be very effective in controlling the insect pest. However, such cultural control practices are not followed by farmers seriously, although, no extra cost is involved in its implementation.
3. *Injudicious use of insecticides*: Insecticides are costliest input in crop protection and injudicious use of it makes the pest control even more costly. These chemicals are useful and cost effective alternative only when applied at right time, at right dose by using right application technology. Most unfortunately, the insecticides in Indian agriculture are the highly misused tools which are mainly responsible for increasing cost of protection.

4. Development of resistance against insecticides: Resistance develops rapidly if most of the pest population is exposed to the specific insecticide, if the insect can multiply quickly or if there is limited immigration of unexposed individuals. Selection pressure for resistance is reduced if part of the pest population is on alternative host plants or other crops which are not treated with the chemical. Migration of these susceptible individuals may be important in delaying the onset of resistance, thus emphasizing the need to minimize drift of chemical beyond the treatment area to other host plants. Unfortunately, the user is often tempted to increase either the dosage, or frequency of application, or both, when resistance is suspected, and this merely aggravates the situation and increases the cost of protection.

B. Strategies for cost effective insect pest management

Cost effective management of insect pest is possible only when agricultural ecosystem is planned after proper understanding and control measures are applied after cost/benefit and benefit risk analysis.

1. Proper understanding of pest and IPM

Proper understanding of pest is prerequisite of any pest management programme, as in many instances, the farmers have been found applying fungicides or other chemicals for the control of insects. Also, in so many conditions the farmers are themselves responsible for elevation of insect to pest level. Under such conditions the farmers should be ready to answer following questions:

What is a pest? Any form of plant or animal life or any pathogenic agent, injurious or potentially injurious to plants, plant products, livestock or man are known as pest.

What is an insect pest? Insects which feed on crop plants and cause economic damage directly or indirectly by transmitting the diseases are known as insect pest.

Should all the insects present in the crop field be treated a pest? Never. Many of the insects present in crop field are beneficial to it as they control harmful insects or pathogens. Even those who damage the plant are not always a pest and their presence in the field as non-economic population is essential for successful integrated pest management as they serve as food of parasites and predators.

When insects become a pest? When they get most suitable food over a long period of time, the climatic conditions become most favorable for their growth and development and when their natural enemies are killed due to various reasons.

Do we have any role in helping insects or pathogens to become a pest? And how? Yes. By cultivating most susceptible variety of crop over long area for longer duration, enhancing vegetative growth of susceptible variety by excessive use of fertilizers or by killing natural enemies of pest by using insecticides injudiciously.

Do all the insect pests cause economic damage all the time, every where? No. They cause economic loss when their population exceeds a certain level - Economic Injury Level.

Do all the pests have equal pest status? No. Depending on the level of damage they may be classified as:

Key or major pest: These pests occur perennially and cause serious and persistent economic damage in absence of effective control measures. E.g. Brown plant hopper, Stem borers in rice.

Minor pest: These pest cause economic damage only under certain circumstances in their local environment, e.g. Leaf folder, Gall midge.

Occasional pest: In certain years and in certain locality they build up their population and achieve pest status, e.g. Army worm, Ear-cutting caterpillar.

Potential pest: They have potential to develop into major pest depending upon circumstances, e.g. Leaf folder, Gall midge, Whorl maggot.

Migrant pest: Move from one zone to another, e.g. Locust.

Is it necessary to eradicate all the pests from the field? No. A non economic population of the pest should always be maintained in the field so that natural enemy of the pest could survive and multiply itself. With the eradication of the pest the natural enemy of the pest may also be eliminated from the field.

How to control insect pests? Insect pests may be controlled by:

- Manipulation of agronomic practices - Cultural control
- Causing mechanical injury to pest or killing them mechanically - Mechanical control
- Manipulation of physical factors - Physical control
- Introduction and enhancement of natural enemies - Biological control
- Application of pesticides - Chemical control
- Legislation for pest control related activities - Legislative control

Is chemical control always necessary to reduce crop losses? No. Chemical control should be adopted only when the population of the pest or their damage reaches to Economic Threshold Level even after implementation of non-insecticidal control measures.

Are other non-pesticidal methods of control effective? If applied timely, they are highly effective against concerned pest.

Why should we adopt IPM?

- To avoid the collapse of control system based on single technology.
- To avoid pest control to enter exploitation, crisis and disaster phase

Exploitation phase : The pest control programme is dependent solely on chemical pesticides which are exploited to the maximum resulting in high yield.

Crisis phase: After many years in exploitation phase and heavy use of pesticides, pesticide resistance, pest resurgence and out break of secondary pest occur which increases the production cost?

Disaster phase: The use of pesticide increases the cost of production to the point where crop could no longer be grown profitably. High residue of pesticide in the food makes the produce unacceptable to consumers.

Currently most control is in exploitation phase and integrated pest management should be quickly adopted to avoid the crisis and disaster phase.

2. Proper understanding of area specific agro-ecosystem

Crop: Cropping pattern of the area, seasonal history of the crop, growth stages of crop, agronomic practices

Pests: Taxonomic category and common name of the pest: Insect, fungi, bacteria and mycoplasma like organisms, virus and viroides, nematodes, protozoans, algae, phaenerogamic parasites, rodent, molluscs, weeds

Status of the pest: Key or major pest, Minor pest, Occasional pest, Potential pest, Migrant pest

Economic injury and economic threshold level

Economic Injury Level: It is the lowest pest population density/disease index that will cause economic damage

Economic Threshold Level: It is the pest density/disease index at which control measures should be applied to prevent an increasing pest population/ disease from reaching the Economic Injury Level.

Seasonal and life history of pests: In which month they appear in the crop? In which stage they infest the crop? How they damage the crop? Where they lay the egg and in how many days egg hatch to larva? Where and how the adult or larvae feed? Where the larvae pupate? When the adult emerge and mate? In how many days the pests complete its development? How many generation of the pest are there in that area? Do they remain active through out the year? Where they hibernate/survive during the off season?

Abiotic and biotic conditions favorable to pest: Alternate hosts of the pests susceptibility of main host, shelter of pest, susceptible stages of the pest, temperature, relative humidity, rain fall.

Natural enemies of the pest: Taxonomic category of natural enemy, seasonal and life history of natural enemy, susceptible stages in the life history of natural enemy, alternative host and artificial food of natural enemy, super or hyper parasites of natural enemy, climatic conditions unfavorable to natural enemy, susceptibility of natural enemies to pesticides.

3. IPM Tools

Cultural methods: These comprise regular farm operations, which are so performed as to destroy insects or to prevent them from causing injury. Practices such as tilling and cultivation of soil, use of clean seed, regulating irrigation, use of resistant varieties, controlling crop growth with fertilizers, clean cultivation, cropping scheme and trap crop, pruning and thinning, time of sowing and harvesting and destruction of crop residue are helpful in reducing pest population and damage.

Mechanical methods: These comprise various mechanical devices to kill the pest. Hand picking, use of hand nets and bag nets, beating and hooking, shaking and jarring, sieving and winnowing, trenching field and use of mechanical traps are highly effective in small fields.

Physical methods: Application of physical factors such as heat, cold or moisture to the disadvantage of pest is also useful in management of storage pests.

Biological control: This constitute a deliberate attempt to use natural enemies either by introducing new species in the environment of the pest or by increasing the effectiveness of those already present by conservation and enhancement. It is implemented by preservation of inactive stages, avoidance of harmful cultural practices, and maintenance of diversity of agro-ecosystem, providing natural food, artificial food supplement and shelter, control of honeydew feeding ants and protection from pesticides

Chemical control: Use of different chemicals such as insecticide, attractants, repellants, sterilants, growth inhibitors for killing or reducing insect population.

Genetic method: Propagation and release of sterile or genetically incompatible insect to check breeding.

Regulatory method: Legislation to control pest activity and pest control operation by plant and animal quarantine, eradication and suppression program.

4. Implementation of insect control methods after proper evaluation

None of the insect control methods are equally effective in all the agro-ecosystem due to difference in cropping system, climatic conditions, biotic and abiotic stresses and level of resistance. Therefore, before implementation all the available insect pest controlling techniques should be evaluated at farmer or institutional level. Special attention should be given to bio-control agents and bio-pesticides whose performance is highly unstable in different agro-ecosystem.

5. Consolidation of effective control measures in unified program

In the Integrated Pest Management only effective control measures should be integrated in unified programme to achieve maximum yield without any adverse effect on natural enemies or other biotic components of environment.

6. Insecticide Selectivity

Judicious application of insecticide is strongest strategy to make the insect management cost effective and it can be implemented through following ways:

Need based application: Schedule based application of insecticides are very costly and ineffective alternatives in pest control as in most of the conditions the chemical is not present at lethal concentration when the insect pests attack the crop. On the other hand, need based application based on ETL is cost effective alternative in IPM as application coincides with attack and requires less insecticides.

Use of recommended insecticides: All the insecticides are not effective against all the insects on all the crops. Therefore, crop and insect specific recommendations should be followed to effectively control the insect at minimum cost.

Use of new active ingredients: Most of the insect pests have developed resistance or cross resistance against so many chlorinated hydrocarbons, organophosphates, carbamates or synthetic pyrethroids which are being used since long and they are not much effective in most of the conditions. Under such conditions newly developed active ingredients of spinosyn, oxadiazine, pyrazole, nicotinoid, nereistoxin, growth regulators, antibiotic should be preferred to make the control more effective.

Table 1. Active ingredient and formulation of important insecticides

Group	Active ingredient and formulations
Organochlorine	Endosulfan 35 EC
Organophosphorus	Acephate 75 SP, Chlorpyrifos 20 EC, Dichlorvos 76 SC, Dimethoate 30 EC, Oxydemeton-methyl 25 EC, Profenofos 50 EC , Quinalphos 25 EC, Triazophos 40EC
Carbamate	Benfuracarb 40 EC, Carbaryl 50 WP, Carbofuran 3 CG, Methomyl 40 SP, Thiodicarb 75 WP
Pyrethroid	Bifenthrin 10 EC, Cypermethrin 10 EC, Deltamethrin 2.8 EC, Lambda-cyhalothrin 5 CS
Spinosyn	Spinosad 45 EC
Oxadiazine	Indoxacarb 14.5 SC
Pyrazole	Ethiprole 10 SC, Fipronil 0.3 GR, Fipronil 5 SC, Chlorantraniliprole
Nicotinoid	Acetamiprid 20 SP, Clothianidin 50 WDG, Imidacloprid 17.8 SL, Imidacloprid 70 WG, Thiamethoxam 25 WSG
Nereistoxin	Cartap 4 GR, Cartap 50 SP
Growth regulators	Buprofezin 25 SC, Diflubenzuron 25 WP, Lufenuron 5 EC, Novaluron 10EC
Antibiotic	Emamectin benzoate 5 SG
Diamide	Flubendiamide

Use of broad spectrum insecticides: Some active ingredients are very effective against defoliators, borers and sucking pests while others are specific to each type of pest. Broad spectrum insecticides should be selected if crop is attacked by different types of pest.

Use of branded products: Markets in almost all agro-ecosystem are flooded with spurious products the efficacy of which is very low. It is advantageous to use the product of reputed manufacturers even if is costly.

Use of safe products: Products which are toxic to natural enemies and non-target organism should not be used. Newly developed active ingredients are comparatively safer to natural enemies.

Use of proper formulation: Granules are safer as compared to sprays and dust as they contain systemic insecticides. Water dispersable granules are safer as compared to wettable powder.

Use of different active ingredients in successive applications: Repetition of same group of active ingredients in successive application leads to fast development of resistance.

Use at right time: Time of insecticide application is very important in reducing the cost. Seed treatment or application in nursery cost less but provide adequate protection for longer duration. Application at ETL controls the pest at low cost. Late application is more effective to nocturnal pests.

Use at recommended dose: Application of insecticide at lower or higher dosages accelerates the development of resistance and increases the cost of protection. Crop, stage and insect specific recommended dose applied with recommended quantity of water gives maximum coverage and distribution.

Proper application method and equipment: All the pesticide application equipments are not suitable for application of insecticides. To increase the coverage and distribution and check the drifting of insecticide due care should be taken in selection of equipment and nozzles.

Spot treatment: If the infestation is restricted in some pockets spot treatment is less costly alternative.

Application in clear weather: Application of insecticides in bad weather is always disadvantageous as droplets are drifted or washed on windy or rainy days. Endo and exo-drift cause more than 50% loss of insecticide due to which its efficacy is reduced. Repeated applications are required if deposits are washed with in 3 hours.

7. Check or delay the development of resistance

To check or delay the development of resistance different groups of active ingredients should be selected in different applications and insecticides should be applied at recommended dose at right time using right application technology.

Adoption of abovementioned strategy may lead to long lasting and cost effective management of insect pest without any adverse effect on environment and non-target organism.

□□□

18

Novel Indicators to Assess the Impact of Pesticides on Environment

SUMITRA ARORA

We have been witnessing a silent chemical revolution over the past half century. Innumerable chemicals have been manufactured that do not otherwise exist in nature but possess very important and useful properties for the life and work of human beings. Some of these useful substances are now, however, playing havoc with the lives of humans and environment because of their overuse. Pesticides can be potential polluters of water, soil and air. They are hazardous to humans when they are handled, stored, applied and disposed off improperly. For several years, increased attention has been focused on integrated pest management (IPM) programs and alternative methods of pest control to reduce pesticide use in agricultural systems because of these reasons. IPM is a pest management strategy that uses a combination of methods to manage pests without solely relying on chemical pesticides to produce a safe, economic crop. If, however, no other control measure is effective in preventing pest damage, a chemical pesticide is recommended. In past IPM programs, pesticides were generally chosen based on their efficacy or cost rather than on their potential environmental impact. Although some pest management practitioners did take into account the effect of the pesticides on the applicaton or beneficial natural enemies such as predatory mites when making pesticide recommendations, no formal method was available to assist them in making environmentally based pesticide choices. Because there is no easy method to assess pesticide impacts, each individual had to rely primarily on their own judgment to make these decisions. In addition, IPM programs throughout the country use various methods (number of sprays, the amount of active ingredient or formulated product used per acre, dosage equivalents, etc.) to quantify pesticide use and environmental impact to compare different pest management strategies or programs. None of these methods estimates the environmental impact of specific pesticides.

Environmental impact quotient (EIQ)

An assessment of pesticide active ingredient use as well as an assessment of the specific pesticides used via an indicator is known as the *environmental impact quotient* (EIQ). A relative value that estimates the environmental impact of a pesticide, by taking into account toxicity to natural enemies, wildlife, and humans, degree of exposure, aquatic and terrestrial effects, soil chemistry, etc.

The Environmental Impact Quotient (EIQ) of a given pesticide is based on the active ingredient's potential effects on the health of *farm workers* and *consumers* and on potential adverse effects to *non-target aquatic and terrestrial organisms*.

EIQ is only an indicator and therefore does not take into account all environmental issues and impacts. The use of environmental indicators is commonly used by researchers in comparing the environmental impacts of GM and non-GM crops.

Method

The toxicity of the active ingredient of each pesticide and the effect on each environmental factor evaluated, grouped into low, medium, or high toxicity categories and rated on a scale from one to five, with one having a minimal impact on the environment or of a low toxicity and five considered to be highly toxic or having a major negative effect on the environment (Table 1).

The Extension Toxicology Network (EXTOXNET), a collaborative education project of the environmental toxicology and pesticide education departments of Cornell University, Michigan State University, Oregon State University, and the University of California, was the primary source used in developing the database (Hotchkiss *et al.* 1989). EXTOXNET conveys pesticide-related information on the health and environmental effects of approximately 100 pesticides.

An equation was developed based on the three principal components (Figure 1) of agricultural production systems:

- A farm worker component
- A consumer component, and
- An ecological component

Each component in the equation is given equal weight in the final analysis, but within each component, individual factors are weighted differently. Coefficients used in the equation to give additional weight to individual factors are also based on a one to five scale.

Table 1. The rating system used to develop the environmental impact quotient of pesticides (EIQ) model.

Mode of Action	Toxicity to Fish-96 hr LC50
non-systemic- 1	> 10 ppm - 1
all herbicides - 1	1-10 ppm - 3
systemic - 3	< 1 ppm – 5
Acute Dermal LD50 for Rabbits/Rats(m&/kg)	Toxicity to Birds-8 day LC50
>2000 - 1	> 1000 ppm - 1
200 - 2000 - 3	100-1000 ppm - 3
0 – 200 - 5	1-100 ppm – 5
Long-Term Health Effects	Toxicity to Bees
little or none - 1	relatively nontoxic - 1
possible- 3	moderately toxic - 3
definite - 5	highly toxic – 5
Plant Surface Residue Half-life	Toxicity to Beneficials
l-2 weeks- 1	low impact- 1
2-4 weeks- 3	moderate impact - 3
> 4 weeks - 5	severe impact - 5
pre-emergent herbicides - l	
post-emergent herbicides – 3	
Soil Residue Half-life	Groundwater and Runoff Potential
Tl/2 <30 days - 1	small - 1
Tl/2=30-100 days - 3	medium - 3
Tl/2 >100 days - 5	large -5

Note: l = least toxic or least harmful, 5 = most toxic or harmful

In order to further organize and simplify the data, a model was developed called the environmental impact quotient of pesticides (EIQ). This model reduces the environmental impact information to a single value. To accomplish this, an equation was developed based on the three principal components of agricultural production systems: a farm worker component, a consumer component, and an ecological component. Each component in the equation is given equal weight in the final analysis, but within each component, individual factors are weighted differently. Coefficients used in the equation to give additional weight to individual factors are also based on a one to five scale. Factors carrying the most weight are multiplied by five, medium-impact factors are multiplied by three, and those factors considered to have the least impact are multiplied by one.

The EIQ Equation

The formula for determining the EIQ value of individual pesticides is listed below and is the average of the farm worker, consumer, and ecological components.

$$EIQ=\{C[(DT*5)+(DT*P)]+[(C*((S+P)/2)*SY)+(L)]+[(F*R)+(D*((S+P)/2)*3)+(Z*P*3)+(B*P*5)]\}/3$$

DT = dermal toxicity, C = chronic toxicity, SY = systemicity, F = fish toxicity, L = leaching potential, R = surface loss potential, D = bird toxicity, S = soil half-life, Z = bee toxicity, B = beneficial arthropod toxicity, P = plant surface half-life.

EIQ field use rating

Once an EIQ value has been established for the active ingredient of each pesticide, field use calculations can begin. To accurately compare pesticides and pest management strategies, the dose, the formulation or percent active ingredient of the product, and the frequency of application of each pesticide need to be determined. To account for different formulations of the same active ingredient and different use patterns, a simple equation called the EIQ Field Use Rating was developed. This rating is calculated by multiplying the EIQ value for the specific chemical obtained in the tables by the percent active ingredient in the formulation by the rate per acre used (usually in pints or pounds of formulated product).

EIQ Field Use Rating = EIQ x % active ingredient x Rate

With this method, comparisons of environmental impact between pesticides and different pest management programs can be made. For example, if several pesticides can be used against a particular pest, which pesticide is the least toxic choice?

Table 2 shows an example comparing the environmental impact of three insecticides: carbaryl (Sevin 50WP), endosulfan (Thiodan 50WP), and azinphos-methyl (Guthion 35WP). Although carbaryl has a lower EIQ (22.6) than endosulfan (40.5) or azinphos-methyl (43.1), it may take more of it to provide equivalent control. In this situation, Guthion would have the lowest EIQ Field Use Rating (33 .2) and would be the least toxic choice. Thiodan (60.8) would be the second choice and Sevin (67.8) would be the last.

return to A Method to Measure the Environmental Impact of Pesticides.

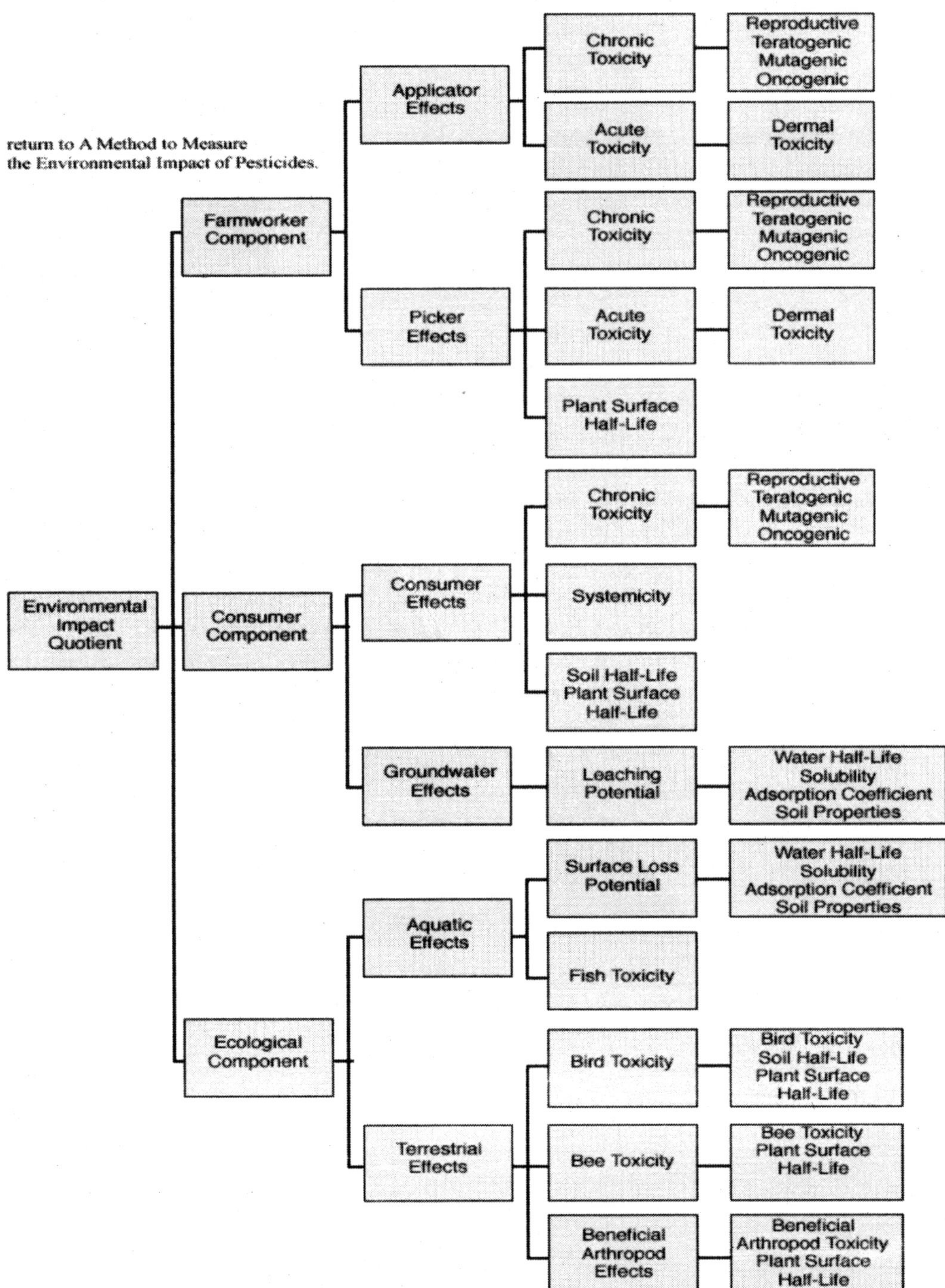

Fig. 1. A diagram showing the individual environmental factors that were evaluated in developing the environmental impact quotient of pesticides (EIQ) model.

By applying the EIQ Field Use Rating, comparisons can be made between different pest management strategies or programs. To compare different pest management programs, EIQ Field Use Ratings and number of applications throughout the season are determined for each pesticide. and these values are then summed to determine the total seasonal environmental impact of the particular strategy. Table 3 compares the theoretical environmental impact of several different pest management approaches that have been used in research projects to grow 'Red Delicious' apples in New York. In this example, a traditional pest management approach to growing 'Red Delicious' apples that does not rely heavily on pest monitoring methods would result in a total theoretical environmental impact of 938 due to pesticides. An IPM approach that incorporates pest monitoring methods, biological control, and least toxic pesticides would have an environmental impact of only 167. The organic pest management approach, which uses only naturally occurring pesticides, would have a theoretical environmental impact of 1,799 according to the model. The environmental impact of the latter approach is so much larger than the other strategies primarily due to the larger quantities of sulfur required and more frequent applications needed to provide the same level of control of apple scab in this variety. By using the EIQ model, it becomes possible for IPM practitioners to rapidly estimate the environmental impact of different pesticides and pest management programs before they are applied, resulting in more environmentally sensitive pest management programs being implemented.

Table 2. EIQ field use rating of different insecticides

Material	EIQ	ai	Rate	EIQ field use rating
Sevin 50WP (carbaryl)	22.6	0.50	6.0	67.8
Thiodan 50WP (endosulfan)	40.5	0.50	3.0	60.8
Guthion 35WP (azinphos-methyl)	43.1	0.35	2.2	33.2

Table 3. Theoretical environmental impact of different pest management strategies used to grow 'Red Delicious' apples in New York.

Traditional Pest Management Strategy					
Material	**EIQ**	**ai**	**Dose**	**Applications**	**Total**
Rubigan EC (Fenarimol)	27.3	0.120.	6	4	8
Captan 50WP	28.6	0.503.	0	6	257
Dursban 50WP	52.8	0.503.	0	2	158
Thiodan 50WP (endosulfan)	40.5	0.503.	0	2	61
Guthion 35WP (Azinphos-methyl)	43.1	0.352.	2	2	66

Contd...

Cygon 4E (dimethoate)	74.0	0.432.	0	3	191	
Omite 6EC (Propargite)	42.7	0.682.	0	2	116	
Kelthane 35WP (dicofol)	29.9	0.354.	5	1	47	
Sevin 50WP (carbaryl)	22.6	0.501.	0	3	34	
Total Environmental Impact	**938**					
Integrated Pest Management (IPM) Strategy						
Nova 40WP (Hydrogen (< 5%) in Ethylene)		41.20	.40	0.3	4	20
Captan 50WP	28.6	0.503.	0	1	43	
Dipel 2X	13.5	0.061.	5	3	4	
Sevin 50WP	22.6	0.503.	0	1	34	
Guthion 35WP	43.1	0.352.	2	2	66	
Total Environmental Impact	**167**					
Organic Pest Management Strategy						
Sulfur	45.5	0.906		7	1720	
Rotenone/pyrethrin	25.5	0.041	2	6	73	
Ryania	55.3	0.001	58	2	6	
Total Environmental Impact	**1720**					

Conclusion

IPM practitioners and growers can incorporate environmental effects, along with efficacy and cost, into the pesticide decision-making process. IPM programs can also use the EIQ model as another method to measure the environmental impact of different pest management and pesticide programs. As newer biorational pesticides are marketed with lower EIQ values and more emphasis is placed on biologically based IPM practices, the EIQ field use ratings will continue to decrease. Eventually these ratings may approach zero, resulting in an environmentally neutral or benign agricultural production system.

Reference

Hotchkiss, B. E., J. W. Gillett, M. A. Kamrin, J. W. Witt, and A. Craigmill.1989. EXTOXNET, Extension Toxicology Network. A Pesticide Information Project of Cooperative Extension Offices of Cornell University, The University of California, Michigan State University and Oregon State University, Cornell University, Ithaca, NY.

□□□

19

Management of Crop Pests Through Indigenous Technical Knowledge (ITKs)

A.K. KANOJIA

Agriculture in every region is being practiced to provide food for growing population and it also occupies a dominant position in India's economic structure. Plants are cultivated with different kinds of package of practices and farmers are after their crops since the preparation of fields to sowing of seeds to final harvesting and storing the produce. There are several other living organisms grow simultaneously with the development of plants feeding on plant parts. Consequently, the farmers in his different ways to obtain good harvest from his crops eventually sees that prominent part is lost due to infestation and damage by pests. It is estimated that one-third of world's food crop is destroyed by these pests annually and nearly 30% of the potential food production in India is lost due to insect pests, plant pathogens, weeds, rodents, birds and in storage, the use of pesticide is vital. In their endeavor to safeguard their crops from pests, farmers use pesticides/fungicides indiscriminately during growth in field or in post harvest storage conditions. This indiscriminate use of pesticides has led to serious environment concerns. Widespread environmental contamination, detection of residues far from their site of application, toxicity to non-target insect species, potentiality to pollute water, soil and air, are becoming day to day problems. Application of high inputs in agriculture, in terms of chemical fertilizers and pesticides has endangered the sustainability of production system. Therefore, organic compounds and their formulations have become the leading pesticides world over now-a-days. They are less hazardous, environmentally sound, nature friendly and economically feasible and are required in much smaller quantities than inorganic chemicals to obtain the same effect. There are indigenously ways available and adopted by the farmers which may check harmful substances from going into the environment.

Indigenous knowledge, which entirely depends on natural resources and having depth of experience and knowledge, is an important means of crop husbandry through which insect pests and diseases are managed knowingly or inadvertently. Indigenous Technical Knowledge (ITK) is an important bank-source of knowledge conserved by different cultural groups followed through the centuries in different agro-climatic conditions in the country. The cotton based cropping system is one of the systems where such indigenous people are engaged in practicing ITKs controlling and managing pests and diseases of crop plants. Since cotton crop consumes huge amount of pesticides, it attracts the attention of everybody concerned with the development of agriculture towards the consequences of vicious cycles of chemical inputs, which are not only polluting our environment and posing economical burden on farmers but also killing natural enemies of the pests. This very reason has prompted the authors to focus on the indigenous knowledge in this cropping system. Indigenous knowledge of agriculture, is the result of farmers' thousands years of experience with nature. Indigenous practices are known to the farmers and are helpful in maintaining and enhancing the quality of the environment. With the dissemination of modern practices the indigenous practices have started to loose their ground and have been eroded to a large extent (Joshi and Singh, 2006). It is the information base for a society, which facilitates communication and decision making. Indigenous information systems are dynamic and are continually influenced by internal creativity and experimentation as well as by contact with external systems (Flavier *et al.*, 1995).

Biological control of plant pests and pathogens continues to inspire research and development in many fields. Plant pathogens are just one class of targets of biological control, which also is designed to limit other pests such as insects, parasitic nematodes, and weeds (Gardener 2002). Targeting the major plant pathogen through plant based botanicals-ITKs as biological control may help in a big way.

The information given in this chapter is based on the farmers' knowledge and their experience with the components and application methodology of ITKs and these are in practice by groups of farmers on a large scale at some places and in restricted areas by limited number of farmers at other places.

However, it needs not to be adopted as such and proper scientific validation for field application in different agro-ecological regions is sought for fullest utilization of the given useful tactics in the area of plant protection in cotton based cropping systems.

What ITK is

Indigenous Technical knowledge (ITK) is also called as Indigenous Traditional Knowledge or Indigenous Know how. It is the local knowledge - knowledge that is unique to a given culture or society (Warner 1991). It contrasts with the international knowledge system generated by universities, research institutions and private firms. It is the basis for local-level decision making in agriculture, health care, food preparation, education, natural-resource management, and a host of other activities in rural communities (Warren 1991). Such knowledge is the consequence of practical engagement in everyday life, and is constantly reinforced by experience and trial and error. It is, as Hunn (1993) neatly puts it, 'tested in the rigorous laboratory of survival'. It is really the product of many generations of intelligent reasoning and time tested, having the confidence of the indigenous people. This knowledge is empirical rather than theoretical knowledge. It is shared to a much greater degree than other forms of knowledge, including global science. This is why; it is sometimes called 'people's science', an appellation which also arises from its generation in contexts of everyday production. However, its distribution is still segmentary, that is socially clustered (Hobart 1993). The knowledge usually goes silently side by side without any external intrusion and users evolve their own way of using the available resources by adopting appropriate methodology for preparation and application. Confidence and for doing better than other farming fellows in the same settings of ecological environment energizes the innovators in acquiring and selecting the specific knowledge based technical skill. Such propounders are reluctant to share their knowledge as it is believed that science based knowledge is advanced and valid and what these fellows are practicing is outdated and not relevant to modern know-how. This kind of attitude hinders the natural development for transfer of the available knowledge and even farmers to farmers technology transfer does not take place in many of the cases. Sincere efforts are to be put forward for natural blending of traditional with modern in general and farmers technical knowledge in crop protection with the latest synthesized packages of pest management. "Indigenous knowledge is the knowledge held by people, no matter how recent their arrival, who happen to live in a certain place. Generally the expression "indigenous knowledge" is used to make reference to knowledge held by indigenous people, where "indigenous" stands for aboriginal, native or autochthonous (Heyd 1995).

Cotton, the white gold is a highly valued commercial crop and is prone to pest attack at all the stages of growth. Though it has been reported that about 162 insect pests attack on cotton in India, only a dozen are major and half of them are key production constraints which cause losses to the extent of 30-80%. Jassids, *Amarasca biguttula biguttula* Ishida; Thrips, *Thrips tabaci* (L.); Aphids, *Aphis gossypii* (Glover) and Whiteflies, *Bemisia tabaci* (Genn.) are the important sucking pests start to de-sap the cotton crop at seedling stage and cause heavy losses (Udikeri *et al.*, 2007). Crop loss estimated at Rs406 crore; 6 per cent of farmland hit. About six per cent of the cotton crop has been affected. The production losses are about 13,67,130 tonnes.

This chapter throws light on some useful Indigenous Knowledge (IKs) which are in practice by the farmers in India to manage insect pests and diseases of cotton crop. These tactics are not only cheaper and evolved by incorporating local and easily available material but environment friendly certainly. Such tactics are of important value and meaningful to provide alternative means in plant protection in the country and elsewhere as scientific community is in search of alternative ways to manage and control pest problem.

Useful and promising ITKs practiced by farmers

Control of pests in Cotton (i) Mechanical control measures include cotton seed treatment with cow dung resulted in delineating of the seed (fibre free seed), followed by identification and removal of pink boll worm infested seeds and hand collection, destruction of larvae and infested plant parts leads to reduction in insect pest population. (ii) Cultural control measures includes field sanitations - clean cultivation during the entire cropping seasons which results in the reduction of insect fauna and weed flora in the filed and intercropping or mixed cropping of cotton with sorghum/maize/greengram/ blackgram for augmentation and conservation of natural enemies of major insect pests. (iii) Application of botanical pesticides include aqueous decoction of *neem* leaf *(Azadirachta indica)* or neem seed kernel or tobacco leaf *(Tobaccum solanum)* or alcoholic extracts of *neem* seed kernel against pest complex of cotton. (iv) Alternatively for pest management, application of ash water in which leather is soaked by cobbler against sucking pest or fermented butter milk or mixture of fermented farm animal urine + distillates of *neem* seed kernel or application of kerosene oil is practiced against insect pest complex of cotton in West Nimar, Madhya Pradesh (Paliwal et al., 2002).

Non-pesticidal management technologies include deep ploughing in summer, use of bonfires/light traps, nipping of the terminal buds to destroy eggs and larvae of spotted bollworm, selecting tolerant varieties for sucking pest complex, spraying of tobacco decoction, spraying of 5% neem seed kernel suspension, chilli garlic spray and spraying of cattle dung and urine, bird perches, collection and destruction of affected bolls. The farmers who practices this traditional knowledge gained more income as compared to the farmers who treated with pesticides (non NPM) in Warangal district of Andhra Pradesh. The farmers cultivating cotton in Warangal district with high pesticide load are prone to risk of negative returns as compared to the traditional practices. Farmers in Jalgaon/Jalna spent more money for pesticides which enhanced cost of production while traditional NPM technologies reduced the cost of production by 25% (Chary *et al.* 2000).

Sufali (*Mundulea suberosa*) to check aphids

Sufali (*Mundulea suberosa*) leaves and neem leaves are crushed to obtain extract. Cow urine and water are added to the obtained extract in an earthen pot. This pot is placed underground in a dark place by digging a pit to accommodate the pot for 7-8 days. Then it is taken out and the fermented formulation so obtained is applied as spray @ 1ltr/10ltr water on cotton, chickpea, pigeon pea, and vegetables crops. It is believed that this practice is able to check the attack of sucking pests. Materials required for this tactis are: Sufali (*Mundulea suberosa*) leaves 1 kg; Cow urine 5 ltr; Neem leaves extract 2.5 ltr; Water 10 ltr. (Kanojia *et al.* 2005).

Rodent control by Parsara plant

To get rid of rats from the field, farmers in Ahmadnagar district of M.S. practice insertion of 10-12 inches long fresh pieces of Jatropha locally known as Parsara. Holes of rat are searched in the field for putting the pieces of that plant. Its smell emanating from those pieces seems to irritate rats and repel them from the field.

Calotropis and castor oil against aphids

Leaves of *Calotropis* plant are immersed in water channel during irrigation to minimize aphid infestation practiced in Banaskantha, Gujarat (Thakor 1991), while a little quantity of castor oil is added during irrigation in the water channel for reducing aphid population in Ahemdabad, Gujarat (Patel 1991a).

Control of pest complex in cotton

Helicoverpa, spotted bollworm and aphid are generally found in cotton. Aphids are also common in brinjal, chilli and Lucerne in Amreli district of Saurashtra. A mixture was prepared from dried tobacco leaves (250 g) *hirakasi* (300 g) and citric acid by boiling in 1-2 litres of water and filtered. Approximately 250 ml of the decoction was mixed in 15 litres (one pump full) water and sprayed over crop. Depending upon the growth and density of the crop, total quantity required varied; however three to four pumps were enough for cotton crop. After a week decoction was sprayed again. It could control pest complex in cotton and the growth of the plants also boosted in Amreli, Gujarat (Kalyanbhai 1995).

Control of whitefly

Whitefly (*Bemisia tabaci*) is usually found on the dorsal surface of the leaves. The nymphs are small and flat with visible beaks that suck the plant sap. To get rid of these flies, papers pasted with castor oil/grease are hung at five or six places in the cotton field. Then air is blown using a sprayer over the crop. The disturbed adult flies (white fly) come out of the plant and stick to the oily paper. By this method 90% of white fly can be successfully reduced in Pondicherry (Balasubramanian 2000-2001).

Calotropis for control of caterpillars

Caterpillar infestation can severely damage a cotton crop. Reportedly the latex of *Akda (Calotropis gigantia,* when diluted with 15 parts water and sprayed on the crop, effectively controls the pest within three days. The new growth after treatment is also free from infestation in Bhavnagar, Gujarat (Jani 1992).

Control of green worm by dhatura

Of all the pests occurring in the cotton crop, greenworm (*Heliothis* spp.) is one of the most harmful pests. Leaves and stem of Dhatura *(Datura metel)* are used to control green worm. About 250 to 300 g of Dhatura' leaves along with stem are dipped in 1 litre of lukewarm water. After cooling down, 250 to 300 grams of the solution is mixed with 15 litres of water and sprayed on the crop. Pests perish within a period of six to seven hours. Spraying the mixture when the crop is of one-month-old yields better results in Amreli district, Gujarat (Kaachhadiya 1998).

Okra (*Ablemoschu s esculentus)* as a sacrifice crop

Use of okra (*Ablemoschu s esculentus)* as a sacrifice crop for pest control in cotton. Farmers grow 2-3 lines of okra plant surrounding the cotton field. Cotton is more susceptible to insect pests like spotted bollworms (*Earias vitella),* worm (*Platyedra gossypiella),* Jassid (*Amarasca biguttula*) etc. Farmers believe that pests prefer okra plant as compared to cotton plant and attack them first. Pest is controlled by simply eradicating/destroying the okra plants attacked by insect and pests in Surendranagar district, Gujarat (Patel 1991 g).

Cow urine-neem pesticide

Neem leaves 2.5 kg. are put in 10 ltr cow urine and kept in an earthen pot underground for 21 days. Fermented solution is taken out after 21 days and it is filtered. The filtrate is then boiled in a copper utensil vigorously so as to reduce to one fourth. It is applied as spray on cotton, chickpea, pigeon pea, tomato, brinjal and soybean crops @ 100 ml/10 ltrs water for insect control. Its results are claimed to be very promising on insect pests and diseases of the crops, besides it provides many nutrients through leaves and acts as a tonic to the plants (Kanojia *et al.* 2005).

Salt with manure to increase cotton yield

Increasing yield of Cotton by application of salt mixed with manure is claimed by taking 20 to 25 kg of common salt and mixing with about 10 cart loads of natural manure. It is then applied in farm during *Jyeshtha-Vaishakh* (May). It has bee observed to be more effective if the mixture is applied uniformly in all the furrows. 10 maunds of cotton/*bigha* is received (1/3 of acre). It is believed that the manure increases the crop's resistance to pests. This knowledge is from Surendranagar in Gujarat (Sanghani 1998).

Nerium arali for white fly

White fly control in cotton by seed extract of *Nerium arali (Nerium oleander)is practiced by* pulverizing and soaking its seeds in water overnight and filtering. This filtrate is diluted in water and sprayed on cotton fields. This practice provides 70% control in Thiruvadagam (Murugesan 1994).

Dry sowing of cotton and pigeonpea

For getting advantage of early seeding in *Kharif* season, farmers resort to dry seeding of cotton (local variety), pigeonpea and also rice. Seeding is generally done 10-12 days before the anticipated date of onset of monsoon. In this practice, dry seeded crops give the advantage of 8-10 days earlier sowing than normal. When sufficient rains are received, the dry seeded grains start germinating. In dry seeding, however, there is a risk of re-sowing in the event of receipt of inadequate rainfall. Otherwise it gives 10% more yield. Dry seeding of cotton and pigeonpea is in practice in Dhule, Nandurbar and Jalgaon districts of Maharashtra. Dry seeded rice is taken in Taloda and Akkalkuwa tehsils of Nandurbar district in Maharashtra. This knowledge as been opined by Directorate of Extension Education, Mahatma Phule Krishi Vidyapeeth, Rahuri, Ahmednagar, Maharashtra.

Sufali (*Mundulea suberosa)* to control borers

To control the borers of tomato and chili there is a practice to make a paste of *Mundulea suberosa* (*Sufali*) leaves of 3 kg which is mixed with 10 ltr water in an earthen pot and kept overnight. The formulation so prepared is applied next evening @ 200 ml/15 ltr water on standing crop. Materials required for this ITK are: leaves of *Sufali* 3 kg; water 10 ltrs. The *sufali* (*Mundulea suberosa)* plants are locally available at the elevated lands in and around Ahmadnagar during rainy season and it is said by the farmers that the tribal community of the region practiced the same leaves to kill fish by taking the extract of the leaves and mixing with the balls of edible flour. This knowledge inspired the villagers to use the leaves in plant protection (Kanojia *et al.* 2005).

Raising of sorghum as mixed crop

Raising of sorghum as mixed crop in cotton as bird percher is adopted in Jalgaon, Nadurbar, Dhule and Ahmednagar districts of Maharashtra. Farmers raise sorghum as a mixed crop scattered in cotton fields. The grain of sorghum attracts the birds and served as a perch for the birds to reach the insects of cotton plants. Director of Extension Education, Mahatma Phule Krishi Vidyapeeth, Rahuri, Ahmednagar (Maharashtra) 413 722.

Spraying of mixture of garlic extract and monocrotophos to control bollworm in cotton

Farmers of Sindhkheda and shirpur tehsils of Dhule district and Kopargaon tehsil in Ahmednagar district are using this practice. Garlic bulbs are crushed and soaked in water to draw extract. This extract is added to monocrotophos and it is sprayed on cotton to control cotton bollworm, explored by Mahatma Pule Krishi Vidyapeeth, Rahuri, Ahmednagar, Maharastra.

Marigold to manage Helicoverpa *in cotton*

Marigold plants are sown as trap crop around cotton plants which attract *Helicoverpa*. It prefers these plants and gets attracted to the plants. It starts laying eggs on the top of the marigold flowers. Farmers pluck such flower and sell those in the market. Thus, practicing this knowledge reduces pest load.

Garlic and *chilli* extract to control pests

Use of garlic and *chilli* extract to control pests in cotton is in practiced by using 500g garlic and red *chillies*. The extract is added to 100 litres water, mixed with 200g soap and sprayed on cotton to control sucking as well as other pests. This practice is in vogue in Dhule and Ahmednagar districts of Maharastra (Inventory of Indigenous Knowledge 2003).

Mixture of ash, cow urine, kerosene and soap against aphids and jassids

Dusting of ash, spraying of cow urine and kerosene+soap mixture for control of aphids, jassids and whitefly on vegetable crops and cotton. The tribal farmers of Nandurbar and Ahmednagar districts of Maharashtra treat the seeds of vegetable crops by dusting ash and spraying of cow urine. It is very rarely applied on *chilli*, brinjal and cucumber to control aphids and jassids. Some farmers spray cow urine in cotton for control of aphids and jassids. In tribal areas some farmers also spray kerosene + soap on cotton crop to control whitefly attack. This is an economic practice and the cost involved is much less as compared to chemical insecticides, explored by Mahatma Pule Krishi Vidyapeeth, Rahuri, Ahmednagar, Maharastra.

White fly control in cotton by jaggery solution

Some farmers use gur (jaggery) solution to control white fly and for this they take 1 kg gur and dissolve it in 10-12 ltrs of water and filter through a thick cotton cloth. Approximately 5-6 ltrs solution is sprinkled in one acre with the help of sprayer pump gently on the foliage. The tiny insects of white fly stuck on the leaf surface due to stickiness of the solution. All the stuck eggs of white fly are killed on the leaf surface which increase the possibility of occurrence of Madhiya disease. Hence they prefer to use this practice once only. This practice can achieve 40-50 percent control of white fly. This technique is used in the months of November-December in Bharuch, Gujarat (Ahmedabadi 1992).

Control of bollworm in cotton by extract of *mahua* and tamarind

Bollworms like spotted bollworm and pink bollworm of cotton crop attack cotton plants frequently. These pests can be controlled by making a stock solution of 500ml taking equal amount of *mahua* (*Madhuca indica*) and tamarind (*Tamarindus indica*) bark. This solution is filled into a pump of 15 ltr capacity and sprayed on the infested crop. If the attack is severe then the same solution is sprayed once again after a week. If it is used more than twice then there are chances of root burning (Kodarvi 1991). This practice has been used for over 30 years in Banaskantha in Gujarat, however, this need to be used judiciously.

Control of *Spodoptera litura* and *H. armigera*

To combat the menace of *Spodoptera litura* and *H. armigera,* farmers use 4 kg of *Kuvarpathu* (*Aloe vera*), 50 ml of neem oil and 500 g of tobacco powder by boiling together these ingredients in 20 lts water for 3-4 hours to make a 5-ltr solution.Then 50 g of aritha (*Sapindus emarginatus*) powder is added to the solution and mixed thoroughly. It is applied @ 100-150 ml/15 lts water on the crop and it is practiced in Northern Gujarat region (Patel 2001).

Fungal disease control in cotton and chillies

Fungal disease control in chillies, cotton and also lemon fruits is controlled by spraying a mixture, locally prepared called Panchagavya. The mixture is prepared by mixing 5 parts milk, 2 parts curd, ½ part ghee, 2 parts cow urine, ½ part cowdung and 10 parts water. The mixture is kept for 7 days before it is sprayed in field where fungal disease is predominant. Spray of the mixture also controls leaf shedding and enhances flower setting. Ten farmers of Periakovilankulam village follow this practice for the last three years, opined and evidenced by Shri. S. Murugesan Periakovilankulam Sankarankovil, Tirunelveli, Tamil Nadu.

Ratooning of *Kharif* cotton

Some farmers in pimpri village of Dhule Districts in Maharashtra have started taking ratoon crop of Kharif cotton. The farmers having irrigation facilities take the ratoon. It is taken after 3-4 pickings are done. The crop is provided with chemical fertilizers 2-3 irrigations are given till February – March. This practice is followed to obtain additional returns from the same crop. An additional 10-15 q/ha is obtained from the same crop if it is properly managed. The practice has been in use for 8-10 years. About 10-15 farmers are practicing this technology. This has been opined by Prof. P. S. Patil, Training Organiser, Krishi Vijayan Kendra, Dhule, Maharashtra.

Shaking of pigeon pea plants for managing H. armigera

Shaking of pigeon pea plants and mass collection of *Helicoverpa* for their destruction is practiced to control population of *Helicoverpa* (Fig.2).

Fig. 2.

Control of cutworm in cotton

Citrullus grandiflora (periya kumuttikai), *Cissus quadragularis* (perandi kodi), Tecoma stans (yellow arali seeds) and *Azadirachta indica* (neem leaves/ seeds) are ground and pasties prepared. The paste is allowed to ferment as such for 10 days. Paste from *Sesbania grandiflora* is made separately and fermented. After 10 days, the fermented pastes are filtered though a thin cloth, diluted with required quantity of water and sprayed over the foliage. This controls the cutworms in cotton very efficiently with no side effect evidenced by Shri S. Pidhai Muniandi Koil Street, Jayanangalam Periyakulam, Theni, Tamil Nadu.

Jaggary to check pod borers and pod sucking bugs in pigeonpea

To check pod borers and pod sucking bugs in pigeon pea, many farmers in Nanded district of M.S. dissolve 1kg jaggary in 40 liters of water and use the formulation as spray on pigeon pea plants when they observe eggs of pod borer and pod sucking bugs. The sweet taste of the formulation remains on plants for around one week, which is quite able to attract big sized ants, which prey on these eggs. Thus, the visiting ants check the population of pod borers and pod sucking bugs naturally (Fig. 3). The formulation is sufficient for an area of 1 acre and according to the density of eggs more jaggary may be added in water and vice versa. (Kanojia *et al.* 2002 and Kanojia *et al.* 2005).

Fig. 3.

Suggested methodology for validation of ITKs

There are some ways to be followed while dealing with indigenous knowledge and the broad lines of such ways have been given as suggested methodologies, these steps are very handy and beneficial for scientifically validation of the associated and claimed potentials of ITKs. The given steps are self explanatory and need not be explained in detail:

- Exploration of the concerned ITKs
- Documentation
- Selection of crop
- Lab validation of crude formulation against major insect pests and diseases of the concerned crop
- Dose calculation
- Field validation at experimental plots of crude formulation including pest monitoring using pheromone traps and agronomical practices

- Lab validation of organic extracts
- Economics of the formulation
- Validation at farmer's field
- Large scale validation
- Recommendations for incorporation in national programmes for pest management
- Commercialization of the formulation
- Benefits to farming communities, nations, environment etc

Validation of the explored ITKs

This has also been suggested and emphasized in previous pages of this chapter that validation of ITKs in laboratory before taking them to fields for the claimed potential regarding their efficacies is important. It involves selection of target insect pests and disease causing damages to the concerned crop. In the following example laboratory testing of one ITK has been explained and such testing opens new avenues for further quest. In this example validation of ITK in laboratory against *Rhizoctonia solani* and *Rhizoctonia bataticola,* is presented.

An insect control decoction, which is prepared from cow urine and Neem leaves by putting neem leaves 25kg in cow urine 10ltrs and kept in an earthen pot underground for 21 days. Fermented solution is then boiled in a copper utensil vigorously to make one fourth. This formulation has been taken for the mentioned test organisms. Its efficacy shows that it can be used for seed treatment of cotton & other crops to protect against seed rots and seedling mortality (Kanojia *et al.* 2005).

Test organism were: Rhizoctina solani and *Rhizoctonia bataticola*

Treatment : Control 0 ml/ltr PDA (Potato dextrose agar)
10 ml/ltr PDA
25 ml/ltr PDA
50 ml/ltr PDA
100 ml/ltr PDA

Through this testing it was found that at 1% concentration the formulation started checking the growth of *Rhizoctina solani* and controlled it cent percent at 2.5% (Fig. 4), while in the case of *Rhizoctonia bataticola,* cent percent inhibition was observed at 5% (Fig. 5).

Formulation at 0.0 , 0.2, 0.5, 1.0, 2.5, 5.0 and 10.0 per cent conc. in PDA against *Rhizoctonia bataticola*

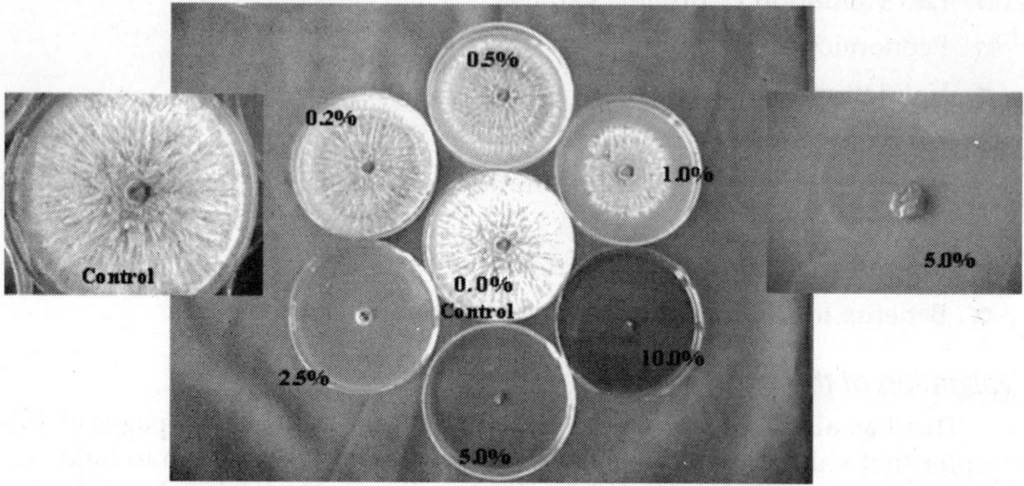

Formulation at 0.0 , 0.2, 0.5, 1.0, 2.5, 5.0 and 10.0 per cent conc. in PDA against Rhizoctonia bataticola

Fig. 4.

Conclusion

This chapter on the discussed indigenous technical knowledge is an effort to provide such information to facilitate scientists, extension workers, planners, farmers and administrators for harnessing the potential by utilizing suitable ITKs as valuable components in the national IPM programme for the sustainability of Indian agriculture and alleviation of rural poverty. It is expected to enhance knowledge base and stir up a light of interest in simple and small tips of indigenous origin for managing in pests and provide new dimension in the generation of IPM technologies at national level abridging the knowledge and the communication gaps between researchers and community involved in pest management in cotton based cropping system.

An escape from rich indigenous knowledge systems may not be spared and wise by treating them outdated and unscientific just because being practiced by the resource poor and under privileged farmers belonging to the underdeveloped regions. Since the ITKs are a Basket of sustainable technological options rather than fixed packages and therefore these may be promoted to strengthen the ongoing pest management programmes in the country and else where creating an environment of respect for local people engaged in practicing indigenous knowledge using sound & thorough knowledge of local inhabitants with locally and easily available resources and home made preparation and reducing the social distance by interacting frequently for greater involvement and further improvement protecting the agricultural environment. The available information on indigenous technical knowledge in managing and controlling pests of crops must be explored and documented.

References

Ahemdabadi, D.K. 1992. White fly control. *Honey Bee* 3(1): 7

Balasubramanian, K. 2000-2001. Beyond clinical pesticides in cotton. *Honey Bee* 11(4) and 12(1):21.

Chary, M.S. 2000. Traditional innovations for sustainable pest management. Proceeding of National Symposium on Sustainable Agriculture Production: Lesson from Traditional Technologies. Jan-30-31, Vigyan Bhawan, New Delhi, pp 24-29.

Flavier, J. M. 1995. The regional programme for the promotion of indigenous knowledge in Asia, pp 479-487. In Warren, D.M., slikkerveer, L.J. and Brokensha, D.(eds.). The Cultural Dimensionof Development: Indigenous Knowledge Systems. London: Intermediate Technology Publication.

Hobart, M. (ed). 1993. An anthropological critique of development: The growth of ignorance. London: Routledge.

Inventory of Indigenous Knowledge. 2003. Use of *chilli* and garlic extract for *helicoverpa* management in cotton. Pest and Disease-code1804. Inventory of Indigenous Knowledge. Pp 156, ICAR, New Delhi.

Jani, B . 1992. Calotropis gigantia- caterpillars in cotton. *Honey Bee* 3(3&4):17.

Joshi1, C. P. and Singh, B. B., 2006. Indigenous Agricultural Knowledge in Kumaon hills of Uttaranchal. Indian Journal of Traditional Knowledge.

Kaachhadiya, D.D. 1998. To control green worm (*Heliothis* spp.) in cotton crop. *Honey Bee* 9(3):15-16.

Kalyanbhai, K.R. 1995. Control of cotton and chili pest complex. *Honey Bee* 6(1):13.

Kanojia A.K., R.V. Singh, Surender Kumar, O.M. Bambawale and T.P. Trivedi. 2001. Exploration, Documentation and Mapping of ITKs in pest management-Paper presented in the Seminar on Vedic Agriculture held at Rajasthan Agriculture University, ARS, Durgapura, Jaipur on August. 10, 2001 organized by Asian Agri-History Foundation, Udaipur (Raj.) and published in the Abstr. pp 28-29.

Kanojia, A.K., Singh, R.V., Bambawale, O.M. and Trivedi, T.P. 2002. Exploration, documentation and mapping of indigenous technical knowledge in pest management in cotton based cropping system in Central India. Annual Report 2001-02, NCIPM, New Delhi, pp 14-15.

Kanojia, A.K., Amerika Singh, R.V. Singh and Surender Kumar. 2001. Indigenous Technical Knowledge in Integrated Pest management. Paper presented in the Seminar on Vedic Agriculture held at Rajasthan Agriculture University, ARS, Durgapura, Jaipur on August. 10, 2001 organised by Asian Agri-History Foundation, Udaipur (Raj.) and published in the Abstr. pp 30-32.

Kanojia, A.K., Singh, R.V., Bambawale, O.M. and Trivedi, T.P., 2005. Explored Indigenous Technical Knowledge in Pest Management. Tech. Bulletin, National Centre for Integrated Pest Management, New Delhi.Pp3-5, 17-18 & 28.

Knowledge and Policy: The International Journal of Knowledge Transfer and Utilization, Spring 1995, Vol. 8, Number 1, pp. 63-73.

Kodarvi, A.D., 1991. Tamarind and *Mahua* scare pests of cotton. Honey Bee 10(3): 14.

McSpadden Gardener, B. B., and Fravel, D. R. 2002. Biological control of plant pathogens: Research, commercialization, and application in the USA. Online. Plant Health Progress doi:10.1094/PHP-2002-0510-01-RV.

Murugesan, P. 1994. White fly control in cotton. *Honey Bee* 5(1):15.

Paliwal, M., Das, S.B. and Veda O.P. 2002. Indigenous technology for effective pest management of cotton-adopted by tribal farmers of West Nimar, Madhya Pradesh. International Seminar on Traditional Knowledge, Health and Environment, 23-24 February, 2002 at Bhubaneshwar. Abs p 42.

Patel, M.H. 1991. Plant Protection measures-aphids-castor oil. *Honey Bee* 2(1):16.

Patel, R.K. 2001. Controlling worms in crop. *Honey Bee* 12(2): 9

Patel, U.T. 1991a. Cotton pest complex – spraying of botanical pesticide. *Honey Bee* 2(1):15.

Sanghani, A. T. 1998. Increasing crop yield. *Honey Bee* 9(1):15.

Thakor, D.D. 1991. Aphids- leaves and twigs of *Calotropis* in irrigation channel. *Honey Bee* 2(1):16.

Thomas Heyd. 1995. Indigenous Knowledge, Emancipation and Alienation

Udikeri, S.S., Patil, S.B., Krishnanaik, L., Rachappa, V., Farook Nimbal and Guruprasad, G.S., 2007. Poncho 600 FS – A New Seed Dressing Formulation for Sucking Pest Management in Cotton. *Karnataka J. Agric. Sci.*, 20 (1): 51-53.

Warner, K., 1991. Shifting Cultivators: Local Technical Knowledge and natural Resource management in the Humid Tropics. FTP. Community Forestry Note 8. FAO. Rome.

Warren D.M., 1991. Using Indigenous Knowledge in Agricultural Development. The World Bank, Washington DC.

□□□

20

De-contamination Methods for Pesticide Use in Commercial Agriculture

SUMITRA ARORA

The pesticides and allied agrochemicals have become an integral component of modern agriculture and public health, but their excessive and injudicious use has raised several environmental issues. Despite the fact that the consumption of pesticides in India is still very low, about 0.5 kg/ha of pesticides against 6.60 and 12.0 kg/ha in Korea and Japan, respectively, there has been a widespread contamination of food commodities with pesticide residues, basically due to non-judicious use of pesticides. In India, 51% of food commodities are contaminated with pesticide residues and out of these, 20% have pesticide residues above maximum residue level on a worldwide basis. The extensive application of pesticides creates an environmental risk due to the total pesticide residues, identified in air, water, soil, foods and human tissues (Dikshith, 1991). Besides contaminating agricultural commodities, their transport may increase their level beyond acceptable limits and this way they may have an impact on the delicate equilibrium of our environment. Some pesticides (particularly organochlorine compounds) were banned after evidence of their toxicity, persistence and bioaccumulation in the environment and their ability to concentrate in the food chain. Organophosphate compounds are considered to degrade either biologically or chemically in plants, animals and soils, more rapidly depending on their formulation, method of application, climate and soil types and conditions (Khan, 1980) but are much more toxic than other insecticides. Municipal and industrial treatment plants, effluents, discharge from factories, accidental spillage during transportation or storage, deliberate pesticide dumping and leakage from waste disposal sites also contribute substantially to pollution. While use of pesticides has risen significantly, crop losses to pests have not declined correspondingly.

The concept of poison or toxicant implies interaction between a living organism and a chemical substance, which cause pathological changes to appear in it. Acute poisoning of an organism by a pesticide occurs when the pesticide acts at once, whereas chronic poisoning of an organism is the result of the repeated action of relatively small amounts of a pesticide and manifests itself in slowly developing the malfunctioning of a normal vital activity. The entry of a poisonous substance into an organism calls forth the response in the form of protective reactions that restrict the toxic action of the poison. These reactions include excretion of the substance (as such or in modified form) from the organism, deposition of it in the tissues, decomposition of the poison into simpler substances with their excretion or inclusion into the general process of metabolism. Although there are many methods, this chapter highlights some of the chemical and physical methods to decontaminate and detoxify the agricultural commodities so as to reduce the environmental hazards.

Hazards of pesticides use

Pesticides are by nature toxic to one or more life forms and it was realized at an early stage by many users that there was a certain risk attached to their use. Pesticides are tissue degradative, relatively stable, as well as toxic or carcinogenic in nature. Generally, pesticides damage the liver and the nervous system. Tumor formation has also been reported in the liver. Besides, certain pesticides are found to disturb the enzymatic activities of the body which leads to different types of diseases.

Pesticides kill millions of fish, birds, bees, and other plant pollinators each year, disrupting agricultural ecosystems and making it harder for farmers to manage their crops.

A study of farm workers exposed to pesticides in grape gardens in India found a 6 fold increase in genetic damage and adverse reproductive outcomes. It has been found that 44% of pregnancies of exposed farm workers resulted in miscarriage or stillbirth. A high rate of sterility was also found (Rita *et al*, 1985; Yoder *et al*, 1973). Organophosphate insecticides have been found to inhibit the activity levels of acetylcholinesterase, succinate dehydrogenase, isocitrate dehydrogenase, pyruvate dehydrogenase, lactate dehydrogenase and cytochrome-c-oxidase and cause accumulation of acetylcholine in the heptopancreas and muscle tissues (Reddy and Rao, 1991). Paraoxon, a component of organophosphates, has lethal effects due to blocking of the neurotransmitter acetylcholine. The blood enzyme paraoxoase breaks down this toxic chemical and different people have genetic differences that result in producing different amounts of PON1, thus having different susceptibilities to pesticides.

Fungicides are an especially dangerous category of pesticides. Fruits and vegetables from Florida are especially dangerous since the humid conditions lead to more mold and fungus, resulting in higher levels of fungicide use. Many fresh fruits and vegetables are sprayed as many as 20 times per year with 5 or more active ingredients (Baslain, 1987; Science News, 1994). Benomyl, vinclozin, ethylene dioxide, and the dithiocarbamate fungicides (ETU) have been shown to be male reproductive toxins in animal studies, as well as causing birth defects.

Scientists from the Environmental Protection Agency say there is "sufficient evidence" to conclude that the country's most widely used pesticide, atrazine, is an endocrine disrupting chemical and causes sexual abnormality in frogs (Hayes, 2003). Six studies involving three species of frogs showed a variety of defects, including frogs with both multiple testes and multiple ovaries, when exposed to the chemical. Two papers suggest that low levels of exposure to atrazine, as low as one part per 10 billion in the water, could cause tadpoles to develop into frogs with both male and female sexual organs. Atrazine has been banned in seven European countries.

Laboratory studies show that a variety of organochlorine, organophosphate, carbamate, and metal-based pesticides (such as those based on arsenic, copper, or mercury) can suppress the immune system of mammals. For example, exposure to the organochlorines aldrin and dieldrin reduces mouse resistance to viral infection, while DDT decreases antibody production in mammal and bird species. The organophosphate parathion delays antibody production and suppresses T-cell response in cell cultures, while chronic low-dose exposures of the commonly used organophosphate malathion can depress several different immune responses. Many solvents, inert ingredients, and contaminants that are part of pesticide formulations can also suppress immune responses in laboratory tests.

Therefore, it is necessary to have awareness as to how to detoxify the pesticides and reduce harmful effects of these pesticides on the environment by decontamination of various components of the environment or commodities (food, feed, soil, water, equipments etc.) with which they might have come in contact and in which they persist in undesirable amounts.

Detoxification and decontamination

Detoxification of pesticide molecules using chemical methods involves acid or base hydrolysis, oxidation, reduction, chlorinolysis, adsorption, fixation, solvent extraction etc. It has been reviewed long back in 1979. The chapter attempts to consolidate the recent methods, while retaining and quoting certain key features of the previous illustrations and attempts. Besides defining severity of the problem, details are given about methods for detoxification of pesticides including chemical industrial methods. Decontamination of edible commodities as well as spray equipment is also documented.

Detoxification and decontamination are two different terms. Detoxification is breaking or modification of the pesticide in order to reduce its toxicity or to reduce the hazard due to residues. While the first term is usually employed before pesticide disposal, decontamination is a process for removal or segregation (Roy, 2003) of toxicant from an agricultural commodity for example by washing. Some detoxification chemical methods are listed below:

Hydrolysis (acid or base)

Most pesticides are lipophilic and are not excreted in an unchanged form. Due to hydrolysis, lipophilic substances transform into hydrophilic compounds and the reaction products can penetrate only slightly through the membranes to the vitally important centres and are excreted from the organism at a faster rate. Though, toxicity in general decreases in such processes, but exceptions are also known. For example, acetyldiptrex gets hydrolyzed under the influence of enzymes to more toxic trichlorfon.

Oxidation and reduction

Compounds, which resist hydrolysis, prefer oxidation as the main route of their metabolism in organisms. For example b-oxidation mechanism in living organisms consists of step-wise decomposition of fragments of the hydrocarbon chain with an even number of carbon atoms to acetic acid. Phenoxybutyric acid is oxidized to more phytotoxic phenoxyacetic acid in plants. The oxidation of side bonds of cyclic and heterocyclic compounds occurs when direct hydroxylation of the ring is hampered. More polar and less toxic products thus formed accelerate ring decomposition processes.

Chemical industrial method

The detoxification methods, which the firms producing or formulating pesticides, generally utilize involves:

- Optimization of the chemical reactions involved in the production of the pesticides, not only increases yield of the product but also decreases the demand on the air and water pollution control system

- The chemicals with low water solubility can be extracted using an organic solvent
- Neutralization by addition of concentrated solution of acid or base or by precipitation and coagulation
- Wet scrubbing of gaseous pesticide containing effluents. The effluents are either sent through trapping or detoxifying solution and then through an activated charcoal filter or they are scrubbed in towers, where scrubbing liquid, generally water, is sprayed concurrently to the upward flowing gas

Fixation and chlorinolysis

Chemical fixation of toxic wastes involves reaction of at least two reactants with the hazardous molecule from waste to form a chemically and mechanically stable, cross-linked, three dimensional matrix. This immobilized waste is fairly resistant to leaching and the fixed material can be deposited in a landfill (Newman, 1976).

Pesticide decontamination of equipments

One of the sources of high-pesticide-content polluted waters is the rinsing of used pesticide spray containers and their disposal. The main characteristics of this wastewater are its extreme toxicity, low volume and well-defined location. Suitable treatment, is, therefore, required to prevent this pollution, which is easier than cleaning up the environment afterwards. Malato *et al.* (1999) used a pre-industrial solar TiO_2-photocatalytic treatment for this purpose. Commercial pesticides (Lufenuron and Propamocarb) are used to demonstrate the practical application of the treatment. The photo-catalytic pilot plant used for these tests has 6 CPC modules (collector surface 8.9 m^2, photo reactor volume 108 L, total plant volume 247 L) connected in series and mounted on a platform inclined 37°. The results are compared with phenol degradation under the same experimental conditions, and relative photonic efficiency (îr) is calculated to facilitate comparison with other experimental devices. Total Organic Carbon (TOC) analysis is employed to confirm the mineralization of the contaminants. Relative photonic efficiency related to TOC is also introduced (îrTOC). The pesticides were completely photo degraded, but mineralization is slow with only illuminated TiO_2. The use of an additional oxidant (peroxydisulphate) enhances the degradation rate by 5 (propamocarb) and 9 (lufenuron), compared to TiO_2 alone.

Physical methods

The right amount of pesticide needed should be bought from market for the season. The quantity of diluted pesticide required for the job should be calculated. All the pesticides should be mixed in accordance with label instructions. Pesticide residues that cling to empty containers can be dangerous to humans and the environment. Improper disposal can result in serious hazards to humans (children) pets, wildlife and the environment (water, soil), or legal problems. After mixing, loading or applying a pesticide, you should:

- Wash pesticide equipment, return to designated area
- Safely store or dispose off all pesticides and other chemicals that you used
- Be sure that work site presents no hazards to humans or environment (never leave site unattended until clean)
- Triple-rinse or pressure-rinse empty containers (plastic, glass or metal) as soon as they are empty
- While cleaning a pesticide-contaminated equipment, wear appropriate personal protective equipment and clothing
- Wash yourself and clean personal protective clothing
- Always clean pesticide mixing, loading, application equipment as soon as you are finished using them (pesticides left in equipment can cause corrosion, clog nozzles, and contaminate future applications)
- Make record of pesticide applications while your memory is fresh

It is illegal to burn or bury any pesticide container in North Carolina or dump such containers in any unauthorized location. Pesticide containers that have been triple-rinsed or pressure-rinsed usually can be disposed of as trash in a sanitary landfill (unless prohibited by label or state/local laws).

Personal cleanup

- Wash the outside of gloves before removing them
- Carefully peel off protective clothing to avoid getting pesticide on your skin
- Put protective clothing in a plastic bag until they can be cleaned or disposed of properly
- Remove additional work clothing carefully; keep them separate from other clothing until they are laundered
- Wash face, hands, forearms, etc. with mild soap and water right away; wash entire body in shower at the end of the work day

Pesticide waste

- Excess pesticide and rinsates that can not be used must be disposed off as hazardous waste

- Other pesticide wastes include contaminated soil, spill cleanup materials, and personal protective equipment or clothing that can not be cleaned and reused.

Pesticide spill management

A spill is any accidental release of a pesticide. You should know how and be prepared to respond immediately to a pesticide spill. Never leave a pesticide spill unattended. If you can not manage a spill by yourself, get assistance as soon as possible. Three C's for spills: Control, Contain, Clean up.

Controlling a pesticide spill

- Protect yourself by putting on appropriate personal protective equipment before contacting spill (also be aware of pesticide vapours)
- Stop the source of the spill (if a small container is leaking, put it in the larger chemical-resistant container)
- Protect other persons from the spill by isólating the spill, warn others of the spill, and keep people out of reach of the pesticide's drift or fumes
- Stay at the spill site until another knowledgeable and correctly protected person arrives
- Confine the spill in as small an area as possible (small spills can be contained with containment "snakes"; larger spills may require digging soil dike)
- Protect water sources (ponds, creeks, wells, etc.) from spill and runoff
- Absorb any liquid pesticide spills with absorbent materials such as spill pillows, sand, vermiculite, sawdust, clay, kitty litter, shredded newspaper, etc.
- Cover dry pesticide spills (dusts, powders, granules) with sweeping compound, plastic covering or light mist of water.

Methods for decontamination of food commodities

Decontamination means reduction of contamination present due to pesticide residues in/on edible or other commodity without spoiling the commodity. The decontamination methods are generally washing, cooking, boiling and peeling, but involvement of chemicals is on the rise. Easiest example is that of washing the fruits. Simple washing is effective only for the decontamination of fresh samples, for aged samples, special methods are still needed. In decontamination, the pesticide is present as a residue in a complex biological matrix, so the residue is to be inactivated with as little change to bulk of the sample as possible and the decontamination procedures must also be economically feasible. Due to the diversity in chemical properties of pesticides and their formulated products and range of quantity and composition of pesticide wastes which must be treated, no single treatment system can be universally applied. The pesticide on the fruit or vegetable sample may be dislodged by any of the following physical method-

- Washing/rubbing removes up to 60-70% of the pesticide
- Treatment with saline water or vinegar solution can dislodge up to 70-75% of the pesticide (Dixit *et al.*, 1984; Jacob and Verma, 1989)
- Chemical methods like cooking, baking, frying or boiling can bring about 70-80% reductions of residues. In some cases, where the pesticide is present inside the tissues, only boiling can decontaminate the pesticide to the extent of safe limit. Common processing procedures for the removal of different insecticides from the vegetables is given in Table 1.
- Application, technology, quality, dose, formulation and method of application of pesticide should be taken care of - use right doses, of right pesticides at right time
- By need based application - Doing regular monitoring of the insect pests and beneficial in the field. Applying chemical pesticides when its use is needed
- Spraying pesticides in the evening when the pollinator's population is less in the field
- Using alternative options to pesticides - Use of botanical pesticides and bio-pesticides at appropriate stage
- Using protective clothing while applying pesticides
- Giving training to pesticide dealers and users
- Health care and gender voice should be taken in to consideration
- First aid and medicinal treatment in case of pesticide poisoning and accidents
- Awareness on precautions, protective clothing and safe handling of pesticides

It has been observed that long-term, low-dose exposure of pesticides are increasingly linked to human health effects such as immune-suppression, hormone disruption, diminished intelligence, reproductive abnormalities, and cancer. In this light, problems of pesticide safety, regulation of pesticide use, use of biotechnology, and bio-pesticides, integrated pest management (IPM), where natural and biological controls replace some pesticide use, and use of pesticides obtained from natural plant sources such as neem extracts are some of the future strategies for minimizing human exposure to pesticides. It would be more ecologically and economically effective in controlling pests.

Table 1. Removal of insecticides by common processing procedure in vegetables

Vegetable	Insecticide	Washing with 2% salt water and cooking for 15 minutes (Percentage lost)	Washing and steam cooking for 15 minutes (Percentage lost)
Beans	Malathion	60	69
	Monochrotophos	42	47
	Carbaryl	58	69
Chillies	Monochrotophos	28	30
	Quinalphos	22	29
Tomatoes	Carbaryl	67	75
	Monochrotophos	31	32
	Quinalphos	29	30

References

Baslain B.J., in Harrowsmith, Feb 1987.

Dikshith. 1991. Toxicology of pesticides in animal. CRC Press Boca Raton.

Dixit, A.K., M.D. Awasthi and S.K. Handa. 1984. Decontamination of insecticide residues from cowpea. *Pesticides* 43-44.

Hayes T. B. 2002. In : Proceedings of the National Academy of Science, April 2002, and T.B. Hayes, Nature , Oct 2002; & U.S. EPA, Atrazine linked to amphibian abnormalities, Environmental Health Perspectives, June 17, 2003.

Jacob, S. and S. Verma. 1989. Persistence and decontamination of malathion on cauliflower. *Indian J. Ent* **51** (4): 377-83.

Khan SU. 1980. Pesticides in the soil environment. Elsevier Sci. Publishing Co., Amsterdam, The Netherlands.

Malato S., J. Blanco, C. Richter, B. Milow and M.I. Maldonado. 1999. Solar photocatalytic decontamination of wastewater from the rinsing of pesticide containers. Proceedings of the 9th *Solar PACES International Symposium on Solar Thermal Concentrating Technologies*. J. Phys. IV France 09 (1999) Pr3-277-Pr3-282

Newman JH. 1976. Fixing hazardous wastes. *Processing* **22** (6): 41.

Reddy MS, Rao KV. Phosphamidon, methylparathion and dichlorvos impact on tissue oxidative metabolism in penaeid prawn, Metapenaeus monoceros. *Biochem Int* 1991. 23(3):439-47

Rita P. *et al.*, 1985. "Monitoring of Workers Cccupationally Exposed to Pesticides in Grape Gardens", Environmental Research, Vol 44.

Roy A.K. 2003. Mycological problems of crude herbal drugs- overview and challenges. *Indian Phytopathology* 56 (1): 1-13.

Science News July 1994. "Another Emasculating Pesticide Found", Vol 146, p16.

Yoder J.M. *et al.*, 1973. "Lymphocyte Chromosome Analysis of Agricultural Workers in Occupational Exposure to Pesticides:, Mut. Res., Vol 21.

□□□

21

Agronomic Practices for Management of Insect-pests in Sugarcane

RAJENDER KUMAR AND SMRITI SHARMA

Sugarcane is a long duration crop of 10-18 months and therefore is liable to be attacked by a number of insect pests which according to an estimate, decline the sugarcane production by 20.0 per cent (Dhaliwal *et al* 2004). Hence, to minimize these losses, management of insect-pest below the economic threshold levels is of great significance. Insects are major biotic constraints in sugarcane crop production particularly in subtropics that provides a conducive environment for insect pests built up. The physical characteristics of sugarcane crop limits the use of chemicals against the insect pests once the canopy has closed, spraying of the insecticides at cane formation stage and thereafter. The practice of ratooning further limits the opportunities for soil application of chemicals against root feeders.

The magnitude of damage caused by a particular insect varies in different sugarcane growing areas. In general, some insects like top borer, shoot borer, stalk borer, pyrilla, whitefly and black bug are major problem and quite destructive in sub-tropical sugarcane growing areas whereas, others prefer tropical growing areas of India. The increased vulnerability of most of the high sugared/ high yielding genotypes and the consequent unscrupulous use of chemicals have brought imbalances in biotic system resulting into development of resistance against insecticides, loss of bio-diversity (parasites, predators), resurgence of pests and secondary pest outbreaks apart from environmental pollution, contamination of food chain and pose various health hazards to human beings and animals. A paradigm shift in pest control strategies from unilateral chemical approach to non-chemical methods such as biological control, ecological management or host plant resistance has therefore become imperative (Jaipal 2000).

Adoption of various agronomic practices like planting of insect free cane setts, altering date of planting, earthing up, trash mulching, detrashing of dry foliage, removal and destruction of infested plant parts, intercropping, destruction of water shoots/ late shoots, harvesting close to surface or deep harvesting, stubble shaving, trash burning, balanced fertilization and timely irrigations for pest management in sugarcane are well established (Table 1). These practices are not only very effective but also environment friendly. As the studies on these aspects for management of insect pests have been done in isolation at different locations, so an effort has been made to review the significance of these practices for the management of sugarcane insect pests thereby minimizing the yield losses caused by them.

Sugarcane insect pests can be classified into three categories according to their nature of damage and feeding habits:

1. Borer complex- Early shoot borer, top borer, stalk borer, root borer, Gurdaspur borer and internode borer
2. Sucking pests - Pyrilla, black bug, whitefly, mite, thrips and scale insect
3. Soil pests - Termites and white grubs

As mentioned above, all the pests have specific plant parts where they feed upon, like borers feed on young plants and later feed on the cane stalk. Similarly, termites feed on stalks, roots and underground portion of the plant. Sucking pests like pyrilla, whitefly, thrips and mites feed on leaves while black bug prefers to feed hidden under leaf sheath and leaf whorl. Mealy bug and scale insects prefer to feed on cane stalks (Table 1). Most of these pests infest both plant as well as ratoon crop of sugarcane. However, incidence of black bug in particular and the incidence of all other pests in general, is higher in ratoon crop. These pests individually also cause great losses to the crop (Table 2).

Brief information of various insect pests regarding their importance, active phases, preferred crop stages and the mechanism of damage along with symptoms and cultural practices for their management are as follows.

Table 1. Extent of adoption of cultural aspect of integrated pest management strategies by sugarcane growers

S. No.	Cultural practice for pest control	Sugarcane growers Frequency	%
1	**Management of early shoot borer in sugarcane**		
1 a)	Sugarcane plots free from early shoot borer	78	43.33
1 b)	Sugarcane plots with early shoot borer problem (n=102)	102	56.66
	i) Practiced earthing up	43	42.15
	ii) Practiced mulching operations	38	37.25
	iii) Biological methods	4	3.92
	iv) Chemical methods	68	66.66
2	**Management of termites**		
2 a)	Sugarcane plots free from termites	128	71.11
2 b)	Sugarcane plots with termite problem (n=52)	52	28.88
	i) Frequently irrigate the crop	45	86.53
	ii) Catch the termite queen and destroy it	04	7.69
3	Management of wooly aphid incidence		
3 a)	Sugarcane plots free from wooly aphid	18	10.00
3 b)	Sugarcane plots with wooly aphid problem (n=162)	162	90.00
	i) Wide row/ paired row spacing	35	21.60
	ii) Judicious use of nitrogen and irrigation	18	11.11
4	**Management of scale insect**		
4 a)	Sugarcane plots free from scale insect	124	68.88
4 b)	Sugarcane plots with scale insect problem (n=56)	56	31.11
	i) Detrashing lower leaves leaving 8 leaves in the top	38	67.85

Number of growers sampled (n) = 180 *Source:* Marradi *et al* (2007)

Table 2. Extent of losses due to different insect pests in sugarcane

Insect- pest	Per cent loss Cane yield	Sugar recovery
Early shoot borer	22 to 33	1.7 - 3.07
Internode borer	34.88	0.2 - 4.1
Top borer	21-37	1.7 - 3.07
Stalk borer	Up to 33	0.1 - 0.8
Gurdaspur borer	5-15	0.3 - 2.90
Root borer	35	1.5 - 2.5
Scale insect	32.60	brix loss 16.20
Black bug	31.60	2.0-3.0
Pyrilla	14.70	1.0-1.5
Whitefly	80	5.0-6.0
White grub (H)	100	Complete drying
White grub (L)	33	4.5
Termite	22.27	-

Source : http://dacnet.nic.in/Sugarcane/PestManage.htm

Borer complex

Importance

Among insect-pests of sugarcane, borers are the most damaging, which cause a great loss to cane and sugar yield. Borer complex includes top borer, early shoot borer, stalk borer, root borer, Gurdaspur borer and internode borer. Sugarcane borers are widely distributed and serious pests, causing shoots death at young stage and damage stalks in grown up canes. The extent of losses caused by these borers ranged from 5 to 37 and 0.1 to 4.1 per cent in terms of cane yield and sugar recovery, respectively (Table 2). Feng (2003) has also reported the cane yield loss up to 20 per cent, and sugar loss of 0.5 per cent. Borers occur severely in all sugarcane-growing regions and even several species of borers can occur simultaneously in the same field with their overlapping generations. The control of these borers is one of the most difficult tasks faced in the cultivation of sugarcane.

Nature of damage and symptoms

Important insect pests of sugarcane and their specific area of damage are given in table 3.

Early shoot borer (*Chilo infuscatellus*)

Early shoot borer is widely distributed in all sugarcane growing areas in India which infests the crop during early growth stages. In case of serious infestation, its larvae can damage 60-70 per cent shoots resulting in gappy stand of the crop. Its incidence is more pronounced in areas where drought and hot weather prevails during early stages of the crop growth. Its larvae cause maximum damage during pre-monsoon period (April- June) and decreases rapidly when crop attains the cane formation stage. During winters, being adverse conditions for survival, the pest as full-grown larvae hibernates in the stubbles. After hatching, the larva reaches the plant base and bore into the shoot (about 2 cm below soil surface) to feed on the central leaves of shoot. Infested shoots typically produce dead hearts which can be easily pulled out emitting foul smell. But after cane formation, no such symptoms appear and attack cannot be identified from the appearance of the canes. The pest causes maximum damage to late sown plant crop, but its damage is comparatively low in autumn planted and ratoon crop.

Top Borer (*Scirpophaga excerptalis*)

Top borer infestation on sugarcane can be seen through out India but its economic effects are more pronounced in sub-tropical sugarcane growing regions. It is unique among the moth borers as it infests sugarcane during all the stages of crop growth. Top borer remains active from March to October as it prefers moderate temperature and high humidity and then hibernates as a full-grown larva in the cane top during winter. Hibernating larvae pupate in end February and subsequently moths emergence occur in April. The first brood appears in the beginning of March on ratoon and autumn crop. Its incidence is generally higher than second brood owing to the favourable weather conditions. This brood however does not cause economic damage as enhanced tillering compensates the shoots killed by the pest. The second brood appears during hot and dry weather conditions in mid of May, which results in lesser egg laying and high mortality of young larvae. The third brood appears in the end of June or beginning of July coinciding with the monsoon. Due to favourable weather conditions, the incidence is quite high and causes maximum damage to the crop. By the time fourth brood appears, in the end of August, the crop completes almost 70-80 per cent of its growth and therefore losses by this brood are not so high.

Table 3. Major insect pests of sugarcane and their specific area of damage

Insect - pest	Part of sugarcane plant
Termites and white grub	Cut ends of setts and stubbles and roots and underground portion of cane
Shoot borer, root borer and stalk borer	Young crop (shoots), plant and ratoon
Root borer (most of the borers aestivate and hibernate in stubbles)	Stubbles and under ground portion of cane
Top borer, Gurdaspur borer, stalk borer, mealy bug and scale insect	Cane portion
Black bug, whitefly, red mite, pyrilla, thrips, webbing mite	Leaf blade
Black bug	Leaf sheath

Source: (Madan 2001)

The young larvae bore into the mid rib of the leaf from lower side, mining their way to the base. From there, they enter the spindle, feed on the growing point and soft portion of the cane. The central leaf of the cane top dries up and turns dark. The other typical symptoms are the shot holes in the leaf, white or red streaks on the upper side of the leaf midrib and bunchy top from July onwards (Anonymous 2009).

Stalk Borer (*Chilo auricilius*)

In the early eighties, stalk borer spread all over India and its incidence went on increasing by the end of eighties and later on stabilized during nineties. The pest breeds from February to October and remains active throughout the year. The eggs are laid in clusters on the leaf sheath or on the underside of the leaves. The young larvae feed under the leaf sheath or on mid ribs for about a week and thereafter bore into the stalks by making circular holes in the rind. Dead hearts are produced in young plants however, after the cane formation stage, the attack can not be easily detected. Entrance or exit holes on the infested canes can be seen only after stripping the canes. The larva can cause damage up to three nodes and the cane may be attacked at several places. Pest incidence remains low during pre-monsoon period, multiplies rapidly from July to October and keeps on damaging the crop during winters. Heavy rains with intermittent prolonged dry spells coupled with high temperature, lodging of canes, water logging and heavy manuring are conducive factors for its multiplication. Severe infestation of the pest adversely affects both the cane yield as well as sugar recovery.

Gurdaspur Borer (*Acigona steniellus*)

Gurdaspur borer is not as serious problem as other borers under present scenario however, it was once considered the most destructive pest of sugarcane in north-west India and Pakistan. It prefers humid weather, 30-38 °C temperature and waterlogged conditions so remains active from July to October and passes the winter and early part of summer in diapause as full grown larvae in sugarcane stubble and has no alternate host. All the larvae hatching out of one egg cluster enter the same stalk close to the nearest bud and feed gregariously there for 8-10 days. After this, they come out and enter the adjoining canes, singly or sometimes in two's. Due to the galleries made by these larvae in the cane, top of the cane starts drying. The pest causes withering of the central leaves (notably fifth leaf) followed by total drying up of the tops. On shaking, the plant breaks from the place of attack. In September, the larvae start migration from the upper part by coming out and re-entering the cane near the base to diapause.

Root Borer (*Emmalocera depressella*)

Root borer generally infests the young shoots, though it is known to infest grown up crops as well. Its infestation during early stages of crop growth is marked with the formation of dead hearts but in grown up canes no external symptoms are visible except yellowing of leaves. Therefore, the cane is to be dug out to detect its infestation. Unlike early shoot borer, dead hearts cannot be easily pulled out and some times in the process, the plant gets pulled out with the larva either hanging out at the broken end of the shoot or remaining partly outside. Though the plants damaged by the pest can form millable canes but results in about 10 per cent reduction in yield and up to 0.3 units loss in sucrose.

Internode borer (*Chilo (Saccharíphagus) indicus*)

Internode borer is a major pest of sugarcane in peninsular India causing considerable reduction in tonnage and sucrose content of cane. It has also been reported from certain pockets in north India, Madhya Pradesh and Orissa. The pest remains active throughout the crop season under south Indian conditions. Newly hatched caterpillars prefer the top portion of the cane for their early feeding. Soon after the emergence they feed by scraping the leaf spindle and subsequently, they bore into the tender canes. The attack of the pest in young shoots is more commonly found in the late-planted crops and in the ratoons. In advanced stages of crop growth, the damage can be observed only by removing leaf sheath. Fresh excreta can also be seen sticking outside the larval holes on the internodes. Grown up larva is highly migratory in habit and often transfers itself from one internode or cane to another.

Agronomic Management Practices

Planting Time

Time of planting plays an important role in preventing the damage caused by the borers. Early planting in the month of February and early March is very helpful and non-monetory way in checking serious damage caused by shoot-borer under north Indian conditions. As in early-planted crops, tillering phase initiated ahead of the time of borer infestation which appears in April. Similarly, autumn crop escapes the damage as well developed shoot and root can withstand the attack (Table 4). In contrary, dense growth helps in survival of other insect pests by providing shelter from ravages of weather and better availability of food. Based on various studies on this aspect, the following conclusions have been made:

i. Autumn planted crop suffers more seriously from ravages of stalk borer than the spring planted one. This may be attributed to better growth of the autumn planted crop due to longer period of growth and advancement of cane formation.

ii. In the areas where stalk borer infestation is high, autumn planting of sugarcane may be avoided and to check the further build up of pest population heavily infested crop should not be ratooned. (Khan *et al* 2006; Duhra and Suri 2002).

iii. Autumn planting generally protects the crop against shoot borer, Gurdaspur borer, top borer, termites and white grubs in sub-tropical India (Kalra 1963, Murthy 1953, Varasi 1954).

iv. In Andhra Pradesh, the higher incidence of the top borer has been reported during January to April in the November planting, February to March in December planting and during May in the January planting in case of *Eksali* crop. The incidence of top borer decreased with the advancement of plantings, being highest in November planting and lowest in February planting. The higher incidence in November planting may be due to the fact that the tillering completes by March- April coinciding with higher activity of the pest and due to the dense growth, the young larvae emerging from the egg masses during this period have a greater chance of survival and to infest the cane crop (Khan *et al* 1965).

Table 4. Cultural practices for management of different insect pests in sugarcane

Name of practice	Time of operation	Crop stage	Effective against
Timely irrigation	Pre monsoon (40th day of planting till end of June)	Formative phase	Shoot borer
Timely urea application	With 1st, 3rd and 5th irrigation	Formative phase	Shoot borer
Earthing up	May and June	Formative phase	Stalk borer, Shoot borer
Mechanical removal/ killing of damaging stages of insect pests	First fortnight of July (every 4th day)	Grand growth phase	Top borer
Propping of cane stalks and liberation of *Trichogramma chilonis*	Monsoon and post monsoon (July to October) at 10 days interval	Grand growth phase	Stalk borer
Crop residues removal and foliar application of urea (2.5%)	Before sprouting of ratoon crop	Formative phase	Black bug

(*Source:* Jaipal 2000)

v. In case of *Adsali* planting of sugarcane, the activity of the top borer was higher during January to February in the June planting, November to January in July planting and December to March in the August planting. However, the incidence decreases in the September planting. (Khan *et al* 1965). Studies have also reported that *Adsali* planting can effectively prevent the severe attack of shoot borer in Maharashtra (Kalra 1963, Murthy 1953, Varasi 1954).

vi. To escape the crop from damage by the pest, early planting has been advocated to get the crop past the susceptible stage. In north India, planting early i.e., before the mid of March has been found very effective in lowering the incidence of shoot borer (Karnatak, 2006).

Planting method, seed rate and seed quality

Insects' preference to crop changed with different planting methods and seed rate because of variation in crop geometry, plant population, crop growth and input requirements. Studies conducted at various places on this aspect in minimizing the borers' incidence are summarized below:

i. The higher incidence of stalk borer but lesser top borer incidence in ring (pit) and trench planting method against conventional planting method has been reported at Indian Institute of Sugarcane Research, Lucknow. Greater stalk borer incidence in ring planting method may be due to dense canopy and luxuriant crop growth.

ii. Sugarcane raised through spaced transplanting technique recorded lower incidence of stalk borer and top borer as compared to that of normal flat planting (Chandra and Singh 2007).

iii. Deep trench (40 cm wide) planting of sugarcane at Anakapalle recorded only 19 percent incidence of early shoot borer against 34.2 per cent when planted in shallow furrows. Therefore, sugarcane planting in deep trenches can be useful in reducing the attack of early shoot borer (Rao *et al* 2008).

iv. The incidence of shoot borer reduced considerably with higher sett (seed) rate. The experiment conducted at the Sugarcane Research Station, Cuddalore concluded that the cumulative incidence of shoot borer reduced progressively with a higher number of setts. The mean shoot borer incidence was found to decrease in all the higher sett rate plantings. Among the different sett rates economic advantage in terms of yield and benefit was more conspicuous when 30 per cent hike in sett rate over the conventional rate. The mean reduction of shoot borer varied from 3.80 to the highest of 39.16 per cent among the different sett rates tested as against the conventional sett rate. The gross cost-benefit ratio was 2.17 under 30 per cent increased sett rate. The technique of using higher seed rate, being a cultural and ecofriendly one could be specifically advocated to late planted sugarcane (March to June) and shoot borer endemic locations (Rajendran 2005).

v. Use of insect free setts helps in reduction of attack of stalk borer. Therefore, cane from infested field should not be used for seed purpose (Anonymous 2009).

Irrigation

Time and frequency of irrigation (irrigation interval) has been considered economical, effective and safe pest management technique especially against early shoot borer and top borer. Pests preferring hot and dry conditions can be minimized by adopting proper scheduling of irrigation during hot summer months which also prevent moisture stress to the crop. The changes in the microclimate of the crop due to irrigations can also be beneficial to the pest or crop according to the nature of damage by the pest.

i. The incidence of top borer can be managed to a considerable level by avoiding irrigations during the peak periods of moth emergence. (Rao *et al* 2008).

ii. In case of shoot borer, frequent irrigations during April to June lower down the incidence, as the pest prefer hot and dry conditions. The incidence of shoot borer decreased with 8-10 days irrigation interval whereas it increased with increasing irrigation interval to12-14 days (Duhra and Suri, 2002; Singla and Duhra 1992).

iii. Closer irrigations during hot summer months were found essential for the management of shoot and root borer under drought conditions to avoid heavy build up of these pests. (Varma and Tanwar, 1993).

iv. Judicious use of irrigation preferably with trench planting method avoid lodging thus reduced the incidence of stalk borer in sugarcane (Duhra and Suri, 2002).

v. Various cultural practices have been evaluated for the control of root borer. It was observed that first irrigation applied 35-40 days after planting and subsequent irrigations during summer at 10 days interval reduced root borer incidence besides improving the tiller counts, millable canes and cane yield (Singh *et al* 1997).

Intercropping

Intercropping is an important component in sugarcane cultivation to get higher returns. In addition to this, the practice is also useful for management of insets as reported by several workers. The mechanism behind altering the incidence is variation in the host location process, repelling effects, dense foliage, changes in microclimate and faster crop growth with the adoption of various intercrops. A variety of cereals, oilseeds, pulses and other crops grown as intercrop with sugarcane in India influence the insect pests incidence in sugarcane ((Kailasam, 1994; Verma *et al.*, 1984) in positive, neutral and negative manner. Some of the research findings are concluded as:

i. Intercropping of onion, methi, saunff and ajwain can reduce the infestation of top borer. Similarly, intercropping of transplanted gobhi sarson, sunflower, transplanted maize resulted in lesser incidence of top borer as compared to sugarcane sole crop (Karnatak 2006; Singla *et al* 1994).

ii. Growing twelve feet arhar around cane fields and coriander as an intercrop with sugarcane reduced root and top borers damage (Khanna 1942, Misra and Hora 1982).

iii. Intercropping of spice crops like ajvain, garlic and saunf in sugarcane were found to be the most effective in reducing the incidence of the top borer. The reduction in incidence was due to repelling effects of these intercrops (Varun *et al* 1990).

iv. Pulses like black gram, cowpea, green gram and soybean as intercrops were also found effective in lowering the incidence of top and early shoot borers (Srikanth *et al* 2000).

v. During summer, intercropping is much helpful in managing and / or reducing the incidence of early shoot borer. In a study it has been concluded that sugarcane with moong as an intercrop reduced shoot borer incidence (9.50%) to a great extent affecting maximum of 63.16 per cent reduction over 25.79 percent in the sole crop of sugarcane. Sunhemp in sugarcane appeared to be the next best companion crop (10.25%) reducing early shoot borer incidence by 60.26 percent. Intercropping also gave higher crop yield of about 12 percent over the sole crop of sugar cane. The lowest incidence of shoot borer in moong and sunhemp companion crops might be due to changes in the microclimate and faster crop growth, which were unfavorable to the pests. The non-host crops might have also acted as a physical barrier preventing initial build up of borer population. (Thirumurugan and Koodalingam 2005).

vi. Intercrops like wheat, potato and coriander reduced the incidence of sugarcane top borer (Tiwari and Om Prakash, 1980). Field and vegetable crops like radish and turnip reduced its first brood and spice crops like garlic and fennel *(Foeniculurn vulgare)* reduced its first three broods (Verma *et al.*, 1981).

vii. Companion cropping of sugarcane during summer may also prove to be helpful in reducing the incidence of root borer in the early stages of crop growth. Intercrops such as mentha, sorghum and turmeric have been found to reduce the incidence of root borer as compared to sugarcane sole crop (Sardana 2001).

Trash Mulching

Trash mulching in sugarcane during early stages of crop growth helps to conserve soil moisture during hot summer months and also changes the microclimate of the crop, thus decreases the incidence of pre-monsoon pests. Therefore, trash mulching plays an important role in reducing borers infestation in sugarcane and has been recommended as an alternative for chemical control of sugarcane shoot borer. It has also been reported that sugarcane trash proved better with minimum cane damage and maximum yield. But on the other hand, some times it can increase termite incidence as the organic matter content increases in the field.

i. Shoot borer infestation reduced significantly with trash mulching and incidence of egg parasitoid, *Trichogramma chilonis* comparatively increased with this practice. However, other crop residues such as wheat straw mulch can also increase cane yield but sugarcane trash was found to be more effective, easily available and economical (Rana *et al* 2007).

ii. Besides conserving moisture during summer months, trash mulching also protected the crop from the damage by shoot borers (Parthasarthy 1959, Gill 1963, Rajani 1966). The reduction in shoot borer incidence due to trash mulching could be related more to the increase in sheath moisture (Avasthy and Tiwari, 1986) than changes in organic matter content of soil.

iii. Spreading of a thick blanket soon after germination greatly helps in suppression of shoot borer. However, this practice may increase the incidence of termite, armyworm and rat damage in areas where these pests are prevalent.

Earthing up and propping

Earthing up during formative phase helps the crop escape pest attack and also improve crop vigour in terms of cane height and number of millable canes, which ultimately contributes to higher cane yield than that obtained with insecticide treatment (Jaipal 2000).

i. Two to three light earthing ups during April and May covered the exit holes therefore, adults could not come out for egg laying which resulted in declining of the shoot borer population (Duhra and Suri 2002, Gupta 1945, Khan 1951).

ii. Three earthings at 8, 10 and 12 weeks after planting was found to reduce the incidence of shoot borer significantly over no earthing (Malik 1982).

iii. A light earthing up usually at 45 days after planting was effective to suppress both root and shoot borer population (Khan et al 2008).

iv. Planting the crop preferably by trench method, earthing up the crop in end June and propping up the canes in end August has been reported to control stalk borer infestation effectively. According to Jaipal (2000) innundative releases of egg parasitoid during July-October and the crop production practices like earthing up and propping of cane stalks could curtail stalk borer infestation adequately (Table 5).

v. In addition to early shoot borer and stalk borer, earthing up in May to June has also great impact on the larval population of root borer. Two earthing up in May and June have shown maximum reduction in the population of root borer larvae under north Indian conditions like Haryana (Singh *et al* 1997).

Table 5. Effect of various practices for insect pest management in sugarcane

Cultivars	Per cent decrease as of untreated in the incidence of insect pest		
	Timing irrigation and application of urea (Shoot borer; 90 DAP)	**Earthing up, propping and *T chilonis* (Stalk borer)**	**Crop residue removal and stubble shaving (Black bug)**
Co J 64	- 59.5	- 40.9	- 56.6
Co 7717	- 55.6	- 33.4	- 43.8
CoS 767	- 56.3	- 47.7	- 65.7
Co 1148	- 61.1	- 36.2	- 69.4
CoH 92	- 61.7	- 33.9	- 65.1
CoH 99	- 68.2	- 48.7	- 59.7
CoH 108	- 46.8	- 27.2	- 47.3
CoS 8436	- 66.8	- 54.1	--

(*Source:* Jaipal 2000)

Detrashing

Detrashing of lower leaves after cane formation or after grand growth phase helps in lowering the incidence of many pests like pyrilla, stalk borer, scale insect and internode borer. These pests lay eggs on the dried foliage of the crop so by removal of dried leaves, the pest population can be put under check and ultimately the pest load can be decreased.

i. Detrashing in the months of October and November resulted in reduction of damage caused by stalk borer (Madan 2001, Varma and Mitra 1981).

ii. Removal of dry leaves and burning of the trash with a view to destroy the larvae and pupate attached with the leaf-sheaths, proved helpful and recommended for controlling the internode borer. (www.krishiworld.com).

iii. Detrashing at fifth, seventh and ninth month's stages of the crop growth reduced internode borer damage. Similarly, removal of only lower leaves during fifth and ninth month stages of crop growth was also found effective in reducing the activity of stalk borer, internode borer, scale insects, mealy bugs, whitefly and pyrilla (David 1986; Khan *et al* 2008).

Rouging, Destruction of water shoots and Stubbles

Removal of the infested plants, susceptible forms like water shoots, late shoots and stubbles which harbour insects through the process of rouging and destruction helps in preventing further attack. Destruction of stubbles being a part of field sanitation reduces the pest population by suppressing the seasonal carry over of the pest.

i. Collection and destruction of egg clusters and rouging of affected sugarcane tops from March to June before moth emergence, helped in lowering the incidence of top borer. Removal of dead hearts of first and second brood lowered down the population level of third brood. In addition, heavily infested crops should be harvested before mid February (Rao *et al* 2008; Karnatak 2006).

ii. The timely removal of top borer infested shoots, its egg masses and adults reduced the incidence of most damaging third brood alone by over 50 percent. Apart from being completely safe to the environment, the technique involved 3-4 times lesser (Rs 150-250/ha) expenditure than the chemical treatment (Rs. 1000-1500/ha) and found practicable in areas where labour is plentiful and inexpensive and on small family holdings (Jaipal 2000).

iii. The population of stalk borer can be suppressed by burning the trash after harvest and by removing and destruction of late water shoots during February and March as these late water shoots contribute considerably in the multiplication of the pest. This practice also improved tillering. Ratooning in the infested fields should be avoided (Atwal and Dhaliwal 2005).

iv. The practice of destruction of water shoots/ late shoots protected the crop from ravages of stalk borer and internode borer (Varma and Tiwari 1994).

v. As at the initial stages of infestation of Gurdaspur borer, the larvae fed gregariously in the top portion of cane hence rouging of the canes showing withered tops in the afternoon every week from June to September was found very effective for controlling the pest. Also, ratooning of the infested field should be avoided and such fields should be ploughed up to destroy the stubbles as in the case of stalk borer (Atwal and Dhaliwal 2005).

vi. The larval stage of Gurdaspur borer remains hidden in sugarcane stubble from December to the end of May or even beginning of June and can be easily destroyed during this period by ploughing up the harvested cane fields. In case of sporadic infestation, only the infested stubbles may be dug up and destroyed. These can be easily located immediately after harvest by the presence of a round tunnel-hole on the cut surface. The uprooted stubble should then be collected and burnt (www.krishiworld.com).

vii. Deep ploughing must be done after the harvest of the crop. As the larva of root borer remains hidden in the crop stubbles so these must be collected and burnt. Ratoon crop should be avoided in the previously infested fields. Plants with dead hearts should also be uprooted and destroyed at the initial growth stage (Karnatak 2006).

viii. During the pre-monsoon period, cutting and destruction of root borer infested shoots below the ground surface reduced the pest incidence. Also, ploughing up of the infested fields and burning at the stubbles can further suppress the pest build up. If the ratoon crop is to be kept, harvest the cane close to/ below the soil surface (deep harvesting) in order to kill the larvae of root borer which also provided protection against scale insect. However, it has been advocated that heavily infested fields with root borer should not be ratooned (Atwal and Dhaliwal 2005; Cheema 1952; Varma 1981).

Balanced use of nitrogen

Fertilizers when applied at appropriate doses induce tolerance in plants by changing their physiology and plants become unsuitable as host for the insect fauna. However, at high nitrogen budget, insects usually respond with an increase in survival along with faster rate of development due to increased succulence (Butter and Kumar 1994).

i. In areas prone to stalk borer, internode borer and pyrilla, excessive dose of nitrogenous fertilizers should be avoided. Removal of lower leaves during fifth and ninth month stages of crop growth was found effective in reducing the activity of internode borer, stalk borer, scale insect, mealy bug, whitefly and pyrilla (Khan *et al* 2006).

ii. Stalk borer incidence increased with the application of nitrogenous fertilizers which might be due to higher succulency of the stalks (Sukhija *et al* 1994).

iii. Addition of organic matter, lime and sulphur reduced the magnitude of infestation of the root borer (Kundo *et al* 1994)

Sucking Pests

Importance

Apart from borers, sucking pests like pyrilla, whitefly, black bug, mite, thrips and scale insect pose a great threat to sugarcane. Besides losses in yield, the major impact of sucking pests is degradation of juice quality and reduction in sugar recovery. The losses caused by sucking pests have been estimated from 2 to 34, 30 to 40 and 40 per cent in sucrose content due to pyrilla, whitefly, and scale insect, respectively (Singh and Madan 2002). Excretion of sugary fluid like material called honeydew also results in indirect losses by some of these pests. Honeydew is a sweet, sticky liquid that is excreted by sucking insects as they ingest large quantities of sap from the plant which cannot completely utilize so, insect assimilates what it needs and excretes the rest as "honeydew." Sooty moulds grow on surfaces where honeydew deposits accumulate that cause losses indirectly by inhibiting or reducing sunlight penetration. Without adequate sunlight, the plant's ability to carry on photosynthesis is reduced, which results in stunted plant growth and prematurely leaf senescence.

Nature of damage and symptoms

Pyrilla (*Pyrilla perpusilla*)

Sugarcane pyrilla or leafhopper is a sporadic pest and appears in destructive form with a periodicity of 5-8 years. Apart from sugarcane, it feeds on wheat, barley, maize, jowar, bajra and other grasses. It remains active throughout the year, migrating from one crop to another for fresh food but assumes a serious form during August to November. The incidence of this pest is particularly high in a luxuriant crop and in the interior of the field. Both the nymphs and adults suck the cell sap from the leaves, which turn pale and later on dry up. If not managed timely the crop suffers heavy losses as the canes shrivel up and even die in case of severe infestation. The pest exudes honeydew, which falls on the lower leaves where fungus develops rendering the foliage unfit for fodder. Cane quality in terms of brix, sucrose and purity decreases whereas undesirable non-sugar contents increase which results in the loss of commercial cane sugar and it becomes difficult to prepare gur from affected canes. Severe infestation of pyrilla can cause 28 per cent yield losses and up to 50 per cent loss in CCS.

Whitefly (*Aleurolobus barodensis*)

The whitefly is found throughout India, predominant under waterlogged conditions especially in ratoon crop, which usually remain neglected and unmanaged. Nymphs cause damage by sucking sap from leaves. The damaged crop looks pale and the leaves turn black owing to development of sooty mould. The general vitality of the plants is affected and sugar recovery from the infested crop also gets reduced.

Black Bug (*Cavelerius excavatus*)

Black bug is again a sporadic pest of ratoon crop particularly. It is widely distributed through out India and Pakistan. It has also been found feeding on other crops like rice, maize and grasses. It was first observed damaging sugarcane crop in Punjab during late 1930s. as the pest prefers dry weather conditions, infestation is high during pre-monsoon period. However, with the onset of monsoon its infestation gets drastically reduced. Both the adults and nymphs suck cell sap from the central whorl of the young plants and leaf sheath in grown up plants. The affected crop appears yellow with the leaves showing deep reddish brown spots in patches. This pest adversely affects the cane yield and do not cause much harm to commercial cane sugar.

Mite (*Oligonychus indicus*)

Like black bug, mite cause damage to the crop during pre-monsoon period. Mite also prefers hot and dry conditions and the pest remains active on baru (*Sorghum halepense*) through out the year and migrates to sugarcane in April, multiply and damage the crop up to end of June. Both nymphs and adults suck cell sap with their stylet like chelicerae, on the lower side of the leaves under a fine web, resulting in typical red spots. Female spins a web on the underside of a leaf in which it lays eggs. If not controlled in time, it spreads to the whole field and the crop gives a burnt appearance.

Thrips (*Baliothrips serratus*)

Thrips thrive well in hot and dry conditions and remain active from March to July, but its population is high during April - June. Nymphs and adults rasp the leaf surface and suck the oozing cell sap, resulting in drying and rolling of leaf tips. The pest is particularly more problematic for plant crop than that of ratoon crop. Life cycle of the pest is completed in 3-4 weeks during active period and it has many generations in a year.

Scale insect (*Melanaspis glomerata*)

Scale insect usually feeds and establishes on internodes covered with leaf sheath. The leaves of infested canes show signs of tip drying and pale green colour however with continued infestation these leaves finally turn yellow. Desapping also leads to non-opening of the leaves which, ultimately get dried. Nodal region of the cane is infested more than the internodal region. Infested crop shows poor vigour, stunted growth, shriveled canes and drastic reduction in internodal length and ultimately cane dries up. Such canes when slit open appear brownish red. The yield loss could range from negligible to total crop failure and the juice quality deteriorates.

Agronomic Management Practices

Use of insect free seed

Infested seed itself acts as source for spreading and build up of the pest population so, use of insect free seed helps in preventing the initial population build up of the pest and also restricts the spread of the pest from one area to another through infested seed material.

i. Scale insect spreads through seed cane, therefore seed from the infested areas should not be used. As this insect cause drying up of canes thereby led to poor germination therefore, hot water treatment of cane setts at the time of planting is essential for its control in sugarcane (Madan 2001;Atwal and Dhaliwal 2005).

ii. The infestation of seed borne pests like mealy bugs and scale insect can be successfully minimized by planting pest free setts in the endemic areas (Varma and Tiwari 1994).

Irrigation

Water stress in both the ways i.e. excess moisture or water logging and droughts alters the pests' preference to a particular crop. In case of whitefly, water logging causes it's severe out break therefore, the losses can be prevented by draining off excess water from the water logged fields. In contrary to this, summer droughts and dry spells during monsoon season also favour buildup of this pest. Frequent irrigations especially during pre monsoon period were found useful in avoiding the crop losses caused by thrips (Duhra and Suri 2002).

Detrashing

The concept behind detrashing of lower leaves for the management of insect pests is to disturb their life cycle indirectly. The incidence of the pests like pyrilla, scale insect and stalk borer can be lowered by this practice as these pests lay eggs on dry foliage.

i. Removal of lower leaves during fifth and ninth month stages of crop growth helped in reducing the activity of mealy bugs, whitefly and pyrilla (Khan *et al* 2006, Gupta 1948).

ii. In the whitefly endemic pockets, practice of detrashing of lower leaves bearing whitefly puparia reduced the incidence (Anonymous 1991).

iii. Detrashing of 4-5 lower leaves at 4-5 internode stage reduced the scale insect establishment (Khanna 1957, Venkateshwar Rao *et al* 1976).

Rouging, Destruction of water shoots and Stubbles

Most of the sucking pests tide over unfavourable conditions by hiding at suitable places so, the objective of these practices is to put a check on their seasonal carry-over, preventing further population build up and incidence by destroying the source or susceptible targets for the pests like late water shoots and stubbles. Therefore, these practices are essential to restrict the population load of the pest on the crop. The findings of some of the studies are summarized as under-

i. Burning of trash after harvest of the crop upto specific period before the pest migrates to nearby crop kills the carry over population of pyrilla (Gupta 1948), black bug (Gupta and Avasthy 1954) and scale insect to some extent (Subramanyam 1976).

ii. The incidence of black bug drastically reduced by removal of crop residue (Table 4) and stubble shaving (Jaipal 2000). Deep harvesting of cane fields, which are to be ratooned, provided protection against scale insect damage. Invariably shaving of stubbles resulted in the reduction of scale insect population, at the same time this operation also helped in getting uniform and vigorous tillering (Cheema 1952; Varma 1981; Prabhakar Rao *et al* 1976)

Balanced use of nitrogen

Discouraging the factors especially excess dose of nitrogen which favours buildup of sucking pests are important to reduce the infestation.

i. Dense and luxuriant crop due to excess nitrogen application thereby lodging of cane and varieties with broad and succulent leaves increased the incidence of pyrilla and whitefly (Mathur 1941).

ii. The incidence of black bug can be reduced by using recommended dose of fertilizers along with frequent irrigations and timely interculture in ratoon crop during March (Duhra and Suri 2002).

iii. In pyrilla prone areas, excessive dose of nitrogenous fertilizers must be avoided (Khan *et al* 2008).

iv. Since the weak crop with thin stand is attacked more readily than a well-manured and heavy crop, therefore recommended dose of nitrogenous fertilizer should be applied to keep crop healthy for the control of whitefly infestation in sugarcane crop. Infestation can also be further reduced by avoiding water-logged conditions and by removal of lower leaves infested with pupae and nymphs. (Atwal and Dhaliwal 2005, Madan 2001).

Removal of alternate hosts

Alternate host is the one of two species of hosts on which some pests must develop to complete their life cycles or a species of host other than the principal host on which a parasite/ pest/ pathogen can survive. Baru (*Sorghum halepense*), a perennial weed is the alternative host plant from which the mite spreads to the sugarcane crop (Duhra and Suri 2002, Atwal and Dhaliwal 2005). Therefore, the destruction or removal of alternate host like baru in case of mite is very effective and helpful practice in lowering the incidence of certain insect pests in sugarcane.

Soil Pests

Importance

Termites are polyphagous in nature and found throughout the world causing significant yield losses. Its infestation affects germination, initial crop stand and ultimately the cane yield. Germination failure can be up to 60 per cent where as the yield losses of 10-28 per cent have been reported due to this pest. Another serious pest of this group is white grub. The yield losses caused by white grubs are also very high which can be to the extent of 35 t/ ha. However, depending upon the severity of damage, the losses may vary from 30 to 100 per cent or complete failure of the crop (dacnet.nic.in/ Sugarcane/PestManage.htm).

Nature of damage and symptoms

Termites

Termites remain active during April-June and become more serious under prolonged drought conditions particularly in light textured soils where proper irrigation facilities are not available. The damage is more prevalent under situations like rainfed, prolonged dry spells and in stressed crop due to disease or physical damage. Termite infestation usually occurs soon after planting of setts in the field. The cut-ends and the buds of cane setts are the first to be attacked resulting in poor germination and lead to gappy stand of the crop. In case of severe of infestation, germination is so badly affected that it becomes necessary to replant the crop (www.krishiworld.com). During September to December the tips of leaves start drying up and then outer leaves as a whole dry up in grown up cane subsequently, the whole plant withers. The termites are seen inside the stalk, which sometimes reach above middle portion of the stalk. The stalk is filled with soil and can be easily broken. After harvesting, the termites invade stubbles during April – May in ratoon crop and tillering in such infested stubbles is greatly reduced.

White grubs

White grubs damage sugarcane crop by feeding on roots and underground stems. The first symptom is a yellowing (chlorosis) of the leaves. This is usually followed by stunted growth, dense browning, lodging, plant uprooting and death in heavily infested areas. Damage is usually more severe in ratoon crops and is most evident around the edges of a field. Grub damage is also worse on muck soils than sandier soils (Cherry and Nuessley 2008). In case of heavy infestation whole of the crop may be damaged and needs replanting.

Agronomic Management Practices

Planting Time

Autumn planting (September-October) and early spring planting ensure well established root and shoot system which can withstand or escape incidence of soil borne pests This is because of better suitability of environmental conditions for the crop growth than that of insects like termites and white grub. As in case of other pests of sugarcane, incidence and damage caused by soil borne pests also get influenced by planting time in the following way-

i. Besides protecting the crop from shoot, top and Gurdaspur borers, autumn planting also protected the crop against termites and white grubs in sub-tropical India. (Kalra 1963, Murthy 1953, Varasi 1954).

ii. Planting of varieties with a vigorous and prolific root system resulting a luxuriant crop growth was found more suitable to withstand the attack of the pest better by putting forth fresh roots. Early season planted or autumn planted sugarcane crop can develop its root system fairly well by June-July when the young grubs first appear that helped the crop to withstand the pest attack better (Cherry and Nuessly 2008).

Trash Mulching

In areas where these pests are more prevalent trash mulching should be avoided. As this practice increases the organic matter contents of the soil which favour the higher incidence of termite. Spreading of a thick blanket soon after germination greatly helps in suppression of shoot borer. However, this practice may increase termite, armyworm and rat damage in areas where these pests are prevalent (Khan *et al* 2008).

Interculture

Disking of the infested fields, reducing the number of ratoon crops and flooding are some of the well known practices adopted in Florida to control grubs. Out of these, disking was found more effective to kill many grubs and then allows the birds to kill many more by exposing them (Cherry and Nuessley 2008).

Irrigation

Irrigation alone is very important tool in minimizing the incidence of soil pests by creating unfavourable or adverse conditions for the proliferation of these pests. Therefore, the pests preferring hot and dry conditions can be minimized by adopting proper scheduling of irrigation particularly during hot summer months. Some of the studies conducted on this aspect have been concluded as follows

i. Frequent irrigations during pre-monsoon and timely irrigation during post monsoon period controlled the incidence of termites and benefited the crop(Duhra and Suri 2002).

ii. Proper moisture conditions in the field should be maintained especially at early growth stages by frequent irrigations as this practice discouraged infestation of termites otherwise drought conditions could cause heavy infestation of termites and shoot borer (Varma and Tanwar 1993).

iii. Flooding the fields at frequent intervals was found very effective to reduce the damage since the grubs temporarily migrate downward under flooded conditions. Larval and pupal stages of white grub have been reported the easiest stages to kill by flooding in Florida. The practice was found effective under warmer weather conditions with water temperature of 77 °F or more in the flooded field. A continuous flooding for 5 days was found suitable for grub control. The flood water level should be about 2" above the soil surface. Many grubs could come to the soil surface and survive in case of less than 2" of standing water (Cherry 2008).

Conclusion

With the adoption of various agronomic practices, insect-pest burden on sugarcane crop can be minimized to reduce the losses both in terms of cane and sugar yield. The practices like optimum planting time, seed rate, balanced fertilization, irrigation, Intercropping detrashing and many more when adopted judiciously as per the site specific needs create conditions more conducive for crop growth and simultaneously cause hindrance against build up and infestation of various insect-pests. The impact of several such practices has been summarized in Table 6. These practices should be adopted on a large scale to minimize the damage thereby sustaining the productivity for higher returns.

Table 6. Impact of various cultural practices on the major insect pests

S. No.	Practice	Pest	Impact (positive/ negative)
1	Planting Time		
	i) Early planting	Early shoot borer	Negative
	ii) Autumn planting	Early shoot borer, Gurdaspur borer, top borer, termite, white grub	Negative
		Stalk borer	Positive
	iii) *Adsali* and *Eksali*planting	Early shoot borer, top borer	Negative
2	Planting method, seed rate and seed quality		
	i) Pit planting	Stalk borer, pyrilla	Positive
		Top borer	Negative
	ii) Spaced planting	Stalk borer, top borer	Negative
	iii) Deep trench planting	Early shoot borer, Stalk borer	Negative
	iv) Higher seed rate	Early shoot borer	Negative
	v) Insect free seed	Stalk borer, scale insect, mealy bugs	Negative
3	Irrigation		
	i) Avoiding irrigation at the time of moth emergence	Top borer	Negative
	ii)Timely and (or) frequent irrigation	Early shoot borer, root borer, termite, Stalk borer, thrips	Negative
	iii) Water logging	Whitefly	Positive
4	Intercropping	Top borer, root borer, Early shoot borer	Negative
5	Trash Mulching	Early shoot borer	Negative
		Termite, armyworm	Positive
6	Earthing up and propping	Early shoot borer, root borer, Stalk borer	Negative
7	Detrashing	Pyrilla, Stalk borer, scale insect, Internode borer, mealy bug, whitefly	Negative
8	Rouging, Destruction of water shoots and Stubbles	Top borer, Stalk borer, Internode borer, Gurdaspur borer, root borer, Pyrilla, black bug, scale insect	Negative
9	Balanced use of nitrogen	Stalk borer, Internode borer, Pyrilla, black bug, whitefly	Negative
10	Removal of alternate hosts	Mite	Negative
11	Interculture operation and flooding	White grub	Negative

Positive: incidence of the pest increases, Negative: incidence of the pest decreases

References

Anonymous. 1991. Technical report- Entomology. AICRP on sugarcane, Lucknow.

Anonymous. 2009. Package of Practices for Kharif crops PAU, Ludhiana.

Atwal A S and Dhaliwal G S. 2005. Agricultural pests of South Asia and their management. Kalyani publishers, India.

Avasthy P N and Tiwari N K. 1986. The shoot borer, *Chilo infuscatellus* Snellen. In: H. David, S. Easwaramoorthy and R. Jayanthi (Eds) *Sugarcane Entomology in India.* Sugarcane Breeding Institute, Coimbatore pp 69-92.

Butter N S and Kumar R. 2004. Significance of Nutrient and Water Management in IPM. *IPM Mitr* Vol III. pp 36-47.

Chandra J and Singh M R. 2007. Management of Sugarcane insect pests in relation to planting technique. In: Yadav RL and Yadav DV (Eds) Sugarcane planting techniques and crop management, Indian Institute of Sugarcane Research, Lucknow, U.P, India. pp 73-81.

Cheema P S. 1952. Deep harvesting of sugarcane as a measure for exterminating the root borer (*Emmalocua depressella* Swinh.) *Indian J Ent* **14**: 77-79.

Cherry. 2008. White Grubs in Florida Sugarcane. ENY-664 (SC012), EDIS Web site at http://edis.ifas.ufl.edu.

Cherry R A and Nuessly GS. 2008. Insect Management in Sugarcane. ENY-406 (IG065) http://edis.ifas.ufl.edu.

David H. 1986. The internode borer, *Chilo sacchariphagus indicus* (Kapur). In: David H., Easwaramoorthy S. and Jayanthi R. (Eds) *Sugarcane Entomolgy in India.* Sugarcane Breeding Institute Coimbatore, pp 121-34.

Dhaliwal G S, Arora R and Dhawan A K. 2004 Crop losses due to insect pests in Indian Agriculture: An update. *Indian J Ecol* **31**(1): 1-7.

Duhra M S and Suri K S. 2002. Insects and their control. In: Sidhu G S and Waraitch K S *Sugarcane production and protection technologies* pp 85-106.

Feng, Y.X. 2003. Superficial view to the integrated control of sugarcane diseases and pests in Zhangjiang. *Sugarcane and Canesugar*, **27(2)** : 28-29

Gill H S. 1963. Influence of hot weather irrigation and trash mulch on the incidence of certain sugarcane borers and weeds. *Indian J Sug. Res Dev.* 7: 161-63.

Gupta B D. 1945. Effect of early earthing up of sugarcane on the attack of stem borer, Bisettia steniellus Hmpsn *Proc Bien Conf Sugarcane Res Dev Wkrs* **2**: 198-204.

Gupta B D. 1948. Control of sugarcane leaf hopper (*Pyrilla Sp*) in UP. *Indian Fmg* **9**: 401-06.

Gupta B D and Avasthy P N. 1954. Some recommendations for the control of sugarcane pests in India. *Indian Sug* **4**: 387-97. *http://dacnet.nic.in/Sugarcane/PestManage.htm*

Jaipal S. 2000. An IPM module for the management of major insect pests of sugarcane in Indian subtropics. *Sugar Tech* **2(3)**: 1-8.

Kailasam, C. 1994. Evaluation of nitrogen levels and seed rates for short duration sugarcane (Co 8338) intercropped with three soybean cultivars of varying growth habits. Ph.D. Thesis, Tamil Nadu Agricultural University, Coimbatore, India.

Kalra A N. 1963. Control of Gurdaspur borer of sugarcane. *Sugarcane Herald* **5**: 1-3.

Karnatak A K 2006. Insect pest management for minimizing crop losses in sugarcane ecosystem. In: Souvenir: 26th All India workshop of AICRP on sugarcane Oct 16-18, 2006. Pant Nagar pp 96-99.

Khan M A, Mumtaz R, Ahmad T, Kumar S and Maurya R P. 2006. Prospect of bio-intensive management of sugarcane pests. In: *Souvenir*: 26th All India Workship of AICRP on sugarcane Oct 16-18, Pant Nagar. pp 100-104.

Khan M Q. 1951. A note on earthing up as a measure of control of *Argyria sticticrapsis* Hmpsn. the sugarcane stem borer. *Proc Bienn Conf Sugarcane Res Div Wkrs India* **1**: 1-3.

Khan M Q. Krishnamoorthy Rao B H and Krishnamoorthy C. 1965. Incidence of borers in relation to the time of planting of *Adsali* and *Eksali* sugarcane crops in Nizamabad distt (A.P.) *The Indian Sugarcane Journal* **9**(3): 147-52.

Khanna K L. 1942. Cane entomology, 1940-41. In: *A Rep Sugarcane Res Inst. Pusa* (Bihar) pp 111-42.

Khanna K L. 1957. On the coccid, *Melanaspis glomerata* (Green) on sugarcane in Bihar. *Indian J Sugarcane Res Des* **11**: 19-22.

Kundo R, Saha S C, Majid M A and Abdulla M. 1994. Preliminary studies to investigate the possible effects of soil properties on sugarcane root stock berer infestation. *International J Pest Management* **40**: 266-69.

Madan Y P. 2001. Integrated pest management in sugarcane. *Indian sugar* **50**(12): 867-71.

Malik V S. 1982. Evaluation of some important crop management practices for integrated control of shoot (*Chilo infuscatellus* Shell) and top (*Tryporyza nivella* (F) Borer in sugarcane. *Ph.D Thesis,* HAU, Hisar pp.1-160.

Marradi G N, Hirevenkanagoudar LV, Gangadi J and Kunnal L B. 2007. Extent of adoption of integrated pest management practices by sugarcane growers. *Karnataka J Agri Sci* **20(3)**: 564-67.

Mathur R N. 1941. Certain observations on nitrogen nutrition of sugarcane plant in relation to susceptibility to attack of whitefly *Aleurolobus barodensis*. *Proc A Conv Sug Technol Assoc India*. **10**: 45-63.

Misra S C and Hora B S. 1982. Note on the incidence of top borer (*Tryporyza nivella* Fabr) and pink borer (*Sesamia inferens* wlk) on sugarcane grown in companion cropping pattern. *Indian J Agri Sci* **52**: 45-46.

Murthy D V. 1953. Cultural and mechanical methods of control of early shoot borer in sugarcane. *Proc Sug Tech Assoc India* **22**: 108-24.

Parthasarthy S V. 1959. Control of stem borer through trash mulching. *Indian J Sugarcane Res Dev* **3**: 151-52.

Prabhakar Rao C, Sheshagiri Rao C and Venkataswara Rao S. 1976. Sugarcane scale *Melanaspis glomerata* (Green) and its control measures. In: *Proc All India Semi Sugarcane Scale Insect, Nidadavle*, 68-73.

Rajani V G. 1966. Use of cane trash as a cultural measure to control shoot borer attack in sugarcane crop during hot weather in Uttar Pradesh. *Proc All India Conf Sugarcane Res Dev Wkrs* **5**: 613-20.

Rajendran B. 2005. Higher sett rate of sugarcane at planting as a cultural technique for the management of *Chilo infuscatellus* Snell. *Sugar Tech* **7(4)**:165-67.

Rana Z A, Ibrar-ul-Haq, Malik N A and Akhtar A S. 2007. Effect of Trash mulching and *Trichogramma chilonis* (Ishii) on sugarcane borer infestation. *J Agric Res* **45(2)**: 161.

Rao N V, Narsimha Rao C V and Bhavani B. 2008. Management of Sugarcane borers. In: *Sonvenir* 27th Biennial workshop of AICRP on sugarcane 17-19 Oct 2008, pp 49-51.

Sardana H R. 2001. Influence of summer intercropping on the incidence of root borer, *Emmalocera depressella* of Sugarcane. *Indian J Ent* **63(1)**: 49-51.

Singh M and Madan Y P. 2002. Effect of sett dipping and soil application of various insecticides on sugarcane shoot borer. *J Indian Sug* **52**: 105-07.

Singh M, Chillar B S and Madam Y P. 1997. Use of various cultural practices on the control of root borer *Emmolocera depressella* Swinhoe in sugarcane. *Indian J Sugarcane Tech* **12(2)**: 27-32.

Singla M L and Duhra M S. 1992. Effect of seed rate and irrigation interval on the incidence of top borer and shoot borer in sugarcane. *J Insect Sci* **3(2)**: 189-91.

Singla, M.L., Duhra, M.S., Dllaliwal, Z.S. and Sandhu, S.S. 1994. Effects of intercropping in autumn planted sugarcane on the incidence of *Scirpophaga excerptalis* Wlk. and termites. *J. Insect Sci.*, 7 **(2)**: 199 - 201.

Srikanth, J., Easwaramoorthy, S. and Kurup, N.K. 2000. Borer and predator incidence in sugarcane intereropped with pulses. *SugarTech*, **2 (1&2)**: 36-39.

Subramanyam B. 1976. Scale insect on sugarcane, *Melanaspis glomerata* (Green) In: Proc. All India Sem Sugarcane Scale Insect. *Nidadavole* pp 35-41.

Sukhija H S, Chhuneja P K and Grewal R S. 1994. Incidence of sugarcane stalk borer, *Chilo auricilius* dudgeon in relation to application of organic manures and inorganic nutrients. *Indian Sugar* **44(3)**: 179-81.

Thirumurugan A and Koodalingam K. 2005. Management of Borer Complex in Sugarcane Through Companion Cropping under Drought Condition of Palar River Basin Area *Sugar Tech* **7(4)**: 163-164.

Varasi M S. 1954. Influence of climate on the population of sugarcane borers in Hyderabad state. *Proc. Bienn Conf Sugarcane Res Dev wkrs* **2**: 54-55.

Varma A. 1981. Pest complex in ratoons in north India sub-tropical conditions and how to tackle it. *Proc natn sem ratoon management* **1**: 92-95.

Varma A and Mitra K. 1981. Recent researches leading to evolve a suitable control programme for stalk borer *Chilo auricilius* Ddgn. *Proc natn symp stalk borer*. Karnal pp 63-68.

Varma A and Tanwar R K. 1993. Incidence of pests under drought conditions in sugarcane. *Cooperative sugar* **24(12)**: 675-78.

Varma A and Tiwari N K. 1994. Recent concepts in insect pest management in sugarcane. In: Dhaliwal GS and Arora R (Eds) Trends in Agricultural Insect pest management. Pp 238-65.

Varun C L, Singh S and Singh H N. 1990. Effect of intercropping of spices on the incidence of top borer (*Scirpophaga excerptalis* wlk) in sugarcane. *J Indian Sug* **39**: 751-52.

Venkateswara Rao S, Seshagiri Rao C and Krishnamurthy Rao B H. 1976. Mechanical and chemical control of sugarcane scale insect (*Melanaspis Glomerata* Green) *Indian Sug* **27**: 689-93.

Verma, R.S., Kumar, R. and Yadav, R.L. 1984. Status of research in inter-cropping with sugarcane. *Indian Sug. Crops J* 10 (2): 1-6.

Verma, R.S., Motiwale, M.P., Chauhan, R.S. and Tewari, R.K. 1981. Studies on intercropping of spices and tobacco with autumn sugarcane. *Indian Sug* **31 (7)**: 451-56. www.krishiworld.com

□□□

Resource Conservation

22

Resource Conservation in Relation to Commercial Agriculture

V. PRATAP SINGH AND NEETA TRIPATHI

Commercial agriculture is the production of crops for sale, crops intended for widespread distribution (e.g. supermarkets), and any non-food crops such as cotton and tobacco. Commercial agriculture includes livestock production and livestock grazing. Commercial agriculture does not include crops grown for household consumption (e.g backyard garden or from a vegetable garden or fruit tree's)

Commercial farming is a progression from Diversified (sometimes called Mixed) farming, when the landowner acquires a sufficiently large amount of arable land and sufficiently advanced technology (such as hybrid seeds, fertilizers, pesticides, etc.). At this point, it becomes more profitable for the landowner to specialize and focus on one particular crop due to economies of scale. This is further augmented by high levels of technology, which significantly reduces the risk of poor harvests. Another important difference between commercial farming and less-developed forms of agriculture is the new emphasis on capital formation, scientific progress and technological development, as opposed to an emphasis on natural resource utilization that is common to subsistence and diversified agriculture.

Market forces, rising labour scarcity, shifting cropping systems, micro irrigation, new inputs like liquid fertilizers, biofuel crops, growing water scarcity and increasing demand for agricultural land were some of the important factors that were shaping the present day agriculture. The country has moved from subsistence agriculture to 'green revolution agriculture' and now the focus was on sustainable and commercial agriculture and 'market driven agriculture'. Since conservation of natural resources was essential, both sustainable and commercial agriculture has to co-exist in a complementary and balanced manner.

Market gardening

Market gardening is the relatively small-scale production of fruits, vegetables and flowers as cash crops, frequently sold directly to consumers and restaurants. It is distinguishable from other types of farming by the diversity of crops grown on a small area of land, typically, from under one acre (0.4hA) to a few acres, or sometimes in greenhouse. Such a farm is sometimes called a market garden or truck farm.

Market gardening as a business is based on providing a wide range and steady supply of fresh produce through the local growing season . Many different crops and varieties are grown, in contrast with large, industrialized farms, which tend to specialize in high volume production of single crops, a practice known as monoculture. Market gardening also employs more manual labor and gardening techniques, compared to large-scale mechanized farming. Because production is relatively low-volume, sales are often through local fresh produce outlets, such as on-farm stands, farmers' markets, community-supported agriculture subscriptions, restaurants and independent produce stores.

Plantation crops

A plantation is usually a large farm, especially in a tropical or semitropical country, like Brazil or Nicaragua on which cotton, tobacco, lice coffee, sugar cane and the like are cultivated, usually by resident laborers. A plantation is an intentional planting of a crop, on a larger scale, usually for uses other than cereal production or pasture. The term is currently most often used for plantings of trees and shrubs. The term tends also to be used for plantings maintained on economic bases other than that of subsistence farming.

Forestry/ Industrial plantations

Industrial plantations are established to produce a high volume of wood in a short period of time. In southern and southeastern Asia, rubber, oil palm and more recently teak plantations have replaced the natural forest. Industrial plantations are actively managed for the commercial production of forest products. Individual blocks are usually even-aged and often consist of just one or two species. The plants used for the plantation are often genetically improved, e.g. the seeds used may originate from seed orchard. These species can be exotic or indigenous. Industrial plantations are usually large-scale.

Wood production on a tree plantation is generally higher than that of natural forests. While forests managed for wood production commonly yield between 1 and 3 cubic meters per hectare per year, plantations of fast-growing species commonly yield between 20 and 30 cubic meters or more per hectare annually.

Sugar crops

Sugar plantations were highly valued. Sugarcane is still an important crop in India. Sugar plantations also arose in countries such as Barbados and Cuba because of the natural endowments that they had. These natural endowments included soil that was condusive to growing sugar and a high marginal product of labor realized through the increasing number of slaves. These sugar plantations dragged down an economy in the long run as sugar plantations allowed for inequality and low voting rates in any given country. These sugar plantations allowed for there to be a small elite ruling class to have all the power in the country as they overpowered all the slaves and peasants under them.

Conservation practices prevalent in commercial agriculture

Laser land leveling

Precision land leveling is the precursor technology to all RCTs. Uneven fields lead to poor germination and crop stands leading to lower yields and excessive leaching of mobile nutrients (Gupta *et al.*, 2006). Laser assisted precision land leveling saves water in rice-wheat system and improves yield upto 25%. It also reduces the weed problem and increases cultivable area by 3-6% due to elimination of many field bunds, borders and irrigation channels. Laser levelling greatly improves the performance of direct seeded rice (DSR) on raised beds. In Cambodia crop yield increased by 24% by using this technique. Also it reduces the weeds by 40%, reduced weeding time from 21-5 labour-day/ha and improved the water use efficiency (Rickman, 2003).

Use of crop residue/mulch

The multi-crop new generation zero till-seed-cum ferti planters with disc type cultures allow seeding in presence of anchored and loose residues.

In combine harvested field lot of residues retained on field. Before using the various planters such as Turbo seeder, rotary disc drill and double disc planter, the residue must be spread smoothly for efficient use of machine and better crop establishment. Surface retained residues moderate soil temperature, conserves soil moisture, adds organic matter to improve the nutrient water interaction and act as physical barrier to weed emergence. Turbo seeder and rotary disc drill can be used for seeding under conditions of loose as well as anchored crop residues. Double disc planter is lighter in weight and useful for planting wheat and other crops in field having hard stubbles of previous crops such as cotton, pigeonpea, maize and sugarcane.

Potato with mulch

In conventional, tillage-based potato cropping systems, the risk of soil erosion and nitrate leaching can be reduced using the mulch planting technique. The potato beds are prepared well in advance of planting - if potato is to be planted in spring, the beds would be prepared before winter - and seeded with a green manure cover crop. The potato is later planted into the beds which, by then, are covered by the dead mulch of the manure crop. For mechanical planting, planters are equipped with special discs that cut through the mulch and split the potato beds. The mulch protects the soil from erosion during the first weeks of the crop. As the potato plants grow, the reshaping of the beds incorporates the mulch.

Trash farming

Trash farming in sugarcane converts the crop residues, which would regularly be discarded or burned, from a waste commodity into a nutrient source. Three months before harvest, dead leaves are manually removed from the cane stalk (detrashed) and left to decompose on the soil. After harvest, residual sugarcane biomass is again maintained on the field. Through the decomposition process, the trash fixes nitrogen and increases soil organic matter content, reducing application rates of nitrogen fertilizer. Trash farming also enhances weed control, preserves soil moisture, minimizes erosion and also increase sugarcane yields, particularly those of ratoon crops (regrowth of cane after harvest). In Southeast Asia, yields increase on average by 5.8% in the plant crop (initial cane planting) and 21.1% in the first ratoon crop.

Green manuring/brown manuring

Instead of traditional practice of growing green manuring crop and incorporating it in soil before rice transplanting which led to more water consumption and fuel cost, a new technique of co-culture is developed. This comprises of growing of both rice and *Sesbania* crop. *Sesbania* crop is knocked down at 25-30 DAS with 2, 4-D at 500 g ha^{-1}. *Sesbania* surface mulch decomposes very fast to supply N. Moreover, its intercropping is helpful in areas where soil crust formation is a problem for easier emergence of rice seedlings. The seeds of rice and *Sesbania* can be mixed and sown with the same drill/planter. *Sesbania* reduces the need of additional irrigation water for raising in summer before rice when the evaporative demand of water is more as compare to *kharif* season. It also keeps the soil moist for long thereby obviating the need to irrigate rice frequently and reduces the weeds. Studies conducted at Pantnagar, revealed that weed dry weight was reduced when *Sesbania* was used as brown manure (Table 1).

Table 1 : Mean total weed density (No m^{-2}) as influenced by different treatments at various stages of plant growth

Treatments	28 DAS	56 DAS
DSR+Sesbania (broadcasting)+pendimethalin+2, 4-D (0.5kg ha^{-1}) +HW	25	7.5
DSR+Sesbania (row seeding)+pendimethalin+2, 4-D (0.5kg ha^{-1})+HW	43	7
DSR (weedy)	319	206
DSR+Sesbania (row seeding)+pendimethalin+2, 4-D (0.3kg ha^{-1})+HW	31	12

Zero-tillage/Direct seeding

Direct seeding of wheat without tillage has gain popularly in most of the areas. Zero-tillage planting reduced yield that were equal or better than those obtained from conventional tillage practices. Zero till planting is cost effective due to less use of tractor and fuel. The consumption of fuel under ZT was just 10.5 per cent that of CT and the tractor time required for seeding one hectare area under ZT system was also 11.7 per cent compared to CT system. Zero-till takes immediate advantage of residual moisture from the previous rice crop, as well as water use is reduced by about 10 cm - hectare. Production cost of wheat is reduced by Rs. 1500-2000 ha^{-1} by practicing ZT. Also in ZT fields, emergence of crop is earlier than CT and weed growth particularly *Phalaris minor* is much less in ZT.

Direct seeded rice with brown manuring

Direct seeded rice which removes nursery growing, puddling and drudgery, transplanting the young rice seedlings and pre-emergence application of herbicides provides an option to resolve the edaphic conflict and enhance the sustainability of rice-wheat cropping system. Also DSR avoids the problem of seasonal labour requirement for rice nursery raising and transplanting operations and facilitates early wheat planting as it matures 7-10 days earlier than the transplanted crop. For direct seeding there are four popular methods based on the physical conditions of the seedbed and seed. Since the weed management in direct seeded rice is a matter of serious problem, which can be successfully controlled through the·integrated approaches of weed management by the application of herbicides and cultural/ manual/mechanical weeding process. Although herbicides are available in the market, yet the farmers lack awareness and information how to use them correctly and what are the appropriate time of application and volume of water etc.

Surface seeding

This technology performs well where wheat sowing is not possible due to wet field conditions after paddy harvest. In these areas, dry or soaked seed can be broadcasted a few days before or immediately after harvest of rice under wet condition. Wheat sowing in low land rice fields get delayed considerably as excess soil moisture does not permit time for tillage and sowing operations which in turn result in reduction of wheat yield to the tune of 40 to 70 kg/ha/day depending on the extent of delay in sowing. Wheat sowing may be advanced by 15-20 days adopting this technique. Rice straw mulch after seeding ensures better germination in this situation. Generally heavy textured soils are more suitable for surface seeding as compared to light textured soil. This practice doubles the cropping intensity in areas where only a single crop is possible due to wet soil condition following paddy harvest. Cost of wheat production is only 70-75% as this practices need to tillage operation. About Rs. 1200-1500/ha can be saved for land preparation, and finally higher yield can be obtained due to 15-20 days earlier sowing of wheat (Singh *et al.*, 2001).

Furrow irrigated raised bed system

FIRB system not only saves the resources like water, nutrients and labour but also facilitates the greater diversification. Practice of raised bed planting has been emerged with an alternative for crop diversification. The major concern of this system is to enhance the productivity and save the irrigation water. Potential agronomic advantages of raised beds include improved soil structure due to reduced compaction through controlled trafficking and reduced water- logging and timely machinery operations due to better surface drainage. In bed planting water saving compared to flat surfaces was reported upto 26% for wheat and 42% for transplanted rice, with yield increases at the same time of 6.4% for wheat and 6.2% for rice (RWC - CIMMYT, 2003). Irrigation is the major constraint for production of *rabi* pulses. Raised bed planting was found to reduce the crop water requirement and thus increase crop yield. On an average, consumptive water use of chickpea under raised bed system was 11.7% less than flat bed system (Table 2). Crop planted on raised bed extracted relatively more moisture from the deeper soil profile and also recorded higher water use efficiency and water productivity than flat bed system (Pramanik *et al,* 2006)

Table 2. Water use efficiency in chickpea as influenced by raised bed planting

Treatment	Consumptive water use (ha –mm)	Water- use efficiency (kg/ha – mm)
Method of planting		
Raised bed	162	10.37
Flat bed	183	9.50
CD (P=0.05)	1103	-

The water saving through zero tillage and bed planting was higher than the conventional tillage of wheat (Malik *et al,* 2006). In rice-wheat cropping system, zero tillage with small (10%) saving in water but with larger area of acceptance by farmers, the volume of water saving would be much larger compared to technologies which save more water but have less acceptability among farmers.

Conclusion

Thus, it can be concluded that RCT can be crucial for ensuring the sustainable food production in commercial agriculture in India. Production capacity, production efficiency and crop protection are the major pillars supporting the national productivity. The major challenges involved in agriculture are overcoming the traditional tillage practices and better utilization of the conservation practices. In long run no single technology option will sufficient to maintain food security or preserve soil, water and other natural resources. Thus, policies should be implemented to promote efficient use of resources for commercial crop production.

References

Gupta, R.; Jat, M.L.; Singh, S.; Singh, V.P. and Sharma, R.K. 2006. Resource conservation technologies for rice production. *Indian Farming,* 57 (7) : 42-45.

Malik, R.K, Yadav, A and Kamboj, B.R. 2006. Conservation tillage and crop establishment techniques *In: Proceeding of National Symposium on Conservation Agriculture and Environmental* held during 26-28 Oct, 2006 at BHU, Varanasi. pp. 11-16.

Pramanik, S.C, Singh, N.B. and Singh, K.K. 2006. Improving yield and water use efficiency in rabi pulses through raised bed planting technique. *In: Proceeding of National*

Symposium on Conservation Agriculture and Environmental held during 26-28 Oct, 2006 at BHU, Varanasi. pp. 88-89.

Rickman, Joe. 2003. Land levelling : Rice fact sheets. International Rice Research Institute (IRRI).

RWC- CMMYT.2003. Addressing resource conservation issues in rice-wheat systems of South Asia: a resource book. Rice- wheat Consortium for the Indo- Gangetic Plains. International Maize and Wheat Improvement Center. New Delhi, India.305p.

Singh, B.K.; Sinha, R.K.P.; Jha, J.N.; Kumar, A. and Prasad, B. 2001. Surface seeding : New Technology Extension Bulletin-2, Agricultural Research Institute, RAU, Patna, Bihar.

□□□

23

Scope, Status and Limitations of Protected Farming

P.K. SINGH

Protected cultivation practices can be defined as a cropping technique wherein the micro-climate surrounding the plant body is controlled partially/ fully as per the requirement of the plant species grown during their period of growth. With the advancement in agriculture various types of protected cultivation practices suitable for a specific type of agro-climatic zone have emerged. Among these protective cultivation practices, Green house/poly house cum rain shelter is useful for the hill zones. The green house is generally covered by transparent or translucent material such as glass or plastic. The green house covered with simple plastic sheet is termed as poly house. The green house generally reflects back 43% of the net solar radiation incident upon it allowing the transmittance of the "photosynthetically active solar radiation" in the range of 400-700 nm wave length. The sunlight admitted to the green house is absorbed by the crops, floor, and other objects. These objects in turn emit long wave thermal radiation in the infra red region for which the glazing material has lower transparency. As a result the solar energy remains trapped in the green house, thus raising its temperature. This phenomenon is called the "Green house Effect". This condition of natural rise in green house air temperature is utilized in the cold regions to grow crops successfully. However in the summer season due to the above stated phenomenon ventilation and cooling is required to maintain the temperature inside the structure well below 35ºC. The ventilation system can be natural or a forced one. In the forced system, fans are used which draw out 7-9m^3 of air / sec / unit of power consumed and are able to provide two air changes / minute.

After the advent of green revolution in India, more emphasis is laid on the quality of the agricultural product along with the quantity of production to meet the ever-growing food and nutritional requirements. Both these demands can be met when the environment for the plant growth is suitably controlled. The need to protect the crops against unfavorable environmental conditions led to the development of protected agriculture. Greenhouse is the most practical method of achieving the objectives of protected agriculture, where natural environment is modified by using sound engineering principles to achieve optimum plant growth and yield. Poly house cultivation has become an important policy of Indian Agriculture. Our country is self dependent in food grain production but to fulfill the nutritional security, the gap between increasing demand of horticultural produce has to be filled. This gap can not be filled by the traditional horticulture which required large area under horticulture to increase the production for the ever growing population. Green house technology has potential to produce more produce per unit area with increased input use efficiency. Therefore, this problem can be coping up by adopting green /poly house technology for the horticultural production. For example if one lakh hectare area under vegetable cultivation is brought out under poly house cultivation the annual availability of vegetable will be increased by at least 100 lakhs tons. Besides this it will also increase the significant jobs opportunity for the skilled rural men, youths and rural women.

The greenhouse is typically a structure whose roof and sides are transparent or translucent, permitting a sufficient quality and quantity of solar radiation to enter the structure for photosynthesis. It allows the growing of crops independently of the outside climate, since its interior temperature and humidity can be controlled. Poly house is a type of green house with UV stabilized transparent poly film as cladding material. Normally, 600-800 gauge UV stabilized transparent poly film are used for the construction of poly house in which 5-10 ºC temperature can be increased during winters over the prevailing ambient temperature out side. The environmental conditions inside the poly house can be modified as per the crop need. Depending on the precision on environmental control the cost of poly houses increases or decreases.

The cost of the greenhouse is mainly depending on the construction material used to build frames for the greenhouse and the cladding material. Based on the shape of the structure, poly house can be divided into arc, tunnel, modified tunnel and gable shape. However, gable shape and modified tunnel types are most preferred polyhouse for better crop production per unit area compared to other polyhouses. It is economically feasible to go for 500-1000 m^2 size poly houses for the large scale production of cut flowers, vegetables and cash crops.

Historical development of protected farming

The protected farming consist of cultivation of any kind of agricultural produce under protected environment such as green house (glass house, poly house and shade net house). Greenhouses have come a long way since the wealthy Europeans and Americans used them to grow oranges and pineapples in glass structures known as orangeries and pineries. These structures became status symbols as well as a practical way to grow fruits, vegetables and exotic plants. One of the earliest known greenhouses was built around 30 A.D. for the Roman emperor Tiberius. Remember, glass had not been invented so the "Specularium" was painstakingly fabricated from tiny translucent sheets of mica. All this was done to satisfy Tiberius's cravings for cucumbers out of season. It wasn't until 1599 that the first practical greenhouse was designed by Jules Charles a French botanist. It was built in (where else) Leiden Holland and used primarily to grow medicinal tropical plants. One of the favorite plants of the day came from the Tamarind (an Indian date) whose fruit was made into a curative drink.The idea caught on and greenhouses began spreading throughout Europe. The French, who had a love of a new fruit, the orange, began building orangeries to protect the trees from frost. These structures were cumbersome.

During the sixteenth century, several techniques were used to protect agricultural crops against cold. This includes glass lanterns, bell jars, cold frames and hot bed covered with glass. One structure built by Soloman de Caus in 1619 in Heidelburg which held 340 orange trees. It had a removable roof that had to be painstakingly put up and taken down during the frost season.

In seventeen century, low portable wooden frames covered with an oil translucent paper were used to warm the plant environment such as plastic row cover (low tunnel) do today. In Japan, straw mats were used in combination with oil paper to protect crop from severe natural environment. Green house in France and England during the same century were heated by manures and covered by glass panes. The first glass house in 17th century used glass on one side only as sloppy roof. Later in the century glass was used in both sides. Glass houses were used for crops such as melons, grapes, peaches, and strawberries and rarely for vegetable production. It would seem that developers of this new technology kept market profitability in mind. They produced crops, which appeals to the wealthy and privileged; the only people who could afford the luxury fresh fruit produced out of season in protected environment such as green house. Experimenting with angled glass walls and heating flues to improve the efficiency of greenhouses went on throughout the 17th century. New technologies and improved glass led to larger and more elaborate structures that housed plants simply to please the eyes and palates of the European aristocracy. The palace of Versailles was an example of the elaborate efforts of the royalty to build bigger and more spectacular orangeries. The Versailles orangery was more than 500 feet long, 42 feet wide and 45 feet high. It had a southern exposure for light and heat. Despite the elaborateness of these structures, it was the Victorian age in England that ushered in the golden era of the greenhouse. By the end of the mid nineteenth century glass was plentiful and the prohibitive taxes were repealed. The wealthy began competing with each other to build the most elaborate greenhouse, again primarily just to grow citrus fruits and rare flowers. Little thought was given to using the greenhouse for a complete range of food production. The soaring conservatory at Kew Gardens in England is a prime example of the Victorian greenhouse. There is a replica in San Fransisco's Golden Gate Park called the Conservatory of Flowers

In America the first greenhouse on record was built around 1737 by Andrew Faneuil, a wealthy Boston merchant. Like his European predecessors, Faneuil used it primarily to grow fruits. The concept spread slowly, since almost all greenhouses were built for the wealthy. George Washington, perhaps the richest man in America, craved pineapples and ordered a pinery built at Mt. Vernon so he could serve pineapples to his guests. By 1825, greenhouses were becoming increasingly common. Many of the greenhouses were heated by furnace warmed air; some were pit greenhouses built into the earth and heated largely by south facing windows. This is a design that remains highly practical today. Indeed the modern concept of the greenhouse is simple and practical. No longer is it the private domain of the monied class but something anyone interested in gardening can have for relatively little cost. Today a greenhouse can go virtually anywhere there is space; it can be attached to a house, placed in a backyard or perched on a deck, roof or balcony. In addition, greenhouses are becoming more automated, reducing the time and care owners much spend. With new materials available today such as aluminum, pvc piping, fiberglass, acrylic, and polycarbonate panels as well as pressure treated wood ensures the consumer that their greenhouse will stand the test of time. Long gone are the days of shattered glass and overheated/under heated greenhouses.

Protected agriculture was fully established with the introduction of polyethylene after II World War. The first use of polyethylene was used for the construction of green house (later popularized as polyhouse) in 1948 when Professor Emery Mayers Emmert at university of Kentucky used the less expensive material in place of more expensive glass. Dalrymple, 1973 and Jension and malter, 1995 have more details on historical developments.

Present status of protected farming

The greenhouse technology is more than 200 years old. The Europeans were considered the pioneers in this field. Later with the advent of plastics during the Second World War a new phase in the green house technology emerged. Now a days, nearly 90% of the new green houses are being constructed by utilizing Ultra Violet (U.V) stabilized polythene sheets as the glazing material. In India, the technology is still in its nascent stage. The area under green house cultivation as reported by 2007-08 was about 500 ha in India. This figure is quite significant when compared with the total area under green house in the world which was 275,000 ha as reported for the year 1999-2000. Thus it can be safely stated that there exists a vast scope for expansion for green house technology in India.

Application for the green house technology for commercial purpose is however in its infancy in India. The green houses which existed in 1960s were used for commercial purposes but the impetus and the much required thrust came only in the mid 1980's with the emergence of the industries manufacturing U.V. stabilized Low Density Poly Ethlene (L.D.P.E.) and the development of the indigenous technology for low cost green houses. Indian Petro Chemicals Limited (I.P.C.L) was one of the foremost industries operating in collaboration with the agricultural scientists in this field. In 1985, Indian Agricultural Research Institute (I.A.R.I) designed and set up green house at Leh (J&K). Today there are more than 1840 small and medium sized greenhouses at Leh. The Defence Research and Development Organisation (DRDO) has provided adequate support and technological help in this project. Now a days the much needed vegetables are being grown throughout the year in these hostile climates. Production of brinjal, capsicum, tomato and other cucurbits are being grown in the summer months on a large scale, whereas the green leafy vegetables are being grown in the long frozen winter months when the average temperature reaches -30.2°C. Underground greenhouses and soil trenches are also being used on a large scale in these remote areas.

Farmers are also utilizing low and medium cost greenhouses for raising potted plants and seedlings in the nursery. In the northern gangetic plains especially in Punjab, Haryana and Uttar Pradesh, the farmers are using this technology to raise healthy seedlings of high yielding crop varieties so that they can be transplanted early in the fields during the onset of the spring season so as to capture the early markets and thus fetch higher returns. In the north eastern states, especially in Assam, efforts are on to raise vegetable crops in the greenhouse-cum-rain shelter structures during the long south west monsoon periods. In these regions stress is being given for the development of low cost greenhouses using bamboo frame structures as these construction materials are readily available in these regions. The results obtained in this regard from the concerned agricultural universities are encouraging.

The State Seed Corporation of India is utilizing at present the medium and the high cost greenhouse technology in their seed production units so as to maintain higher degree of purity of parental lineage and genotype. A large number of private companies are utilizing at present the high cost multispan greenhouse technology which has provisions for microprocessor controlled environment control system like operation of motorized shading system, on line drip irrigation system, supplementary heating and lighting system and the likes for vegatative propagation of chrysanthemums, roses, carnations and gladiolus etc.

Development of indigenous evaporative cooling system have taken place in recent times to control greenhouse air temperature during the long summer months. Exhaust fans of upto 60 cm diameter and some specialized equipment are being widely manufactured now a days in the country. At the same time great strides have been made to develop advanced cellulose and polyfab nylon cooling pads for usage in these cooling systems. The thermostats, thermocouples, high pressure mist and fogging system etc. and other environmental control equipments are being manufactured now a days in the country itself. Similarly, various brands of glazing material which include the U.V. stabilized polythene sheets, the fibre glass reinforced plastics, the twin wall polycarbonate sheets, etc.are also available in the local market now a days. Micro irrigation system and application procedures for applying the chemical and bio-fertilizers in the liquefied form are available for ready usage in the greenhouse

Benefits of protected cultivation

The benefits which can be derived from the protected cultivation are as follows :

- Environment control allows raising plants anywhere in the world at any time of the year i.e. crops could be grown under the inclement climatic conditions when it would not be otherwise possible to grow under the open field conditions.
- The crop yields are at the maximum level per unit area, per unit volume and per unit input basis.
- The control of the microcosm allows the production of higher quality products which are free from insect attack, pathogens and chemical residue.

- High value and high quality crops could be grown for export markets.
- Income from the small and the marginal land holdings maintained by the farmers can be increased by producing crops meant for the export markets.
- It can be used to generate self employment for the educated rural youth in the farm sector.

Constraints in adoption of protected cultivation

In spite of vast potential of protected farming technology in improving productivity, profitability and quality of the crops, the pace of its adoption is very slow in the country. Efforts to address the following constraints are required:

Lack of suitable structural design : The design of protected farming structures (polyhouse, low tunnels, and shade net house) should be developed and standardized in different agro-climatic conditions for various types of crops. In this direction efforts on very small scale have been made for the potential regions targeted for the protected cultivation is as under:

- Northern and north- western hilly regions.
- High rainfall region of north-east
- Western warm and dry regions
- Northern plain and temperate zone
- South - western moderate climatic zone

Efforts have been made for the development of suitable poly house structures for above mentioned regions. In this regard, Bureau of Indian Standard has developed three standards on polyhouse which are as follow :

- Protected Agriculture Structures IS 1461 : 1997
- Recommendations for Layout, Design and Construction of Green House structures
- Recommendations for green house heating, ventilating and cooling of green houses

The improvement on design, structures and environmental controls are being made with the coverage of more area under poly houses in different climatic conditions of India. Since 1985, efforts have been made to improve the design and structures of poly houses by the plastic development centers located in different climatic regions of India. Besides, plastic development centers, large numbers of private entrepreneurs have contributed for bringing out the suitable structures in various regions of India and as on today more than 5000 ha area is under poly house cultivation for the production of vegetables, cut flowers, elite planting materials and other applications in agriculture.

Indian Agricultural Research institute, New Delhi has designed and developed various poly house structures for different climatic conditions and crop specific poly houses including poly house structure for hilly regions of northern India.

The production technology for the cultivation of tomato, cucumber, capsicum, broccoli, cucurbits, cut-flowers and fruit planting materials have been developed by different precision farming development centers located all across India.

Difficulty in getting suitable cladding material

Presently, quality UV stabilized film is being manufacturing by few manufacturer. To entertain quality demand of lager volume of different types of cladding material significant number of manufacturer should come forward in this sector. Also, the quality cladding materials are not available in the local market. Its availability in local market will help popularization of green house sector.

Less competition among Green- house Manufacturers : Due to less competition among green/shade net house manufacturers the construction cost is higher along with inferior quality of construction.

Lack of local Manufacturers: Local manufacturers will be easily approachable. Any complaint/modification can be attended easily besides lowering down the construction cost.

Lack of easy availability of cooling and heating equipments: Partial control over protected environment for short period shall help in improving the production and accommodation of several desired crops.

Lack of package and practices: Depending upon the green house design and local climate, identification of suitable crops for green house cultivation is essential for different climatic zones. Once, the suitability of crop is established, detailed package of production practices including plant protection need to be developed.

Lack of suitable planting/seed material for protected cultivation: The protected cultivation demands high yielding, indeterminate and long duration specially vegetable crops..The availability of such varieties is limited and its cost is very high.

Lack of training and demonstrations: Potential farmers as well as extension workers should get adequate training for the handling of protected cultivation before they get actually involved in green house cultivation. Field demonstration of technology at progressive farms, in different locations will help its further popularization.

Monitory assistance: Since initial investment in construction of green house is higher, some short of financial assistance will encourage new growers to adopt the technology.

Future potentialities

Research conducted at various locations by the SAUs, PFDCs, ICAR institutions, DRDO and other organizations have confirmed that, there is a vast scope of utilizing the protected cultivation technology for growing many vegetables, flowers and few fruit crops in different parts of the country. The studies conducted by DRDO have proved that even in high altitude areas of Leh and arctic deserts, the green house cultivation can assure fresh supply of vegetables.

A number of opportunities exist for nursery men, commercial crop growers, seed producers and researchers in utilizing greenhouse for their benefit. In temperate and subtropical zones protected farming can easily be used for raising vegetable nursery in late winters which could be transplanted in early spring. This can advance the cropping by one to one and half month and thus, may provide remunerative price to the farmers. In plains green house may be utilized for year round propagation of many tropical and subtropical fruits which could prove a boon for nursery men. There is a demand of quality vegetable nursery in the vegetable growing areas.

Year- round cultivation of selected vegetable crops such as tomato, capsicum and cucumber is possible under protected environment with single/ multi crops in a year, which fetches off season higher price with quality produce and also with lower cost of cultivation and longer crop duration. There is high demand of colored capsicum, parthenocarpic cucumber and cherry, tomato in the hotel industries and export market throughout year at very attractive market price of the produce. India has a vast scope for exporting cut -flowers such as Gerbera, Carnation, Lilium etc. grown under protected environment.

□□□

Alternate Agro-Enterprises

24

Horticulture Scenario in India

J.P. TIWARI

India is the second largest producer of fruits after China, with a production of 44.04 million tonnes of fruits from an area of 3.72 million ha. The agro-climatic conditions of India permit the cultivation of a variety of fruit crops. Fruits have been associated with the Indian civilization since time immemorial, being mentioned as the food for sages, articles of offering to God and for medicinal and nutritional uses. A large variety of fruits are grown in India, of which mango, banana, citrus, guava, grape, pineapple and apple are the major ones. Apart from these, fruits like papaya, sapota, annona, phalsa, jackfruit, ber, pomegranate in tropical and subtropical group and peach, pear, almond, walnut, apricot and strawberry in the temperate group are also grown in a sizeable area. The Horticulture sector, which includes fruits, vegetables, root and tuber crops, floriculture, medicinal and aromatic plants, spices and condiments, plantation crops like cashew nut, coconut, and mushroom, contributes over 24.5% to agricultural GDP from less then 8% area under horticulture. It contributes over 10% of the total agricultural export earnings and supports more than 19% of the labour force. The sector has established its credibility in providing nutritional security to the people besides livelihood to womenfolk and poorest of the poor.

Trend in horticulture development

The emerging trend world wide and also in the country is indicative of a paradigm shift in the dietary needs of the people with rise in the income, which demands for more horticultural area. Since growing of horticultural crops is rewarding to the farmers in terms of returns per unit area, the sector is expected to contribute significantly for food and nutritional security, employment opportunity and poverty alleviation. The decade has witnessed the development of horticulture, which has resulted in appreciable growth and laudable achievements, owing to technological advancement and policy environment.

Past trend in development has been satisfying in terms of technological adoption, production, availability and export of horticultural produce and this trend has been marked as "Golden Revolution". India is not only the second largest producer of fruits and vegetables; it also has first position in several other horticultural crops. Production and export of flowers have increased many fold and the country has a major stake in the global trade of spices. Export of medicinal plants, fruits and vegetables has also exhibited rising trend. Admittedly, the sector is moving dynamic despite numerous challenges and shortcomings and is in crucial phase of development needing initiatives for sustainable development of horticulture.

Table 1. Demand and consumption of horticultural commodities

Commodity	Domestic Consumption popu. 2012 1.19 billion	Loss @ 20%	Export @ 8 %	Processing @ 10 %	Total Demand
			In million tonnes		
Fruits	52.38	16.20	4.19	5.24	78.01
Vegetables	122.23	37.00	9.78	12.22	181.23
Spices *	3.06	1.10	0.61	0.31	5.08
Coconut	8.73	4.00	0.70	0.87	14.3
Cashew nut	0.870	0.34	0.17	0.09	1.47
Others	4.36	1.36	0.35	0.44	6.51
Total	191.63	60.00	15.80	19.16	286.59

* Spices export will be 20%

Source : Planning Commission Documents/DPR of NHM

Potential of processing industry

The present level of annual production of fruits in India is about 45 million tonnes. Hardly 20% of the fruits produced in the country are used for processing for their consumption in the off season and also for export. Hence, there is a vast potential of this industry to develop which will go a long way to improve the national economy. Fruits and vegetables are preserved for a long time by reducing its moisture content. There is very good domestic and export market for the dehydrated fruits and vegetables. Dehydrated fruits and vegetables have good export market in Middle East, Russia and part of European countries. As a whole, there is a good prospect for new entrepreneurs who want to venture into this field.

Table 2. Production of horticultural crops

Commodity	Production (Million tonnes) 2002-03	2006-07	Target 2011-12
Fruits	45.20	59.36	81.00
Vegetables	84.81	115.52	185.00
Spices *	2.88	5.00	5.50
Coconut	8.82	18.00	20.00
Cashewnut	0.50	1.50	1.70
Others	2.20	6.00	6.80
Total	144.41	205.38	300.00

* Spices export will be 20%
Source : Planning Commission Documents/DPR of NHM

Among preserved fruit products, jam, jellies, pickles and squashes, RTS (ready to serve) beverages, wine, vinegar and other alcoholic beverages form an important class of products. Apart from fruit juices, other items like amchoor, aam papad, candies, murabbas, puree, sauces, ketchup and other canned and dehydrated products are also quite popular. Pickles and chutney are used along with other foods to make the lunch or dinner tasty. Market opportunities for the sale of these products are increasing with a considerable rate. The consumption of these products is increasing day by day and is expected to increase in future also. For an example, at world level more than 90% of grapes are processed into wine and raisins and little quantity for juice. However, in India the position is reverse, i.e., only 18-20 per cent are processed into raisins, juice and wine. The demand can be increased by proper advertisement and market management for each item. This industry possesses an excellent future and can allow new entrepreneurs to establish new unit for producing these products.

Ninth Plan intervention for development of fruits

To enhance the development of fruits, the Central Sector Scheme of Integrated Development of Tropical, Temperate and Arid Zone fruits was implemented during the Eighth Plan and was continued in the Ninth Plan with an outlay of Rs. 125 crores. The main thrusts of the programme were Supply of quality planting material, Nursery development, Establishment of tissue culture units, Area expansion, Improving productivity, Upgrading technical know-how, Publicity and media support for spreading awareness among farmers with respect to advancement in horticulture and information about Govt. incentives, Front line demonstrations, Establishing disease forecasting units, Horticulture mechanization, Establishment of leaf analysis units and plant health clinics, On farm handling of fruits, Tackling of emergent requirements, Technology development and application, etc.

Constraints in fruit production and utilization

Major constraints in production of fruit crops are inadequate availability of quality planting material, long gestation period, heavy initial investment, lack of technical know-how resulting in poor quality produce, low density plantations, weak post harvest infrastructure for storage, packaging and marketing, non availability of credit, inadequate extension service and lack of appropriate tools and machinery, inadequate facility for identification of nutrient deficiency and disease forecasting.

The fruits are highly perishable. About 20-40% of different fruits are destroyed due to post harvest losses. This further restricts the availability of fruits to a limited period of time. Mostly in the country, the fruits are consumed as fresh and only 1-2% are processed whereas, in the developed countries a sizeable quantity is processed into a variety of produce.

Strategies to improve production and productivity

- Promotional as well as commercial production for selected crops on cluster basis-project-based approach;
- Strengthening of infrastructure for production of quality planting material;
- Rejuvenation of senile orchards with high yielding cultivars;
- Area expansion under potential crops of the region in clusters;
- Technological intervention including high density planting, INM, IPM, micro irrigation, fertigation, greenhouse cultivation, etc.;
- New gardens through development of wasteland/utilization of bunds/ diversification from other crops/ inter cropping and
- Farmers' HRD through institutional trainings/farm visits.

□□□

25

Agroforestry as an Alternate Land Use System (ALUS)

RAJESH KAUSHAL, SALIL TEWARI AND SUMIT CHATURVEDI

Expanding human requirements and economic activities are putting ever-increasing pressures on land resources. In India, the per capita cropped land and forest land availability is less than 0.2 ha and 0.1 ha, respectively. Due to which there is an acute shortage of timber, fuelwood and fodder. Of the 142.82 m ha of net sown area, 89.82 m ha (63%) is rainfed, where crop production suffers from low yields. Further, continuous land degradation is also a serious threat to our already dwindling natural resources. Out of 329.0 m ha area, 175.0 m ha of the country is under wasteland. The total forest cover as per FSI, 1999 report has been estimated to be 63.73 m ha, out of which 30.70 m ha is degraded forest (Anon, 2003). These degraded lands can be brought under cultivation by adopting alternate land use systems (ALUS). According to National Remote Sensing Agency, cultivable wastelands occupy 32.8 m ha, which could be brought under tree or grass cover through improved technologies.

Alternate land use systems (ALUS)

Any farm enterprise other than crop production is usually referred to as ALUS. Fodder production for raising livestock is part of crop production system and therefore is not a part of ALUS. Tree farming, wood lots, pasture and grassland systems, ley farming and agroforestry systems are different versions of ALUS. (Singh, 1997, Singh and Osman, 1995). ALUS aims at generation of assured income with minimum risk through efficient utilization of available resources and are appropriate for rehabilitating degraded marginal wastelands and saving arable lands from further degradation. These systems can withstand vagaries of climate and can efficiently utilize the resources.

Agroforestry–the principal alternate land use

Agroforestry is collective name of land-use systems and technologies where woody perennials (trees, shrubs, palms, bamboos, etc) are deliberately used on the same land management units as annual agricultural crops and / or animals, in some form of spatial arrangement or temporal sequence. In agroforestry systems there must be both ecological and economical interactions between the different components (Lundgren and Raintree, 1982).

Agroforestry is capable of yielding both wood and food, while at the same time conserving and rehabilitating ecosystems. Agroforestry generates high income and minimises risks in cropping enterprises and is thus considered a potential technology for commercial farming, improving degraded and polluted sites, an opportunity for stabilizing fragile ecosystems. Agroforestry is an appropriate alternate land use system to all classes of land aiming at generation of assured income with minimum risk through efficient utilization of available resources.

Classification of agroforestry systems

Classification of various agroforestry systems has been dealt in detail by Nair (1993). Based on nature of components, three basic types of agroforestry system exist:

- Agrisilviculture system (trees+crops)
- Silvopastoral system (pasture/animals and trees)
- Agrosilvopastoral system (crop+ pasture/animals and trees)

Within agroforestry systems, there can be more than one agroforestry practice. An agroforestry practice usually denotes a distinctive arrangement of components in space and time. Several such practices are involved in the constitution and maintenance of an agroforestry system. Although hundred of agroforestry systems have been recorded, they all constitute 20 agroforestry practices. These include:

- *Agrisilvicultural systems*. Improved shifting cultivation, taungya, alley cropping (hedgerow intercropping), multiple trees on croplands, shade trees for plantation crops, homegardens, mixture of plantation crops, shelterbelts and windbreaks.

- *Silvopastoral systems.* Trees in rangeland and pastures, Fodder banks (multipurpose, leguminous fodder trees and shrubs grown on or around farmlands); living fences of fodder hedges; trees and shrubs on pastures;.
- *Agrosilvopastoral systems.* Tree livestock crop mixtures around the homestead (compound farming); multipurpose trees and hedgerows for browse, green manure, soil conservation; apiculture with trees, aquaforestry and multiple wood lots.

I. Present agroforestry practices in india

The systematic research in agroforestry geared up after establishment of International Council for Research in Agroforestry (ICRAF) in 1977 (which was renamed in 1991 as the International Centre for Research in Agroforestry). In India, organised research in agroforestry was initiated during 1983 with the establishment of All India Coordinated Research Project on Agroforestry by ICAR at 20 centres and later establishment of National Research Centre for Agroforestry at Jhansi in 1988. At present 39 centres of agroforestry (33 regular, 5 voluntary and I PC unit) are working in the country.

Through ICAR centres, a Diagnosis and Design exercise for existing agroforestry practices in India has been conducted. The prominent and priority agroforestry practices include agrislviculture, silvopastoral, agrisilvopastoral, agrisilvihorticulture, agrihorticulture, hortipastoral etc. Some of the prominent agroforestry practices in different regions are presented in table 1.

Table 1. Prominent agroforestry systems in different agroclimatic regions of India

S.No	Agro-climatic zones	Agroforestry Systems*					
		Agri-Silvi-Culture	Agrihorti Silvi Culture	Agri-horti Culture	Horti-Pastoral	Silvi Pastoral	specialized
1.	Western Himalayan Region	4	4	5	4	4	-
2.	Eastern Himalayan Region	5	3	5	4	3	Jhum
3.	Lower Gangetic Plains Region	5	-	3	2	2	Energy
4.	Middle Gangetic Plains Region	4	-	2	-	-	Homestead
5.	Upper Gangetic Plains Region	5	-	5	-	-	Agrihorti-silviculture
6.	Trans-Gengetic planis Region	5	-	5	-	-	Agrihorti-silviculture

Contd...

7.	Eastern Plateau & Hills Region	5	-	4	3	4	Block Plantation
8.	Central Plateau & Hill Region	4	-	5	4	4	Block Plantation
9.	Western Plateau & Hills Region	4	-	5	2	4	-
10.	Southern Plateau & Hills Region	5	-	3	2	4	-
11.	East Coast plains & Hills region	4	5	4	2	2	Home garden/shelt er belt
12.	West Coast Plains & Ghats Region	4	5	4	2	2	Home garden
13.	Gujarat Plains & Hills Regions	4	2	4	3	5	Block plantation
14.	Western Dry Region	5	3	4	-	2	Shelterbelt
15.	The Islands Regions	4	5	4	1	2	Home garden

*No. 1-5 indicate the priority practiced mode1= Least and 5= Highest

II. Traditional agroforestry systems in india

Traditional agroforestry systems in India range from apparently simple forms of shifting cultivation to complex home-gardens, from systems involving sparse stands of trees on farm lands to high-density complex multi storeyed homesteads of humid lowlands, from systems in which trees play a predominantly 'service' role to those in which they provide main commercial product. Most of these are anecdotal but in some enough research efforts have been carried out in recent times (Pathak *et al.*, 2006; Anon, 2003). Some of the traditional agroforestry systems are:

a) Shifting cultivation

It is prevalent in many parts of Africa, Latin America, South-East Asia and Indian subcontinent. In India, it is practiced in Assam, Meghalaya, Jharkhand, Manipur, Orissa, Nagaland, Chattisgarh, Madhya Pradesh, Arunanchal Pradesh, Andhra Pradesh, Mizoram, Tripura, Kerala, West Bengal and Sikkim. In this system, land under natural vegetation (usually forests) is cleared by slash and burn method, cropped with common arable crops for a few years, and then left unattended when natural vegetation regenerates. Traditionally the fallow period is 10-20 years but in recent times due to population pressure, it has been reduced to 2-5 years and the system has degenerated causing serious soil erosion thereby causing decline in productivity. The system is however now being modified using improved fallows, where fast growing nitrogen fixing trees are planted in fallow phase.

These trees are helpful in restoring soil fertility, reducing soil erosion, improving the economy of the farmers which ultimately helps in shortening the fallow phase. Species being used are *Acacia. Albizia, Alnus,. Eryrthrina. Faidherbia. Gliricidia, Inga. Leucaena. Parkinsonia, Pithecellobium, Robinia* and *Sesbania*. Thus intercropping under or between fast-growing trees in a fallow phase is one of the approaches while finding alternative to shifting cultivation. Borathakur et al., (1981) suggested that practice of shifting cultivation can be improved by using the top 1/3 of hill slopes under forest, the middle 1/3 under horticulture and bottom 1/3 under bench terraced cultivation.

b) Taungya *System*

This system was introduced to raise forest plantation but is now recognized as one of important agroforestry practice. This is a modified form of shifting cultivation where trees and agriculture crops are grown together for some period of time. It is prevalent in Uttar Pradesh, Uttarakhand, West Bengal, Kerala and to lesser extent in Maharastra, Andhra, Orrissa, Karnatka and Tamil Nadu. The practice consists of land preparation, tree planting, growing agricultural crops for 1-3 years, until shade becomes too dense, and then moving on to repeat the cycle in a different area. In some cases crops may be grown one year before the trees are planted. A large variety of crops and trees, depending on the soil and climatic conditions, are grown in India. Due to prevailing land shortages. and possibility of growing highly remunerative crops like paddy, tapioca, ginger and turmeric, the farmers are practicing this system. In the high hills of Tamil Nadu, *Eucalyptus globulus, Acacia mearnsii* and *Pinus* species are established after growing crop of potatoes. More recently, cashew plantations in the coastal areas of Andhra Pradesh have been established in combination with groundnut. This method is also used for establishment of poplar plantation with irrigation in *tarai* region of Uttar Pradesh and Uttarakhand.

c) Plantation crops under shade of trees

Tea, coffee, cacao, black pepper, small cardamom, large cardamom are some of the plantation crops grown under shade of trees. Tea (Camilia sinensis) is grown under shade of *A. chinensis, A. odoratissim, A. lebbek, A. procera, Acacia lenticularis, Derris robusta,Grevillea robusta, Acacia spp., Erythrina lithosperma, Indigofera tesmanii, Coffee (Coffea arabica* and *Coffea canephora)* is grown under the shade of *Erythrina lithosperma* as temporary shade while, *Ficus glomerata, F. nervosa, Albizia chinensis, A. lebbek, A moluccana, A. sumatrana, Dalbergia latifolia, Artocarpus integrifolius, Bischofia javanica, Grevillea robusta* are used as permanent shade trees. Cacao is grown under the shade of coconut and areca nut, *and Dipterocarpus macrocarpa (in forest*). Black pepper *(Piper nigrum*) is grown with support from *Erithrina indica, Garuga pinnata, Spondias,*

Mangifera, Gliricidia maculate and *Grevillea robusta.* In home gardens it grows on 27 tree species. Small cardamom (*Elettaria cardamomum*) and large cardamom (*Ammomum subulatum, A. aromaticum*) grow in forests under temporary shade tree of *Mesopsis emini*.. Large cardamom is grown under the shade of natural forest as well under planted shade trees *viz., Alnus nepalensis, Schima wallichii, Cinchona spp., Lagerstroemia spp., Albizia lebbek, A julibrissin, Castanopsis tribuloides, C. hystrix, C. indica, Terminalia myriocarpa, Bischofia javanica* (Tejwani,1994).

d) Home-gardens/homesteads

It is an old-age practice in coastal states particularly in Kerala. Tripura. Assam. North-eastern states and parts of West Bengal and Andaman and Nicobar Islands In this practice, a number of crops including tree, crops are grown with livestock. poultry and/or fish production, mainly for the purpose of meeting the routine basic needs of the farmer. Different components occupy different strata in homegardens. Top most strata is occupied by forest trees, the middle by fruit trees and the ground crops are generally agricultural crops, spices etc. Coconut, arecanut, guava, mango, citrus, tamarind, jack-fruit, papaya, banana, moringa, sesbania, custardapple are the major fruit trees found grown in home-gardens. They provide fruits and vegetables and also support climbers such as betelnut. Pepper, cucurbits, clove and cinnamon. Coffee and cacao are also found in these gardens. Among food crops, cassava, yam, sweet-potato, taro, elephant-yam, and vegetables are grown. Spices includes cardamom, cinnamon, clove, ginger, nutmeg, pepper and turmeric. Pineapple is common fruit grown as ground storey crop. Kapok *(Ceiha pentandm)* is common tree along the boundary providing fibre (Tejwani,1994). Due to large diversity, home gardens can be considered as excellent tools for biodiversity conservation.

e) Winbreaks and shelterbelt

Windbreak or a shelterbelt are a densely planted strip of tall vegetation usually between 50 - 200 feet wide oriented at right angle to the directions of prevailing winds in order to reduce their velocity near the ground. Farmers build shelter-belt by either small dead wood or local vegetation to check wind velocity within safer limits. These are integral part of arid land use systems. *Acacia nilotica, Albizzia lebbek, Azadirachta indica, Tamarix articulate, Acacia* spp, *Casuarina equisetifoli,Parkinsonia aculeate* are the common tree species used as shelterbelts. *Callognum polygonides, Crotalaria burhia,. Leptadenia pyrotehnica* and *Aerva pseudotomentosa* bushes and grasses such as *Cenchrus ciliaris, C. setigerus. Lasiurus sindicus* are also used in windbreaks. Dead wood and shrubs are also grown in checker board system to check the movement of sand dunes. This permanent vegetation helps accumulate sand and check the soil erosion and helps in increasing crop yields along the lines.

f) Scattered trees on farm lands

The practice of growing agricultural crops under scattered trees on farm lands is being practiced for centuries. The density of the trees in this practice varies from 10-50 trees/ha. Some of the practices are very extensive and highly developed in India. For example, growing *Prosopis cineraria* and *Ziziphus* in arid areas; *Acacia nilotica* in Indo-gangetic plains; *Grewia optiva* and other tree species in Himachal Pradesh and Uttarakhand; *Eucalyptus globulus* in southern hills of Tamil Nadu. Trees are mainly used for shade, fodder, fuelwood, fruit, vegetables and medicinal uses and has not much commercial value.

Agroforestry based ALUS for arable lands

Land capability classes I to IV are fit for all agriculture uses including crop production. These lands can be profitably brought under pasture development and forestry/agroforestry practices as per the capability and needs. Some of the agroforestry practices for arable lands are:

a) Hedgerow intercropping

Hedgerow intercropping is an agroforestry practice in which perennial, preferably leguminous trees or shrubs are grown simultaneously with an arable crop. The trees, managed as hedgerows, are grown in wide rows and the crop is planted in the interspace or 'alley' between the tree rows. During the cropping phase the trees are pruned and the prunings used as green manure or mulch on the crop to improve the organic matter status of the soil and to provide nutrients, particularly nitrogen, to the crop. The hedgerows are allowed to grow freely to shade the inter-rows when there are no crops. Alley cropping retains the basic restorative attributes of the bush fallow through nutrient recycling, fertility regeneration and weed suppression and combines these with arable cropping, so that all processes occur concurrently on the same land, allowing the farmer to crop the land for an extended period. Hedges can also be used for fodder production instead of pruning. *Gliricidia sepium, Calliandra calothyrsus, Leucaena leuccephala, Flemingia macrophylla and Erythrina subumbrans* are some of the important species used in hedgerow intercropping. On sloping land, the tree rows act as a physical barrier to soil and water movement, resulting in significant reductions in erosion losses.

b) Boundary plantation

Majority of the boundary plantations are used as shelterbelts and windbreaks, particularly in fruit orchards. In northern parts of India, particularly in Haryana and Punjab, *Eucalyptus* and *Populus* are commonly grown along the field boundaries or bunds; other trees which are found grown as boundary plantations or live hedge include *Acacia nilotica* and *Dalbergia sissoo*. In Bihar. *Dalbergia sissoo* and *Wendlandia exserta* are most common boundry plantations. *Casurina equsitifolia* is extensively planted on field bunds and along sandy coastal areas in Orissa. Farmers of Sikkim grow bamboo all along the irrigation channels. In Andamans, farmers grow *Gliricidia sepium*. *Jatropha* spp., *Ficus*. *Ceiba pentandra*. *Vitex trifolia* and *Erythrina variegata* as livehedges. At many places, succulents like *Agave* and many cactoids are grown as common live-fence.

c) Commercial agrofrestry practices

i) *Poplar based agroforestry*: Poplar (*Populus deltiodes*) is widely grown in Punjab, Haryana, Uttarakhand and Uttar Pradesh. Nearly 15 million poplar plants are raised under agro-forestry plantations in these states every year covering roughly 30,000 ha. Similar area is harvested yielding about 3.6 lakh cubic meters wood of poplars yearly supporting the safety match industry and a large number of veneer and plywood units (Lal, 2004). The tree due to its deciduous nature is compatible with various agriculture crops. Poplar is grown at spacing of 5mx4m. All *rabi* and *kharif* crops can be grown during the first three years except paddy. However, inter-cultivation of sugarcane is preferred for first two years as it is more profitable. Third year onwards cultivation of wheat, cabbage, chilly, tomato, barley, coriander, turmeric, ginger, strawberry, oats, berseem, sarson etc. can be raised throughout the rotation. Poplar can give yield of 150-200 m^3/ha in block plantation and 12-20 m^3/ha in boundary plantation at 8 years rotation under good care. It can give gross income f Rs 6-8 lakhs/ha under block plantation and productivity of 20 m^3 /ha/yr while, income from intercrop is additional.

ii) *Eucalyptus based agroforestry*: Eucalyptus is grown in Punjab, Haryana, Uttarakhand Uttar Pradesh, Gujrat, Andhra Pradesh and many parts of the country to cater the needs of pulpwood, plywood and timber industries Eucalyptus are grown at spacing of 3x2 m^2 or 4x2 m^2. Cotton, chillies, blackgram, greengram, rice and groundnut are taken up as intercrops during the first year of plantation cycle. Intercrops are also taken up during the second and third year of plantation, but the yield reduction observed is up to 70%. Eucalyptus trees are harvested at the end of fourth year for pulpwood and after eigth year for timber/ plywood. A minimum of three coppices are taken up. Intercrops are taken up during the second and third coppice also. Average productivity of 20 to 25 cubic meters/ ha/ year has been achieved under un-irrigated conditions. However, productivity standards have been redefined by many farmers by achieving unprecedented record yield of 50 cubic meters/ ha/year compared to yield of 6 to 10 cubic meters/ha/year from normal seedlings. Because of better soils, adequate irrigation facilities and more progressive farmers in Punjab and Haryana productivity of clonal eucalyptus plantations is likely to establish new records. Farmers can expect net returns of Rs.4 to 4.5 per acre at 10 years rotation from irrigated clonal eucalyptus plantations assuming yield of 200 tones and current price of Rs.2500/ per tones for eucalyptus logs (Lal 2004).

iii) *Subabul based agroforestry* : It is mainly practiced in Prakassam, Krishna, Guntur, Khammam, Rangareddy, Medak and Nizamabad districts of Andhra Pradesh to cater the needs of pulpwood requirement of ITC, J.K. Sirpur, Orient, West coast and A.P. paper mills. For pulpwood spacing of 1x1 m is adopted and no intercrop is taken. This can give net returns of Rs 50,000/ha after three years. With intercrops of blackgram, cotton and tabacoo in first year and limited fodder use, a spacing of 2.5x0.75 m is adopted which can give net returns upto Rs 72,000/ha after three years. For fodder production in animal husbandry, spacing of 1.25x1.25m^2 is followed which can give net returns of Rs 1,68,632/ha after three years. After taking 3 crops (9yrs), 20-25t/ha of root biomass is harvested which is sold in tobacco barns for curing @Rs 800/t (Pathak *et al.,* 2006).

Besides above, *Casuarina, Gmelina arborea, Ailanthus excelsa, Acacia magnum,* Bamboos are also being grown commercially on Large Scale.

d) Agrihorticulture system

In this system, short duration arable crops are raised in the interspaces of fruit trees. Trees such as mango, apple, litchi, almond, walnut, apricot, peach, plum, pear etc can be grown in association with many agriculture crops. In this system, agriculture crops are grown in between the fruit trees. The distance of trees varies from 5 to 7 m. In between the rows of these trees additional row of fast growing timber trees can be grown. These acts as fillers. For example poplar is grown as filler crop in mango orchards. By the times, the fruit trees comes into commercial bearing, fillers are ready to harvest.

III. Agroforestry based ALUS for non-arable lands

Class IV and above soils are termed as non-arable land. These are unfit for crop production. Carrying capacity of these lands is very low. These lands thus require immediate attention for rehabilitation which can be done by adopting suitable grass based agroforestry models and tree farming. Some of the agroforestry based systems for non-arable lands include:

a) Silvipastoral system

The importance of silvipastoral system has now been accepted through out the world for increasing the productivity of degraded/problematic lands and rehabilitation of degraded lands. It helps in checking soil erosion, conserving more in-situ moisture, maintaining higher survival per cent, yielding more fodder and fuel wood biomass. Some of the silvipastoral practices are traditional and well developed. For example, *Prosopis cineraria* and *Zizyphus numularia* in arid zone of Rajasthan where community and village grazing lands are called *orans*. These *orans* have 70-90% of zizyphus. Rest area is shared by *Prospis cineraria, Balanites egyptica, Capparis decidua*. Grasses found include *Lasiurus indicus, Panicum antidotale, Saccharum bengalensis, Cenchrus ciliaris, Dicanthium annulatum, Heteropogon contortus, Cenchrus setigerus, Sporobolus marginatus*. In addition to the above *Acacia tortilis, Prosopis juliflora, Albizia lebbek, Holoptelia integrifolia, Dalbergia sissoo, Anogeissus pendula, Acacia senegal, Butea monosperma Dichrostachys nutans, Ailanthus excelsa, Azadirachta indica, Tamarix articulata, Salvadora oleoides, S persica* are the associated tree species. Fruit trees such as ber, aonla, guava, custard apple, bael, are also grown with grasses particularly in these region.

b) Tree farming

Trees can come up in situation where many agriculture crops fail. Trees can be grown in private and community lands for meeting the diverse need. They are also helpful in checking soil and water erosion.

i) *Short rotation forestry*: Short-rotation forestry is defined as the silvicultural practice under which high-density, sustainable plantations of fast-growing tree species produce woody biomass on agricultural land or on fertile but degraded forest land. For commercial purpose trees are generally grown under high density plantation. In this system trees are grown either as single stems or as coppice systems, with a rotation period of about 5-10 years and with an annual woody production of at least 10 tonnes of dry matter or 25 m^3 per hectare. The biomass produced is used for construction, pulp and paper, fodder and energy. The short rotation plantations are established with unrooted cuttings at approximately 3x3 m spacing. The rotation is about 10 years, and the objective is to produce pulp wood trees of 20 cm in diameter and 18 m tall. Annual yields of 20-30 m^3/ha are obtained. In mini rotation plantations are established with densely planted unrooted cuttings (30x90 cm). One to three year old coppices are harvested repeatedly which can yield 15 tonnes/ha or more of total annual oven dry mass.

ii) *Energy plantations*: An energy plantation is one that is grown purely for plant material for their fuel Energy plantations in India has gained emphasis since the inception of the Clean Development Mechanism (CDM) under the Kyoto Protocol due to various environmental and social benefits.

a) *Woodlots*: In India, *Casuarina equistifolia* is extensively grown in Andhra Pradesh, Tamil Nadu and Karnataka, on degraded lands for fuelwood purpose. The trees are planted very close, usually 1m ×1m and harvested in 7 to 10 years. The clear felling of such plantations yield about 120 tonnes of firewood. After 7 to 10 years, the crop is clearfelled and stumps, which also fetch a good price, are uprooted. Often the land is subsequently brought under agriculture for a few years and then is converted into a *Casuarina* wood lot (Anon, 2003). *Eucalyptus* is grown extensively as wood lots. The species is usually grown at spacing of 2m × 2m but narrower spacings of even 1m × 1m, 1m × 1.5m, 1.5m × 1.5m have also been tried. In general standing biomass varies between 11.9 t/ha in three years old plantation to 196.5 t/ha in nine year old plantations.

b) *Plantation for biodiesel production* : *Jatropha curcas* can grown in a wide variety of agro-climatic conditions from arid (200 mm of rainfall) to high rainfall areas and on lands with thin soil cover to good lands. Seed production ranges from about 0.4 tons per hectare per year to over 12 t /ha. The experience in India and elsewhere shows that a plant density of 2500 per hectare (spacing of 2x2 meters) has been found to be optimal - although in one trial in rainfed areas on poor soils a lower plant density of 1666 has been felt to be more desirable. In suitable plantation Jatropha gives about 2 kgs of seed per tree. If planted in hedges, the reported productivity of Jatropha is from 0.8 kg – 1.0 kg of seed per meter of live fence This is equivalent to seed production of between 2.25 t/ha and 5 t/ha, depending upon whether the soils are poor or average for plantations and between 2.5 t/ha/ annum and 3.5 t/ha/annum for hedges (Anon, 2004).

References

Anonymous. 2003. Report of The Task Force on Greening India for Livelihood Security and Sustainable Development. Planning Commission, New Delhi, Government of India.

Anonymous. 2004. Report of the committee on development of Bio fuels. Planning Commission, New Delhi, Government of India.

Borathakur, D.N.; Parsad, R.N.; Ghosh, Singh S.P.; Awasthi, A; Rai, P.P.;Verma, R.N.A; Datta, H.H; Sachan, J.N. and Singh, M.D. 1981. Agroforestry based farming system as an alternate to jhuming. Proc. Of the Agroforestry Seminar, ICAR, N Delhi.pp. 109-131

Lal, P. 2004. Integerated development of farm forestry plantation and wood based industries. *Indian Forester* 1. 71-78

Nair, P.K.R. 1993. An introduction to Agroforestry. Kluwer Academic Pubklishers. The Netherland. 499p.

Pathak, P.S., Slanki, K.R., Rai, P., Handa, A.K., and Pateria, H.M. 2006. Agroforestry. In: Hanbook of Agriculture. Indian Council of Agriculture Research, N Delhi

Pathak, P.S., Suresh, G., Khan,T.A., and Gupta, S.K. 2006. Subabul in the rural economy of Andhra Pradesh-A case study. Indian Grassland and Fodder Research Institute. Jhansi, U.P. India pp 18-33

Roy, D.R. 1995. Agroforestry-Principle and practices. In Sustainable Development of dry land Agriculture in India (Singh, R.P. eds). Scientific Publishers, Jodhpur 399-413

Singh, R.P. 1997. Alternate land use systems for sustainable development In: Silvipastoral systems in arid and semi arid ecsystems (Eds Yadav, M.S., Singh, M., Sharma, S.K., Tewari, J.C., and Burman, U) CAZRI, Jodhpur, India pp. 249-266

Singh, R.P. and Osman, M. 1995. Alternate land use systems for drylands. In Sustainable Development of dry land Agriculture in India (Ed Singh, R.P) Scientific Publishers, Jodhpur. pp 375-398

Tejwani K.G. 1994. Agroforestry in India. Oxford & IBH Publishing Co. Pvt. Ltd., New Delhi, 233p.

Lundgren and Raintree, 1982. Sustained Agroforestry. In: Agricultural Research and Development-Potential and Challenges in Asia, bnestel (Ed) , ISNAR, The Hague, pp 37-49.

□□□

26

Exploitation of Plantation Crops for Commercial Use

B.C. SAINI

Increasing population, continuing degradation and reduced productivity of our natural ecosystem had occurred side by side in most parts of the developing countries. The time has come now to reckon economic and ecological sustainability a paramount place in our production planning. It is a time for planning of a long-term plan to maintaining a harmonious balance between environment and human needs by introducing economically important plantation crops in degraded forests, marginal lands, cultivable wastelands and fallow lands which will provide not only the local employments but also raw material for industrial development in the rural areas. The plantation crops could help small farmers to earn more per unit of land, water, labour, capital and time. It is, therefore imperative to increase the area under plantation which can not only improve the ecological balance but also provide fuel and other needs through renewable resources of raw material.

Coffee, tea and rubber, are important plantation crops grown under different soil and climatic conditions. These crops can be produced in entirely different environment with proper technologies. Pantnagar scientists were successful in developing technologies for cultivation of coffee and tea crops in sub-tropical climate of *tarai* areas as well as in Uttarakhand hills. Tea and coffee both are shade loving evergreen woody perennial plants with 150 and 50-70 years of life period, respectively, are generally grown as rainfed crops in India. About 5, 00,000 ha of land under tea cultivation produces about 956 million kg of made tea, 200 million kg of which is exported. Coffee is grown on approximately 3, 50,536 ha which produces about 2, 94, 00 tones which is almost 3% of the world coffee. More than 80% of the produce is exported and thus both the crops contribute thousands of crores of rupees of foreign exchange. Rubber is a tropical tree and is a hardy plant. It can be grown on a wide range of soils having adequate drainage. The soil should possess good water holding capacity and pH of 5.5 to 6.5 is suitable for its cultivation. Crop harvesting can be continued for at least 20 years which can be extended up to 28 years by adopting different tapping methods. Rubber plantations are spread over 5.97 lakh hectares in the country with an average productivity of around 1796 kg/ha. Now India has become the third largest rubber producer in the world. Recently rubber tree has caught the attention of farmers that it can also be grown for wood as primary and for latex as a secondary crop.

Introduction

Manifested through different kinds of soil and climatic conditions, India enjoys unique agro-eco diversity scattered throughout the country. From traditional subsistence culture, plantation crops have traveled a long and now these commodities have emerged as a state-of-the-art industry. It is a matter of great strength that in present day competitive agriculture, India is global leader in many plantation crops. It holds rank in production of coconut, arecanut, cashewnut, tea, rubber and accounts for sizeable production of many other crops. Substantiated by involvement of many sole trade oriented organizations like APEDA (Agriculture and Processed Food products Export Development Authority), New Delhi, Trifed (Tribal Coperative Marketing Development Federation) New Delhi, Coffee Board, Bangalore (Karnatka), Tea Board, Kolkata (WB) and Rubber Board, Kottayam (Kerala), plantation crops have proved their industrial attributes especially. Now slogan has also been changed from more production per unit land and time to more production and income per drop of water per unit land per unit time. In the past, one used to talk about the yield gap, but at present more emphasis is given to the income of the farmers.

Tea cultivation

Uttarakhand is one of the first state to pioneer commercial production of tea in India. Before 1992, there were hardly 7 tea estates in Doon Valley having

Himachal Pradesh Krishi Vishvavidalaya, Palampur released the following tea varieties:

1. Kangra Asha
2. Kangra Jwala

Irrigation

Tea is usually a rainfed crop and is grown in the plains as well as 2000m above sea level. Most tea growing areas have a rainy season and dry season. The major countries, where irrigation of tea is a regular practice, India, Bangladesh, Malawi, Zimbabwe, Tanzania and South Africa. The duration of the dry season varies from a few weeks to eight months. Generally most of the states in India experience drought between Novembers to March in North - East India, between January to March in South India and between November -June in North-West India.

In four year old tea plantatins, first plucking was started from March 23. Drip irrigation was given on daily basis. It can be concluded that drip irrigation has resulted in 105 and 205 per cent increase in green leaf yield over the rainfed plots in the months of May and June, respectively.

Manuring of young tea

After planting as soon as the plants have produced two or three new leaves, they can be foliar sprayed with NPK or NK mixture at fortnightly intervals till the plants have produced 4-5 new leaves. Well-established plants can be manured with YTD mixture (NPK 10:5:10 or 10:5:15 depending up on soil available potash status) at rates depending upon the spread of the bush as determined by age. During the first four years of the formative stage, plants are ring manured up to 4 splits while on 5th and 6th year; manures are applied in 2 splits.

Manuring of mature tea

It is important to ensure adequate replenishment of nutrients in the soil removed by harvest.In the Darjeeling hills, the agro climatic conditions are different from those of plains. The cool temperature slows down the organic matter decomposition and the productivity levels are lower as compared to the plain districts. Therefore, the N levels should not exceed 110-120 kg/ha.

Weed management

Weed control in tea is the second most expensive operation. Weeds are the greatest threat for the crop for various reasons and generally herbicides like Paraquat, Dalapan are used as post-emergence and Oxyfluorfen , Atrazine and Diuron as pre emergence.

Tea Research Centre, Pantnagar

Table 1. Green leaf yield under best treatment in different experiments during 2005 and 2006

S. No.	Experiment	Variety	Max. yield (Average) Kg/ha 2005	Maximum yield (average) Kg/ha 2006	Year of Plantation/ age	Remarks
1	Studies on bringing young tea into early production	T-78	1450	2472	2002 (4 Yr)	Different treatments imposed
2	Studies on pruning cycles for mature tea	T-78	7562	10743	1996 (10 Yr)	Light pruning in Dec 2003
3	Effect of plucking systems and plucking intervals on quality and yield of tea	AV-2	7865	8908	1996 (10 Yr)	Light pruning in Dec 2002
4	Effect of plucking standards and plucking intervals on quality and yield of tea	UPASI -9	8119	12988	1996 (10 Yr)	Light pruning in Dec 2003
5	Interaction of tipping levels and plucking systems in clonal variety	T-78	6551	8168	1995 (11 Yr)	Light pruning in Dec 2003
6	Interaction of tipping levels and plucking systems in seed stock variety	BSS-449	5520	7398	1996 (10 Yr)	Light pruning in Dec 2003
7	Effect of foliar feeding of secondary and micronutrients on the growth and yield of matured tea plants	BSS-449	792	3658	1995 (11 Yr)	Light pruning in Dec. 2004
8	Response of soil water regimes and fertilizer doses on quality and yield of tea	T-78	2214	6691	2002 (4 Yr)	LOS in Dec. 2004
9	Effect of inorganic and bio-fertilizers on cultivation of tea under modified climate in an agroforestry system (Pine)	BSS-449	499	4268	1996 (10Yr)	Light pruning in Dec. 2004

Table 2 Year-wise green leaf yields in Kausani region (208 ha)

Year	Total green leaf yield (kg)	Average yield (kg/ha)
2003	80,000	381
2004	1,10,000	524
2005	1,62,000	771
2006	2,00,000	961
2007	1,80,000	865

Source: Uttarakhand Tea Development Board

Coffee cultivation

India's coffee grows between 3000 to 6000 feet above sea level, usually on terraces in the mountainous regions. Coffee is cultivated commercially in the four southern states of Karnataka(approximately 80% of Indian coffee is grown here), Tamil Nadu, Kerala and Andhra Pradesh. Coffee cultivation is confined to the hilly areas of the Western and Eastern Ghats. It is also grown on a limited scale in some non-traditional areas of Orissa, West Bengal, Assam and Madhya Pradesh. However, Pantnagar University has successfully introduced coffee in Uttarakhand by planting *arabica* coffee on normal soils of *tarai*. Research results obtained so far indicate that coffee can be grown successfully in plains as well as hills of Uttarakhand. In the hills where soil pH is more than 5.5 and tea can not be grown without amendments, coffee can be grown.

Propagation

Coffee is generally propagated from seeds, though propagation by cuttings and grafting is also possible. Three-fourth kg (i.e. 750 gm) of quality seeds is enough to plant a hectare. A 50% allowance of seeds must be considered for ungerminated seeds, poor seedlings and for replanting;

Soil characteristics

Deep, friable, porous, rich in organic matter, moisture retentive, slightly acidic, pH 6-6.5, well drained.

Varieties

There are six commercially important varieties and selections of *Coffea arabica* grown in India.

1. Old chiks
2. Coorgs
3. Kents
4. S 288
5. S 795
6. S1934.

Planting time

Field planting can begin when the coffee plants in bags have a minimum of six to eight leaf pairs . Plants should be strong and healthy with no sign of pests or disease. Planting out in the field should be done on cloudy days, from June to August during the wet season. Avoid planting trees when conditions are windy or hot and dry or during the hottest part of the day.

Spacing

The distance between row to row and plant to plant would depend upon the type of planting materials. Recommended spacing for important species of coffee is given in table 3.

Table 3 Recommended spacing for coffee plantation

Species	Spacing
ARABICA	3 x 1 to 3 x 2 m 2 x 2 x 2 x 3 m double row
ROBUSTA	3 x 1.5 to 3 x 3m 2 x 2 x 2 x 4m double row

Climate parameters

Arabica varieties are grown on hills at 800-2000m altitude and *robusta* varieties are planted in the plains.

Table 4. Climatic and physiographic requirements of coffee

Parameters	ARABICA	ROBUSTA
Temperature	15-24°C	20-30°C
Annual rainfall	1600-2500 mm	1000-2000 mm
Relative humidity	70-80%	80-90%
Attitude	1000-1500 m	500-1000 m
Aspect	Northern, eastern and north-eastern	Same
Slopes	Gentle to moderate	Gentle to fairly leveled fields

Shade tree spacing

Erythrina subumbrans (Dadap): 4 x 4 m, *Gliricidia sepium* (Khae Falang): 4.5 x 4 m, *Cassia siamea*: 4.5 x 4 m, *Melia azedarach*: 6 x 6 m, *Paulownia tomentosa*: 6 x 6 m, *Salix alba* (Willow): 6 x 5 m. However, later on shade trees should be thinned down to about 12 x 12 m.

Nutrient management

In India, it was found that every 6,000 kg of ripe coffee cherry (1 tonne of green bean) removes, approximately 40 kg nitrogen (N), 2.2 kg phosphorus (P) and 53 kg potassium (K) from soil. Therefore, it is necessary to replenish the nutrients that are removed from the system. The basic requirements of fertilizers for matured coffee are 160:120:160 kg as $N:P_2O_5:K_2O$ for 1250 kg of the crop and above for both the varieties.

Weed management

Coffee trees are shallow-rooted, which means that most feeder roots are near the surface. Weeds compete for both nutrients and water, so it is essential to keep the area under the canopy of the trees, weed-free. Weeds are a serious problem during the first three years. Generally weeds are controlled manually. Chemical weed control methods are employed where labour is scarce or expensive. Paraquat dichloride (Gramoxone) and glyphosphate (roundup or glycel) are recommended herbicides in India.

Critical periods for irrigation

- At the time of flowering
- Period of berry expansion (7 to 17 weeks after flowering)
- Period of dry matter development when the beans fill and become solid.

Shade Management

Coffee is a shade loving crop and therefore different type of trees are planted e.g. Dadap, Bakain, willow etc. at close spacing (4 to 6 m) in comparison to tea.

Harvesting period in India : Harvesting period of coffee is given in Fig. 2.

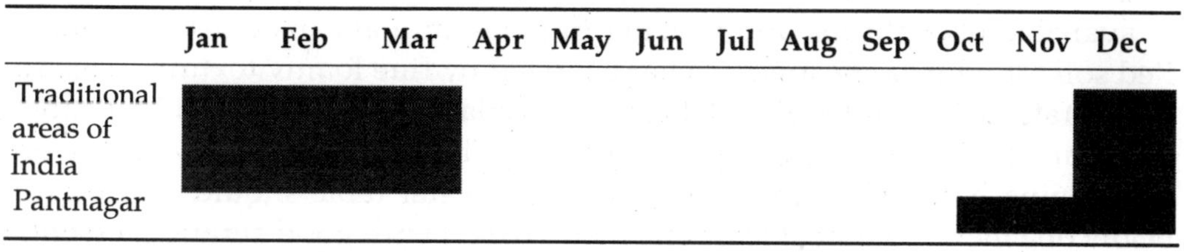

	Jan	Feb	Mar	Apr	May	Jun	Jul	Aug	Sep	Oct	Nov	Dec
Traditional areas of India												
Pantnagar												

Fig. 2. Harvesting time for coffee in Indiae

Table 5. Effect of Salix spacing and fertilizer combination on the four year old coffee yleld under shallow water table condltlons at Pantnagar.

Treatment		Coffee cherries/Bush (g)
Main plot shade tree spacing	S1: 5 X10M	3635
	S2: 5 X 6M	1645
	S0: control	500
	C D at 5%	84
Sub plot fertilizer application	F1: N only	3162
	F2: N + K	3590
	F3: N +P + K	2150
	F4: control	1450
	C D at 5%	205

In Uttarakhand, research work done by Pantnagar scientists has shown that four year old bushes can give maximum yields of about 4-6 kg fresh cherries/ bush planted at 2.5 x 2.5 m. One kg coffee powder can be prepared from one bush whose market value is about Rs.1000/kg.

Rubber cultivation

Natural rubber is grown in the states of Kerala, Tamil Nadu, Tripura, Assam, Megahalaya, Nagaland, Mizoram, Manipur, Goa, Andman & Nicobar Islands and coastal Karnatka. Rubber plantation are spread over 5.97 lakh hectares in the country. India has become the third largest rubber producer in the world with an average productivity of around 1796 kg/ha. During the year 2009-10, the production of natural rubber was projected 8, 61,000 tonnes which was reduced to 8, 59, 000 tonnes because of expected decrease in production due to dry weather. Kerala accounts for 82% of the total area under rubber plantation in India.

Soil

Soil in the rubber tract is generally highly weathered and consists mostly of laterite, lateritic types. Sedimentary types and nonlateritic red and alluvial soils are also seen in some non-traditional areas. The laterite and lateritic soils are mostly very porous, well drained, moderately to highly acidic, deficient in available phosphorus and varying in potassium and magnesium content. Red soil found in some areas is characterized by fine loamy texture. This soil is generally acidic and highly deficient in available phosphorus.Soil for rubber cultivation should have a minimum depth of one metre without any intervening hardpan or impenetrable layer. Water table should also be well below one metre so that at least one metre of soil with good aeration, essential for root penetration is available.Well-drained soil is essential for optimum growth and yield of rubber plants. In marshy areas, owing to poor physical properties and waterlogged conditions growth of rubber is always found to be very poor.

Climatic conditions for optimum growth of rubber tree

- Rainfall of 2000 to 3000 mm evenly distributed without any marked dry season and with 125 to 150 rainy days per annum
- Maximum temperature of about 29°C to 34°C and minimum of about 20°C or more with a monthly mean of 25 to 28°C
- High atmospheric humidity of the order of 80%
- Bright sunshine amounting to about 2000 h per annum at the rate of 6 h per day through all the months

- Absence of strong winds
- Only a few regions in India meet all these requirements. Fortunately rubber can be grown successfully under moderately deviating conditions too.

Propagation

Generally rubber is propagated by seed and bud grafting. However, it is also multiplied by micropropagation.

Planting

1. Seed-at -stake planting- Under *in-situ* planting this method is practiced in which 2-3 germinated seeds may be planted per pit in line or triangle. The plants are thinned out later. 2. *Stump planting-* The seedlings after pruning the stem at height of 45-60 cm from the collar are planted. The leaves appearing on such plants tolerate adverse conditions. Recommended plant density is 420 to 445 trees /ha. Selective thinning should be carried out at in initial of tapping and by 16th year of planting (i.e. 3rd year of tapping) thus stand per hectare may be brought down to 333 trees/ha for both methods at a spacing of 5m x 6m.

Irrigation

Rubber thrives well, if the rainfall is evenly distributed throughout the year and amounts to 1750 mm year. Rubber performs best at 24-35^{0}C provided there is a adequate moisture. Seedlings should be mulched during the late rains or just before the commencement of dry season.

Intercropping

Intercrops should be planted atleast 1.5 M away from plant bases. Intercrops e.g. banana, ginger and turmeric, medicinal plants, tuber crops etc. should be manured as per their recommendations.

Chemical weed control

Weeds can be controlled with the use of herbicides. There are two main types of herbicides, the pre-emergent and post-emergent.

Pre-emergent herbicides

Pre-emergent herbicides are soil-applied chemicals used for killing weeds before they emerge from soil. In newly prepared planting strips, the application of pre-emergent herbicides immediately after planting can check the weed growth for 6-7 weeks.

Table 6. Herbicides, their dosage and dilution

Name of the chemical (Brand name in brackets)	Dose/ha	Water required for application
Diuron (Karmex, Klass)	2.5 kg.	700 litres

Post-emergent Herbicides

These herbicides are applied on the weeds and are used to kill the weeds, which have emerged and are actively growing.

Table 7. Recommended doses and combinations of herbicides

Name of the chemical (Brand name in brackets)	Dose/ha	No. of rounds per year and interval required between rounds	Target weed species	Volume per ha
Paraquat (Gramoxone) and 2,4,D (Fernoxone)	2.25 litres 1.25 kg	4 rounds at an interval of 6-8 weeks	Broad leaf weeds like Chromolaena sp. Boerhhavia spp.	500-600 litres
Glyphosate (Glycel Weed off or Round up)	2 litres	2-3 rounds at 3 months interval and spot application of 0.5 to 2.5 litres Gramoxone per ha	Grass weeds and, overall weed control	400 litres

Mature rubber under tapping

The general recommendation is 30-30-30 NPK kg/ha. This can be supplied through mixing straight fertilisers or through mixtures. The quantity of the same is provided in an earlier session (manuring from 5th year till attainment of maturity for well maintained plantations). Discriminatory fertiliser application based on the results of soil and leaf analysis should be practiced for mature rubber wherever it is possible.

Rubber wood and other by-products : Besides providing raw material for the rubber industry, the rubber tree also provides various by product such as rubber wood, honey and rubber seed oil.

Rubber wood

The continued search for wood led to the utilisation of secondary species of wood like rubber wood, a plantation timber. With over 5,30,000 ha under rubber, India is now the third largest producer of natural rubber in the world. At the age of 22 to 29 years, latex production becomes uneconomic and the trees are then cut and replanted. Thus the rubber plantation is a sustainable source of rubber as well as timber, contributing positively to the environment. In India rubber is predominantly a small holders' crop and over 87% of the rubber is from this sector. A rubber tree from small holdings will have about 0.57 m^3 (20 CFT) of timber and the per ha yield is about 150 m^3 (5295 CFT). The stem wood has timber value of only 60%. Present total availability of rubber wood is estimated as 0.94 million m^3/yr and is estimated to be 1.5 m^3/yr times by the end of the decade.

Physical properties : Most of the strength properties of wood are related to its specific gravity.

Mechanical properties

The mechanical properties determine the ultimate capacity of the wood to bear the forces in bending, tension, compression, shear etc. in the three directions or planes. It is seen that under dynamic conditions in bending, the load bearing capacity increases (Table 8).

Hardness on different surfaces

Hardness of the wood is very important for applications like flooring and decking where abrasion of surface takes place. The table below gives the hardness as measured by indentation of a steel ball. It can be seen that dry rubber wood has better hardness than teak.

Rubber wood is growing in importance, it was originally perceived merely as useful by-product, which was used to dry and smoke rubber and to provide a source of charcoal for local cooking. Now *Havea* is being grown with the preconceived concept of timber harvesting, with the latex being harvested as a secondary consideration. Nevertheless rubber wood continues to grow in importance as a significant by-product from natural rubber production (especially in Europe) as a source of chipboard and pulp for paper making. After completion of economic cycle of latex, rubber plantation yields about 200-250 m^3 of timber / ha. In India about 10% of the rubber wood is value added and this has got tremendous export potential too. Countries like Malayasia and Thailand have achieved significanct progress. India has to utilize wood more effectively for which appropriate infrastructure development is required.

Honey

Rubber is an excellent source of honey. It is obtained from the extra floral nectarines at the tip of the petiole, where the leaflets join. Honey flow period is between the months of Jan. to March. The average production of honey from rubber plantation is about 2500-3000 tonnes/year. However, the potential is about 8000 tonnes/year.

Table 8. Weight and specific gravity of rubber wood

Physical parameter	Rubber Wood		Standard Teak	
	Green	Dry	Green	Dry
Specific Gravity = Wt. oven dry/Vol. at Test	0.521	0.557	0.596	0.604
Moisture content (%)	81.2	12	76.6	12
Weight, kg/m³ at corresponding moisture	944	624	1056	676

Table 9. Hardness rubber wood and teak surfaces

Species and locality	Condition of material (Moisture content, %)	Indentation in Kg			
		Radial	Tangential	End	Average
Rubber Wood	Dry (12%)	549	526	627	567
Teak	Dry (12%)	502	524	488	505

Table 10. Wood working properties

Condition	Sawing		Planing		Boring		Turning	
	Re-sawing	Cross cutting	Ease of planing	Quality of finish	Ease of boring	Quality of finish	Ease of Turning	Quality of finish
Green	Slightly difficult	Easy	Easy	Smooth	Easy	Rough	-	-
Dry	Moderately easy	Easy	Easy	Smooth	Easy	Rough	Easy	Rough

Table 11. Working qualities of wood

Species	Overall performance	Ease of working	Working quality index	Comparative performance in turning
Rubber Wood	194	100	131	101
Teak	100	100	100	100

References

Anonymous. 1989. New Kangra Tea Clones. Tea Experiment Station, HPKV, Palampur

Baggio,A.J., Caramori,P.H., Filho,A.A. and Montoya,L. 1997. productivity of southern Brazalian coffee plantation shaded by different stockings of *Grevillea robusta. Agricultural Systems*.37: 111-120.

Carr, M.K.V.1971. The climatic requirement of the tea plant: A review. *Expt. Agric.* 8:1-14.

Carron, M.P., Enjalric, F., Lardet, L. and Deschamps,A. 1989.. Rubber (H. brasiliensis Muell.Arg.), In: *Biotechnology in Agriculture and Forestry*. Vol.5, Trees II, (ed. Bajaj, Y.P.S.), Springer-Verlag, Beearlin, Heidellberg, pp, 222-245.

Clowes,M. St.J. and Logan, W.J.C.1985. Advances in coffee management and technology in Zimbabwe. Harare, Zimbabwe: Coffee Growers Association.

Farooqi, A.A. and Sreeramu, B.S. 1999. In: *Tropical Horticulture*. Vol. 1. (eds. Bose,T.K., Mitra, S.K., Farooqi,A.A. and Sandhu, M.K.) Naya Prokash Publishers, Kolkata,pp. 521-537.

Graaff, J.De. (Ed.). 1986. The economics of coffee for Agriculture. *Publishing and Documentation, Pudoc wageningen, Netherlands,* pp. 1-24.

Hadfield, W.1974. Shade in north-east India plantation.II. Foliar Illumination and canopy characteristics. *J.Appl. Ecology*. 102-123.

Murty, R.S.R. and Sharma, V.S. 1986. Canopy architecture in tea (*Camellia sinensis* L.). *J.Plant.Crop*. 14: 119-125.

Opeke, L.K.1982. Latex producing plants. In: *Tropical trees* (ed, Opeke, L.K.), pp. 215-240.

Varghese,Y.A. Thulaseedharan,A.and Kumari Jayasree,P. 2001. Rubber. *Biotechnology of Horticultural crops*, Vol.1 (eds. Parthasarthy, V.A., Bose, T.K., Deka, P.C., Das, P., Mitra, S.A. and Mohandas, S.) Naya Prokash Publishers, Kolkata, pp. 630-655.

□□□

27

Scope of Floriculture in Commercial Agriculture

AJIT KUMAR KAPOOR

Over the years, horticulture has emerged as an indispensable part of agriculture offering a wide range of choices to the farmers for crop diversification. This sector contributed about 24.5% of the agricultural GDP from about 8% of the area. Flowers are being cultivated in India from time immemorial. Flowers were cultivated for aesthetic purposes as also for their fragrance, perfumes and extraction of medicines. However, large scale cultivation for commercial purposes and protected cultivation inside polyhouses is relatively new in India. Changing life style of people has led to the commercialization of flower cultivation. The huge demand for flowers coinciding with various occasions with the increasing habit of "Saying it with flowers" during all the occasions has led to growth of market for flowers. According to a report of National Horticulture Board 2010, the total area under flower crops was estimated around 1,67,000 hectares, which included more than 75% area under traditional flowers such as marigold, jasmine, aster, rose, chrysanthemum, tuberose and less than 25% area under modern flowers like carnation, rose, gerbera, chrysanthemum, tulip, gladiolus and anthurium. For the last 12 years (1995-96 to 2007-08), there is 100% increase in area and >230% increase in loose flower production and 480% increase in cut flower production. India's exports of floricultural products are over

Rs. 300 crores and domestic market is over Rs. 500 crores. India's export earnings from floriculture have grown from a mere 1 crore in 1987-88 to over 300 crores in 2008. Yet the country's export is very small in the global trade (0.5%).

Floriculture sector is potential money-spinner for developing countries. As on date flower cultivation has attained industrial status. India produces a variety of flowers including roses and gladioli which are currently in great demand in the international market. The agro climatic conditions are highly suitable for cultivation of high quality flowers in India. Relatively cheap and abundant labour availability favours India with a distinct competitive advantage in terms of prices. The rapidly growing world markets for flowers are holding out adequate inducement for India to expand her production and export of flowers. Moreover, India is an important source of supply of cut flowers for the European market in the winter months when on account of severe cold, the supplies of flowers from Holland and Germany dwindle down considerably. Floriculture today has diversified itself from mere cultivation of roses and chrysanthemum to scientific cultivation of *Lilium*, carnation, tulips, gerbera, *Anthurium*, gladiolus and orchids. Scientific cultivation of flowers today has become a business proposition. After liberalization, the Govt. of India identified floriculture as a sunrise industry and accorded it 100 per cent export oriented status. Owing to steady increase in demand of flowers, floriculture has become one of the important commercial trades in agriculture. Hence commercial floriculture has emerged as hitech activity-taking place under controlled climatic conditions inside greenhouse. Floriculture products mainly consist of cut flowers, pot plants, cut foliage, seeds, bulbs, tubers, rooted cuttings and dried flowers or leaves. The important floricultural crops in the international cut flower trade are rose, carnation, chrysanthemum, gladiolus, gypsophila, liatris, orchids, tulip and lilies. The main exporting countries are Holland, Columbia, Israel, Italy, Spain, Thailand, France and New Zealand. The major importing countries are USA, Japan, UK, France, Holland, Switzerland and Australia. The demand for cut flowers is growing steadily. Germany, Italy, UK, France, Holland and Switzerland contribute nearly 80% of the cut flower consumption in Europe. Rose is the most sought after flower in the Europe followed by carnation, chrysanthemum, orchids and gladioli. Major contributing countries are Holland, Columbia, Israel, Kenya, Spain, Zimbabwe and Italy. India has vast potential for development of horticulture especially floriculture on account of the following distinct advantages:

- Diversified agro climatic and heterogeneous soil conditions
- Plenty of skilled labour at relatively low cost.
- High irrigation potential.

- Round the year radiation.
- Little or no rains in winter.
- Improving domestic market
- Location of the subcontinent to meet the needs of western, eastern and middle east countries.
- Govt. acting as catalytic agent by declaring the sector as "Thrust Area".
- Increasing cost of production in the developed countries.

Diversification from traditional agriculture to growing carnations, gladiolus, lilium, chrysanthemum, alstroemeria, marigold, etc. gave returns over traditional crops as follows:

Annual income from traditional crops: Rs. 40,000 - 50,000 (per hectare)

Income* (Rs. per ha) from floriculture after diversification

- Carnations : Rs. 13,75,000.00
- Gladiolus : Rs. 05,32,500.00
- Lilium : Rs. 11,40,000.00
- Chrysanthemum : Rs. 1,50,500.00
- Alstroemeria : Rs. 17,40,000.00
- Marigold : Rs. 1,25,000.00

* Average of initial 3 years' production

India has a number of advantages for commercial growing of flowers which include:

- Diverse agro-climatic zones and rich floral diversity
- Youth power
- Good soil and abundant water
- Cheap and skilled man power
- Nearness to some potential markets like Japan, Russia, South-East Asia, Middle-East Countries.
- Huge domestic and international demand

Table 1. Area and production of loose flowers in different states/UTs during 2008-09

State/UT	Area (ha) (in 000 ha)	Loose flowers in 000 MT (tonnne)	Cut flowers (lakhs)
Tamil Nadu	29.10	233.70	-
Karnataka	26.00	203.90	5867
Andhra Pradesh	19.50	125.00	3
West Bengal	21.10	52.01	21232
Maharashtra	16.40	89.40	5728
Delhi	5.50	5.70	1038
Haryana	5.50	53.90	929
Uttar Pradesh	13.50	24.30	3467
Gujarat	9.70	49.50	5063
Himachal Pradesh	0.60	03.40	566
Punjab	01.70	82.00	-
Rajasthan	03.40	4.90	-
Madhya Pradesh	03.00	1.80	-
Bihar	00.20	2.30	11
Uttarakhand	00.90	0.60	2056
Jammu and Kashmir	00.10	0.011	20
Orissa	05.70	23.40	-
Others	04.60	31.60	1962
Total	**166.50**	**987.40**	**47942.00**

Source: Indian Horticulture Database - 2009

The liberalization of industrial and trade policy in July 1991 paved the way for the development of export oriented production of cut flowers. The Plants, Fruits and Seeds Order, 1988 known as New Seed Policy had already made it feasible to import planting material of international varieties. The government of India offers tax benefits to new export oriented floriculture companies in the form of income-tax holidays and exemption from certain import duties. Import duties have been reduced on cut flowers, flower seeds, tissue-cultured plants, etc. Setting up of walk in type cold storage has been allowed at the international airports for storage of export produce. Agricultural and Processed Food Products Export Development Authority (APEDA), responsible for export promotion and development of floriculture in India, grants subsidies for establishing cold storage, precooling units, refrigerated vans and greenhouses and air freight subsidy to exports. Eleven-model floriculture centre units and two large centres, 20 tissue culture units have been established by Ministry of agriculture, Govt. of India.

The floriculture industry comprises of

- The florist trade of traditional and contemporary cut flowers and cut foliage, both fresh and dried and value-added products like bouquets, floral baskets, flower arrangements and garlands.
- The plant nursery for propagation and supply of plant material including tissue culture plants, seeds, bulbs, corms and other propagated materials.
- Plant rental service for supply of house plants on annual rent for a specific period.
- Flower based perfume and medicines.

Potential floriculture in commercial agriculture: avenues and scope

1. *Loose flower production*: There is good demand in the domestic market for loose flowers which occupy nearly two-thirds of the total area under flower cultivation in India. This loose flower production sector is still unorganized with lot of scope for its improvement with the involvement of various agencies. The industry is characterized by sale of loose flowers like marigold, jasmine, aster, rose, chrysanthemum, crossandra and tuberose as well as export surplus of the modern cut flowers like carnation, rose, gerbera, chrysanthemum, tulip, gladiolus and anthurium.
2. *Cut flower production*: High-tech floriculture industry is still in its infancy. The corporate sector entered into this sector only in 1990's. As per capita income has gone up, use of modern cut flowers has also increased to a greater extent. The protected cultivation technology for the production of modern cut flowers like carnation, rose, gerbera, chrysanthemum, tulip, gladiolus and anthurium has been adopted from either Holland or Israel. The area under greenhouse cultivation is over 800 hectares in India.
3. *Ornamental foliage plants*: The ornamental foliage plants have a very good scope due to non-availability of ground space in the houses as well as open spaces in the major cities of India. This sector is coming up very well in India. The major production centers of ornamental foliage plants are located around Bangalore, Kolkata, Delhi, Pune, Lucknow, Thiruvanthapuram and Hyderabad. The entrepreneurs supply house plants including flowering plants on rental basis to commercial complexes, homes, offices, banks, shopping malls and other business centers in metropolitan and other major cities of India.

4. *Nursery industry*:

a. *Satellite nurseries*: In this case, small nurseries in the rural areas and at far flung areas are networked through a big nursery in the region for additional income to the growers and to uplift the socio-economic status of the people. The concept of satellite nurseries is well supported by International Fund for Agricultural Development (IFAD).

b. *Plug plant production*: Due to congenial environmental conditions, it is possible to produce large scale production of seedlings of annuals to raise the additional income. Many seasonal flowers can be produced in millions in the highly automated greenhouses for sale to the public and corporate houses.

c. *Corporate landscaping*: The concept of corporate landscaping is developing fast due to increasing aesthetic sense of individuals, industrialization and strict environmental regulations. The increased interest of the individuals in corporate landscaping has led to the blossoming of nurseries, landscape professionals, turf grass industry, etc.

d. *Plant tissue culture*: Plant tissue culture is an important aspect of plant biotechnology. India is making really very good progress in the field of biotechnology. Disease free quality planting material can be produced and multiplied in shorter time and in millions in a relatively very small area as compared to conventional propagation methods. Plant tissue culture has revolutionized the orchid industry globally.

e. *Flower seeds and bulbs*: There is huge demand of flower seeds and bulbs of various flower and bulbous crops throughout India. Large scale production of seeds and bulbs has been taken up in Punjab, Himachal Pradesh, Jammu & Kashmir, Ooty and Bangalore in Karnataka, Uttarakhand, West Bengal and Bihar. Economic liberalization has made it possible to import improved seeds and bulbs of international varieties of flower crops. Farmers of Punjab are even exporting flower seeds to other countries. Indo-American Hybrid Seed Company among other flower companies is exporting hybrid seeds of petunia, nicotiana, geranium and antirrhinum. Developed nations are doing contractual flower seed and bulb production business in collaboration with Indian counterparts due to favourable climatic conditions and availability of cheap labour.

5. *Floral extracts*: Floral extracts like pigments, essential oils, sapogenins, alkaloids, dyes, etc. have huge demand in the national and international market. The major flower crops for floral extract are rose, jasmine, tuberose, marigold, harshringar and vanilla. Damask rose is cultivated for the extraction of essential oil, rose water, attar, pankhuri, otto or

gulkand and concrete extracted from tuberose and jasmine finds application in perfumery and cosmetic industry. The demand for essential oils from flowers is increasing at around 10% per annum and there is great scope for cultivation of essential oil bearing crops for floral extracts.

6. *Pot pourri*: The flowers, herbs and spices have been used since time immemorial to add fragrance to our lives. Pot pourri is a mixture of dried, sweet scented plant parts including flowers, leaves, seeds, stems and roots. The basis of the pot pourri is the aromatic oils found within the plants. Pot pourri constitutes a major share in the export of dry flowers.

7. *Dry flowers*: Dry flowers constitute more than 60% of the total floricultural exports of India and globally the total share contributed by dry flowers is 15%. Any kind of plant and flower can be dried and can be used in making bouquets, greeting cards, flower arrangements, pot pourries etc. This dry flower industry is not well organized and indiscriminate collection of drying material is made from forests and specialized flowers and other plants for drying purpose are not grown in systematic manner in the country. The demand for dry flowers is increasing at the rate of 8-10% and hence there is great scope for the Indian entrepreneurs.

8. *Natural dyes from flowers*: Various plant pigments like anthocyanins, flavonols, carotenoids and xanthophylls are extracted from various flower crops and have manifold applications. In India, large scale cultivation of marigold is done for the extraction of xanthophylls and lutein which is a major portion of xanthophylls is being used in the poultry industry to enhance the egg yolk as well as the colour of meat. Likewise various other pigments extracted from flower crops have huge pharmaceutical and industrial applications and offer a great scope for crop diversification.

9. *Aromatherapy*: The concept of aromatherapy is gaining importance in India and hence hence huge scope exits for commercialization of agriculture through cultivation of plants with known therapeutic uses. Flowers are not only attractive to look, fun to grow, sweet scented but the essential oils from flowers have a variety of therapeutic uses. When the essential oils are used in bath or massaged, they are absorbed through skin and carried to the nervous and muscular system of the body for a healing effect. Aromatherapy is essential for stress relief, calmness and rejuvenation. This influence of aroma on the psyche has led some aromatherapists to practice 'psycho-aromatherapy' where oils are used to enhance mood of an individual.

Conclusion

The commercialization of agriculture/ diversification through floriculture and its products is socially viable and economically sustainable option. Now within floriculture itself, there are a number of options a flower grower can take up. With increase in the demand, production of flowers is expanding from major flower giants like the Netherlands, Columbia, Germany, Israel, Spain, Belgium, Thiland etc. to new growers in Latin America, Africa and Asia. Floriculture industry in India has many challenges, it can grow well with many advantages harnesses through a well planned strategy. Rapid improvement in infrastructural facilities and use of IT in floriculture in the liberalization of market under WTO regimes has bright future for Indian growers for commercialization of agriculture for getting enhanced returns per unit area.

□□□

28

Commercialization of Mushroom Farming

K.P.S. KUSHWAHA

Mushroom farming in India had a humble beginning in lat sixties when a few a progressive farmers in Himanchal Pradesh and Kashipur started growing button mushroom on commercial scale. Notable among the pioneer growers were the Teg's mushrooms. Chail (HP), Saigal Mushroom Farm, Srinagar (Kashipur) and Col. Kak's Mushroom Farm, Srinagar (Kashmir). In seventies, medium scale commercial mushroom growing was taken up by many more mushroom grower was taken up by many more mushroom growers in Kashmir, Himanchal Pradesh, Hilly region of Uttar Pradesh (Now Uttarakhand) and TamilNadu. Mushroom growing took the shape of a cottage industry and village after village was involved in mushroom cultivation in and around areas of Srinagar city. By early eighties button mushroom cultivation was also adopted by farmers in Haryana and UP in area around Delhi as a seasonal crop in winter and Delhi provided a big ready market for the producers. Delhi and Mumbari together form a big market for fresh mushroom scale in India and these two cities have a capacity to utilize 11-12 tons of fresh mushrooms per day. The other cities of importance for fresh mushroom scale are Bangalore, Bhubaneswar Jamshedpur, Kolkata, Chennai, Hyderabad, Goa, Ahmedabad, Lucknow and other such cities with a single health conscious and vegetarian population.

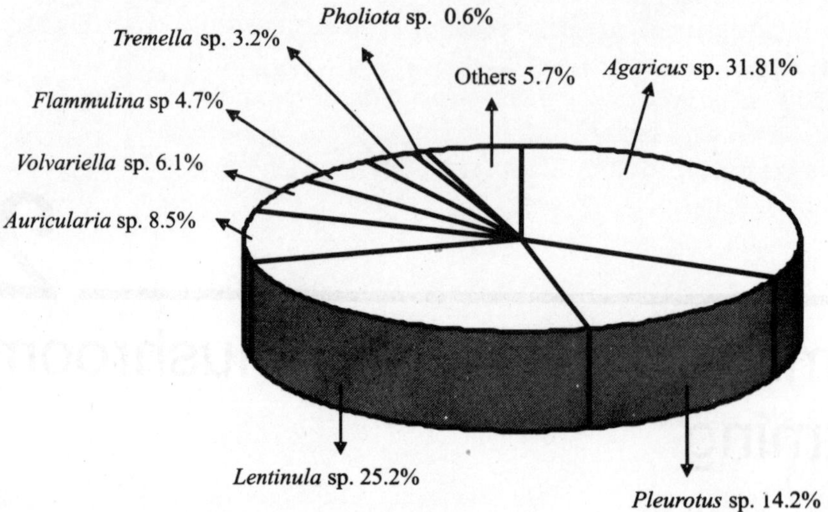

Fig. 1: Share of different type of mushrooms in total world production (6.16 million tons, 1997)

From the early nineties, export oriented units were put up by corporate houses/industrialist in Length and breadth of the country with use of advanced machinery for mushroom growing. In a span of 3-4 year in the 21st century, the mushroom production in India has increased and present mushroom production stand at 70,000 tons.

Production scenario

i) International

Mushrooms are cultivated in more than hundred countries with an estimated total production of 6.16 million tones (1997). Overall, world mushroom production increased more than 12% annually from 1981 to 1997. However, production of *Agaricus* mushroom decreased in the percentage of world total production from 71.6% in 1981 to 31.8% in 1997, even though the actual production of this mushroom increased from 0.9 million tons in 1981 to 1.95 million tons in 1997, an increase of 2.2 times. The reason is that other mushrooms gradually become popular, particularly *Lentinula edodes,* which increased both in percentage of world production, from 14.3 to 25.2%, and in output, from 0.18 million tons to 1.56 million tons, and *Pleurotus* from 2.8 to 14.2% and from 0.035 million tons to 0.87 million tons, respectively. China has become an enormous producer and consumer of cultivated edible mushrooms and also a major producer of medicinal mushrooms. Total production in China in 1997 was 3.91 million tons, which amounted to 63.6% of the total world output. *Lentinula edodes* is a leading cultivated mushroom in China. Its production was 1.39 million tons in 1997, which is equal to 35.6% of the total Chinese mushroom production.

The largest importer of Pre-prepared and preserved mushrooms is Northern America with about 81,000 tones in 1996 followed by EU, importing about 62,000 tones. Most of these supply come from China & East European countries, while as Indian mushroom have an insignificant presence. It is rather interesting to note that USA despite producing about 3,57,000 tones of mushrooms, further imports 1,00,000 tones, thus becoming the largest consumer of mushrooms in the world. For India to, USA is the major market for its mushrooms to concentrate on. Though China has been the largest exporter to USA but due to antidumping duty imposed on China, their presence is likely to suffer, which could be an emerging opportunity for India to increase it's exports to USA. The year -wise supply of mushrooms to USA from 1995-97 is given in Table 1:

Table 1: Mushroom Quantity (M.T.) of Exports, By Leading Countries to USA

Countries	1995	1996	1997
China	34,198	31,840	33,557
Indonesia	16,826	14,210	16,595
Hong Kong	4,146	2,550	2,196
Chile	5,069	4,245	3,026
Taiwan	2,646	2,929	3,011
Mexico	1,754	2,597	2,666
Netherlands	2,464	702	600
India	**2,749**	**2,015**	**4,566**
Others	8,938	8,143	8,091
Total	**78,790**	**69,229**	**74,310**

The demand supply gap for mushrooms in Northern America is likely to go up form 1,26,962 MT 1997 to 2,73,971 MT by the year 2006 the unfortunate situation which can affect Indian supply to some extent is that the American consumers are switching- over from process of fresh mushrooms and it is difficult for India to export mushrooms in fresh form because of the distance.

ii) National

Mushroom farming in India is hardly 3 decades old and with initial lag-phase it has started showing upward trend. From a meager 4000 tones production during 1985, India's total mushroom production reached 30,000 tones in 1995, which has further increased to 40,000 tones in 1996-97 and 52,620 tones in 2000-01. Similarly its productivity has increased from 6kg/q compost in the year 1970 to 22kg/q compost during 1995. However, on some other fronts like diversification of species, adoption of improved technology/ strains and consumption of mushrooms by masses, India has more or less remained static or has shown very sluggish growth. For example, even today, white button mushroom (*Agaricus bisporus*) comprises over 90% of the total mushroom production in our country. Although, artificial cultivation of paddy-straw mushrooms (*Volvariella* spp.) and oyster mushroom (*Pleurotus* spp.) were standardized long ago and a large part of the country having tropical/ subtropical climate are more suitable for cultivation of these two types, yet their adoption by the growers are yet to attain the desired level. This is despite the fact that cultivation of these two types requires much less monetary inputs and their cultivation technology are much simpler than that of white button mushroom. In contrast, China, our great neighbors, could increase the production of oyster mushroom by 400% during late eighties and even today China is the leading producer of that mushroom. India is yet to achieve such a landmark in case of any of these mushrooms, although, it possesses all the necessary inputs for such a breakthrough, viz, favorable climate, abundant agro-wastes, cheap man-power, a largely vegetarian population with rampant protein-malnutrition, etc.

On the technology front, there is a mixed situation in our country due to the fact that both seasonal and climate controlled cultivation of button mushroom are in vogue simultaneously. In most of the Export-oriented farms, improved/imported technology and strains with ample mechanization is being practiced, while in seasonal farms, even the old and outdated technology is not being properly practiced. In between these two extremes, there are some small climate controlled farms, which practice indigenous technology/strains with medium results. Obviously, there is a considerable variation in the productivity as well as quality of mushrooms being produced in these farms, which also reflect on the profitability of these farms. All these have together hampered the growth of mushroom industry in a sound footing in our country, which in turn has arrested the increase in domestic consumption of this nutritious food creating thereby unnecessary marketing problems. States-wise estimates of production of mushroom during 1996-97 and 2000-01 (Table 2) although indicate that most of the States have ventured into this non-traditional farming activity but no effort to develop the domestic marketing consumption of this highly perishable commodity has been made in any of these States either by the private or public sector. Most of the big units have therefore to depend entirely on International market to sell their produce at whatever price and terms they get from their foreign buyers.

Table 2. State wise production of cultivated mushrooms in India

Name of the state	Production (MT)			Type of mushrooms
	1996-97	2000-01	2003-04	
Jammu & Kashmir	400	400	500	*Agaricus*
Punjab	4000	13,000	22500	*Agaricus*
Himanchal	500	600	2500	*Agaricus*
Haryana	4000	5000	6000	*Agaricus*
Uttar Pradesh	4000	2000	3000	*Agaricus/Pleurotus*
Uttarakhand	-----	2640	4050	*Agaricus*
Madhya Pradesh	2000	2500	2800	*Agaricus/Pleurotus*
Rajasthan	50	80	150	*Pleurotus/Calocybe*
Maharastra	7000	7000	7500	*Agaricus/Pleurotus*
Goa	1000	1000	1000	*Agaricus/Pleurotus*
Karnataka	600	700	750	*Agaricus/Pleurotus/Volvariella*
Kerala	100	300	500	*Pleurotus/Volvariella*
Tamil Nadu	10,000	11000	11500	*Agaricus/Pleurotus/Volvariella/ Calocybe*
Andhra Pradesh	2500	3000	3500	*Agaricus/Pleurotus*
Orissa	700	700	800	*Pleurotus/Volvariella*
West Bengal	250	300	400	*Agaricus/Pleurotus*
North Eastern States (Including Sikkim)	1000	1200	1250	*Agaricus/Pleurotus/Volvariella/ Lentinula*
Delhi/Pondichery/ Chandigarh/Others	1000	1200	1300	*Agaricus/Pleurotus*
Total	**40,000***	**52,620****	**70,000*****	

Source: AICMIP Workshop at Solan, 2000*, Coimbatore, 2002** and AICMIP workshop 2004***

Mushroom growing systems

A. Marginal scale

- *Crop Rooms (Huts):* Made up of Sarkanda, Bamboo, Straw and grasses
- *Crop Room/Hut size*: 40′x20′x12-14′ or 35′x18′x10-12′
- *Containers*: Shelves or racks of Bamboo and Sarkanda
- *Composting*: Long method
- *Yield*: 10-12 kg/100kg compost in 8-10 weeks of harvesting

B. Small scale

- *Crop Rooms*: Conversion of old buildings into crop rooms or insulated crop rooms

- *Crop Room size*: 40'x18'x12'or 50'x21'x12'
- *Containers*: 3 to 5 tires bamboo shelves or metallic racks for 10 to 12 kg compost in bags
- *Composting*: Long method/Short method
- *Yield*: 10-15 kg/100kg compost in 8-10 weeks of harvesting

C. Industrial scale

- *Crop Rooms*: Insulated and controlled crop rooms
- *Crop Room*: 48'-100'x18-27'x12-18'
- *Containers*: Metallic racks for bags/ shelves
- *Composting*: Short method
- *Yield*: 18-22 kg/100kg compost in 5-6 weeks of harvesting

Export of mushrooms from India have started looking up only in the nineties due to addition of preserved mushroom in the exportable quota. Prior to that, mostly dried collections from forests comprising of morels/oyster were being exported from India. Table shows that there has been a tremendous increase in mushroom export from 1990-91 to 1997-98 and their value has exceeded 50 crores of rupees. However, the major share (>80%) of the export has gone to USA and there is an urgent need to expand and diversify the area of foreign market for Indian mushroom. In the international market, our main competitor have often been found to be lacking in quality, yet their cost of production being lower, their produce are proving more competitive in the world market. Indian mushroom exporters have therefore to face double challenge-viz. maintenance of good quality and keeping the cost of production competitive.

Nutritional and medicinal values

Today, the popularity of mushrooms is due not only to their culinary value but also to their potential as a source of protein that can enrich human diets especially in some developing countries where meat may be rare and expensive. Mushrooms contain more protein than either fruits or vegetables. They can be eaten, as they are cooked or raw, unlike other protein sources such as soya. Mushrooms are also low in cholesterol. Besides their protein content, mushrooms are also high in certain vitamins such as B, C, D, riboflavin, thiamine, and nicotinic acid. Mushrooms are also a good source of iron, potassium and phosphorus in addition to folic acid, an ingredient known for enriching the bloodstream and prevention deficiencies.

Table 3. Proximate protein content (dry weight) of edible mushrooms as reported by different authors

Species	Protein content (%)
Volvariella volvacea	21.32
Agaricus bisporus	27.8
Pleurotus ostreatus	27.4
Pleurotus florida	37.19
Pleurotus sajor-caju	36.94
Lentinula edodes	17.5
Auricularia auricular-judae	8.1
Flammulina velutipes	21.9

Table 4. Vitamin content of some edible mushrooms

Mushroom	Thiamine	Riboflavin Content (mg/100g air-dried)	Niacin
Agaricus bisporus	1.1	5.0	55.7
Lentinula edodes	7.8	4.9	54.9
Pleurotus florida	0.35	2.97	64.88
Volvariella volvacea	0.32	1.63	47.55
Pleurotus sajor-caju	1.16 - 4.8	-	46.108

Medicinal importance of mushrooms

The invention of the so called "wonder drug" penicillin was a landmark in the field of medicinal uses of fungi. Since then several fungi have been well recognized for their antifungal, antibacterial, antiviral, anti-tumour and many others such properties of pharmacological values. In the recent past a variety of medicinal preparations in form of tablets, capsules and extracts from mushrooms have been produced and marketed. Mushrooms are perhaps the only fungi deliberately and knowingly consumed by human beings and they complement and supplement the human diet with various ingredients not encountered in or deficient in food items of plant and animal origin. Besides, chemical composition makes them suitable for specific group suffering with certain physiological disorders or ailments. Mushrooms are regarded as an ultimate health food, low in calories due to presence of good amount of quality protein, iron, zinc, vitamins, minerals and dietary fibres which protects from digestive ailments and strengthening of the human immune system. Recent investigations have proved the empirical observations of the oriental herbalists that certain mushroom possesses very useful medicinal attributes. In the 1991, the value of world medicinal crops was estimated at 8.5 billion dollars and in the same year 1.2

billion dollars are estimated to have been generated from medicinal products from mushrooms. This was based on the sale value of products from *Coriolus, Ganoderma, Lentinula, Schizophyllum* and other mushrooms.

Although the biggest use of mushroom has traditionally been for reasons of their gastronomic and nutritional appeal. There has always been interest in certain mushroom for their medicinal attributes. Production of medicinal mushroom is now a days increasing over world wide. In the present era a variety of proprietary product based on mushroom nutriceuticals and pharmaceutical have already been produced and marketed. Various mushrooms and their metabolic extract have been reported to protect against cancer, tumor and pathogenic microorganisms. It is suggested that regular consumption of different mushroom varieties not only protects humans from heart trouble but also had medicinal potential for certain ailments.

White Button Mushroom (*Agaricus bisporus*)

Favourable season : Oct. to March (for plains of India)

Required temp. and humidity:14-22^{0}C and 80-85%

Preparation of compost

Unlike other traditional crops soil is not the appropriate substrate for mushroom cultivation. Rather, the substrate for mushroom called compost, is prepared from agro wastes like straw, stem, shoot, apices etc. with organic manure. Mushroom substrate may be simply defined as a lingo-cellulosic material that supports the growth, development and fruiting of mushroom mycelium. This compost is pasteurized by various micro-organisms and at appropriate temperature range. Essential supplement are also added/ supplemented to the compost. The whole process is termed as composting. Generally composting refers to the piling of substrates for a certain period of time and the changes due to the activities of various micro-organisms, which result in a composted substrate that is chemically and physically different from the starting material. The compost provides nutrients, minerals, vitamins and ions required for proper growth of mushroom. This compost supports the growth of only the mycelium of button mushroom and prevents that of other competitive moulds.

Spawning

The process of mixing of the spawn in the compost is known as spawning. Spawn is thoroughly mixed in the compost at the rate of 600-750 gm per 100 kg of compost (0.6 - 0.75%). The spawned compost is filled in tray or polypropylene bags covered with formalin treated news papers. In case of bags, they should be folded at the top and covered up. After spawning, temperature and humidity of crop room should be maintained at 18-22°C and 85-90%, respectively. Water should be sprayed over the covered news papers, walls and floors of the crop room. After 12-14 days of spawning white mycelial growth is seen running the entire length of the tray/bag. This is then covered with casing soil on the surface.

Casing soil

The significance of casing soil is to maintain the moisture content and exchange of gases within the surface of the compost which helps in the proper growth of the mycelium. The pH of the casing soil should be 7.5-7.8 and must be free from any infection or disease. In our country casing soil is prepared from the following ingredients.

Two years old manure + garden soil	3:1
Two year old manure + garden soil	2:1
Two year old manure + spent compost	1:1
Two year old manure + spent compost	2:1
Two year old manure + spent compost	1:2

Harvesting of crop

Pin head initiation takes place after 10-12 days of casing and the fruiting bodies of the mushroom can be harvested for around 50-60 days. The crops should be harvested before the gills open as this may decrease its quality and market value.

Productivity

From 100 kg compost prepared by long method of composting 14-18 kg of mushroom can be obtained. Similarly, 18-20 kg mushroom can be obtained from pasteurized compost (Short Method Compost).

Oyster mushroom (*Pleurotus sajor-caju*)

This mushroom gained importance during the last decade and now several species of Pleurotus are available for commercial production such as: *P.sajor-caju, P.florida, P.sapidus, P.eryngii, P.columbinus, P. cornucopiae, P. flabellatus. P. platypus, P. opuntiae, P. citrinopileatus*

It is now being cultivated in many countries in the subtropical and temperate zones. In China, it is known as abalone mushroom (*P. abalonus or P.cystidiosus*).

Pleurotus spp. can be grown using various agricultural waste materials. The different species of *Pleurotus* grow within a temperature range of 20⁰ to 30⁰ C. *P. sajor-caju* can tolerate temperature up to 30⁰ C although it fruits faster and produces larger mushroom at 25⁰ C. *P. fossulatus* is the so-called low temperature *Pleurotus*, fruiting mostly at 12-20⁰C. The tropical wastes like rice straw, wheat straw, corncobs, dried water hyacinth, sugarcane bagasse, banana leaves, cotton waste or sawdust are used for cultivation. The materials are usually soaked in water chemically sterilized with bavistin (7-10g) and formalin (120-150 ml)/ 100 litre of water for 16-18 hours. Extra water is drained off. The process of spawn making is the same as for *Agaricus* species. *Pleurotus* spawn should be about 15 days old when mycelium has formed complete coating around the grain. The normal rate of spawning in a pasteurized substrate is 2.0-2.5% of the wet substrate. The spawning is usually done thoroughly. Before filling the substrate in polythene bags, holes of about 1 cm diameter are made at 10-15 cm distance all over the surface for free diffusion of gases and heat generated inside. The optimum temperature for growth of *Pleurotus* spp. is 23 ±2⁰ C. Relative humidity in growing room should range between 85-90% during spawn-run. Usually 3 to 4 days after opening the bags, mushroom primordia begin to form. Mature mushrooms become ready for harvesting in another 2 to 3 days. An average biological efficiency (fresh weight of mushrooms harvested divided by dry substrate weight x 100) can range between 70-80% and sometimes even more. To harvest the mushrooms, they are grasped by the stalk and gently twisted and pulled. A knife should not be used. The mushrooms remain fresh for up to 3 to 6 days in a refrigerator/cool place.

Milky mushroom (*Calocybe indica*)

Calocybe indica is an edible white summer mushroom also known as milky mushroom. It can be easily grown in the temperature range of 25-35⁰C. It has moderate protein content and has a good biological efficiency (60-70%) under optimum conditions. Its sporophores have long shelf life. The major advantage is that it can be best fitted in relay cropping when no other mushroom can be grown at higher temperature. *Calocybe indica* has a very good scope for further cultivation and it can replace the other tropical mushrooms like *Pleurotus* spp. and *Volvariella* spp.

Auricularia spp.

Auricularia is world wide in its distribution in temperate and tropical regions. *A. polytricha* is a widely distributed tropical and subtropical species, while *A. auricula* is a temperate species that occurs occasionally in the subtropics. They grew densely on trunks of broad leaf trees of neem, mango, oaks, elm, poplar, banyan, locust-acasia or on dead wood. The fruiting body is gelatinous, elastic, semi-translucent, crop or ear shaped, red-brown in colour and rubbery in texture but becoming hard when dry. In traditional Chinese medicines, the wooden ear is mild and sweet in nature. It activates the blood and stops the pain. It is considered specific for bleeding, especially excessive uterine bleeding, abdominal and tooth pain. It is often used to treat hemorrhoids, to stimulate bowel movements, to help build up energy and to act as an anti-coagulant as well as a source of dietary fibre for the prevention of geriatric disorders.

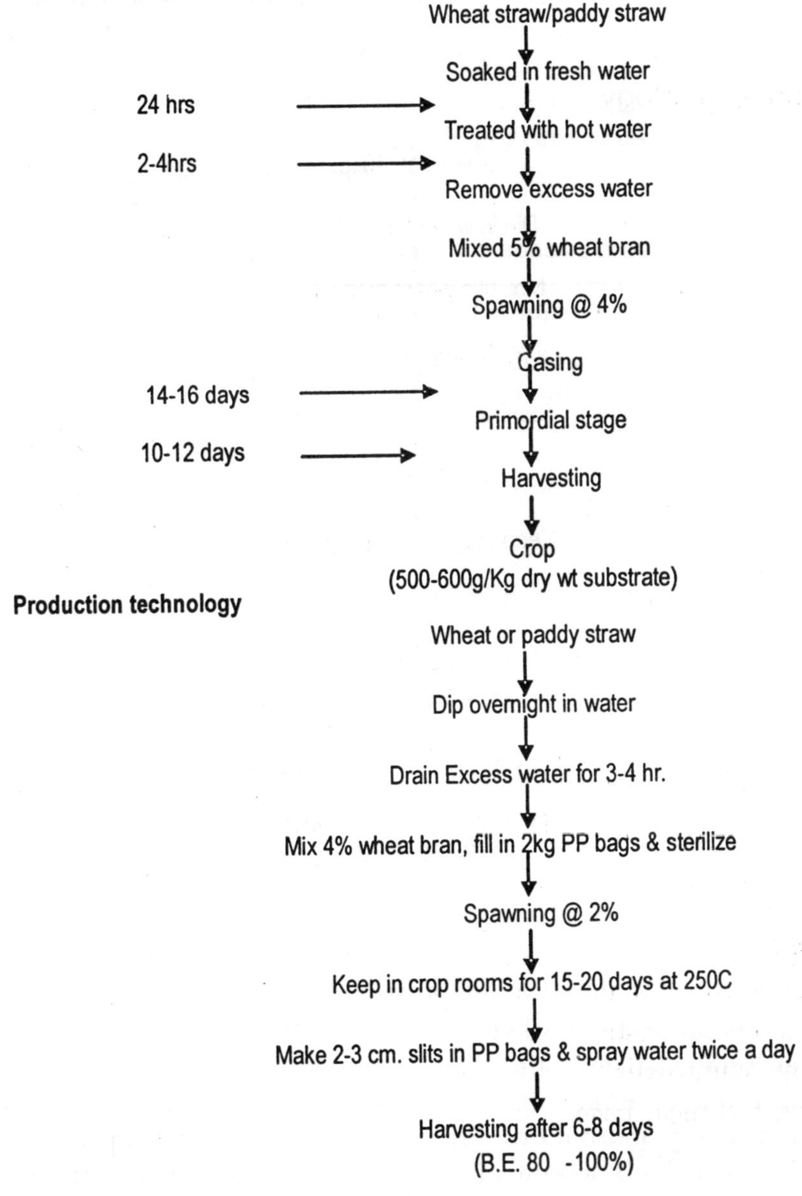

Lentinula spp.

It is the second most important edible mushroom in the world. In 1997, the total production was estimated to be 13,22,000 MT in terms of fresh equivalent weight, which was 25.4% of the total world production of edible mushrooms. This mushroom has been valued for its unique taste and flavour and as a medicinal tonic. This mushroom is capable of generating stamina, curing colds, improving circulation, lowering blood pressure and possessing antitumour and antiviral activities. It has been predicted that within a few years world production of *Lentinula* will surpass the world production of *Agaricus*.

Production technology

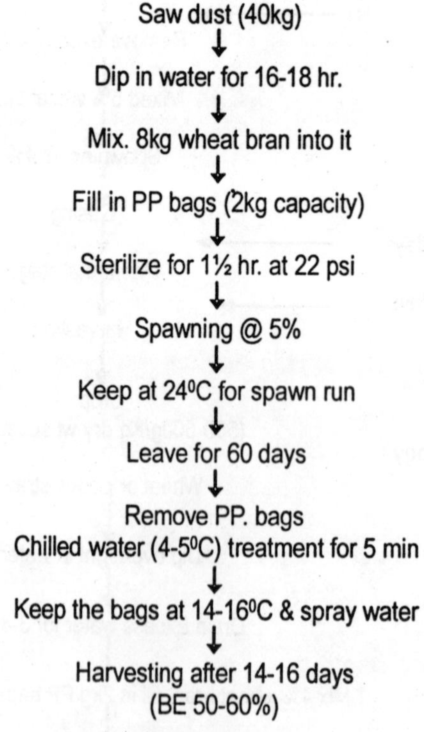

A. Strengths

- Abundant raw material availability.
- Enormous domestic market.
- Cheap manpower.
- Strong technical base.
- Strategic location.

- Climatic zones congenial for mushroom production.
- Established distribution network.
- Govt. support.

B. Weakness

- Much higher cost of Bank borrowings.
- Costly packing material.
- Poor infrastructure i.e. in adequate research, training centres & spawn, compost supply centres.
- Higher cost of energy.
- Inadequate procurement and processing units.
- High transportation costs.
- Improper marketing channels.

C. Opportunities

- Increasing domestic demand.
- Technology absorption achieved.
- Production in other countries declining.
- International trade barriers breaking down.

D. Threats

- China dumping everywhere at cheap rates.
- Demand shifting towards fresh & speciality mushrooms.
- Crash of sea money market & imminent devaluation of Chinese currency.
- Increase in cost of production due to transportation of material for installation of production unit and raw materials for its operation.

□□□

infestation in sugarcane and has been recommended as an alternative for chemical control of sugarcane shoot borer. It has also been reported that sugarcane trash proved better with minimum cost during limited maximum yield. But on the other hand, some times it can increase termite incidence as the organic matter content increases in the field.

i. Shoot borer infestation reduced significantly with trash mulching and incidence of egg parasitoid, *Trichogramma chilonis* comparatively increased with this practice. However, other crop residues such as wheat straw mulch can also increase cane yield but sugarcane trash was found relatively more effective [illegible] (Rana et al [illegible]).

ii. Besides conserving moisture during summer months, trash mulching also protected the crop from the damage by shoot borers (Parthasarathy 1959, Gill 1963, Rajani 1966). The reduction in shoot borer incidence due to trash mulching could be related more to the increase in sheath moisture (Avasthy and Tiwari 1986) than changes in organic matter content of soil.

iii. Spreading of a thick blanket of trash after germination greatly helps in suppression of shoot borer. However, this practice may increase the incidence of termite, armyworm and rat damage in areas where these pests are prevalent.

Earthing up and propping

Earthing up during formative phase helps the crop escape pest attack and also improve crop vigour in terms of cane height and number of millable canes, which ultimately contributes to higher cane yield than that obtained with insecticide treatment (Jaipal 2000).

i. Two to three light earthing ups during April and May covered the exit holes therefore, adults could not come out for egg laying which resulted in declining of the shoot borer population (Duhra and Sur 2002, Gupta 1945, Khan 1951).

ii. Three earthings at 8, 10 and 12 weeks after planting was found to reduce the incidence of shoot borer significantly over no earthing (Malik 1982).

iii. A light earthing up usually at 45 days after planting was effective to suppress root borer and shoot borer population (Khan et al 2008).

iv. Planting the crop preferably by trench method, earthing up the crop in end June and propping up the crop in end August has been reported to control stalk borer infestation effectively. According to Jaipal (2000) inundative releases of egg parasitoid during July-October and the crop

❑❑❑

Live-Stock based Farming Systems

29

Cattle Farm Management

SANJAY KUMAR

Livestock sector is the backbone of India's economy in terms of income, employment, equity sustainability and foreign exchange earnings. By virtue of traditional practice of utilizing crop residues as feed for milch animals, our cows and buffaloes not only produce milk but also produce draft animals that still power our agriculture and give the manure that is necessary to maintain the fertility of our soils. In order to obtain milk form cows and buffaloes in a commercialized manner, one has to attempt to a large sized dairy farm enterprise, whose management differs from peasant farmer' enterprise in many ways. The central theme of the farm management hovers around maintaining productivity; non-negative trend in total factor productivity in long run; natural capital stock over time and meeting the growing demands of present and future generations. India is now proudly ranking as world's largest milk producing nation. When we look at the high producing so called elite animals, we find their presence very negligible in numbers. However, their contribution cannot be ignored. Following strategy may become beneficial in long-term livestock production for the commercial dairy enterprisers of our country.

1. The foundation stock

The foundation stock for dairy enterprise should be so selected that it meets the requirements of age, type and criteria. The stock should be at young to adult stage, to enable the entrepreneur study of certain traits, take decision to replace them easily and at lesser discount. This would provide sufficient time to familiarize the animals with the new environment without causing unnecessary delay in production. They should be true to some breed type. This would ensure coordination of their general characters with production performance and also make ease in selection. They should fall within certain pre-decided selection index. This index can be formulated utilizing the purpose of farming, early age at maturity, regularity of reproduction, resistance to various tropical climatic stress variables, diseases, insects, etc., longevity and desirable production potential (based on the pedigree, relatives and self).

2. The housing

The farm should be located not close and not very far from the market of the produce and near to the team of care takers to enable them supervise the animals easily and frequently. The site should be at elevated and pakka ground for better drainage, cleanliness and durability. The livestock as such do not require costly housing. The floor should be non-slippery, comfortable and durable with slight gradient (say 1:40 in heavy rainfall and 1:100 in arid climates). The walls should be strong and at least 2.20 meter high. They may be replaced partially or wholly with pillars, as per the prevailing climatic conditions. In coastal areas, the wall should be of honeycomb pattern for easy air circulation. The roof should be light, strong, weather proof and be placed at good height (4-5 meters at highest point). This height would, in most cases, eliminate installation of electric fans. The slope should be 22° to 33°. The materials should preferably be thatch or asbestos cement corrugated sheets, fixed on iron pipes/ angle truss. The need for sunlight and natural ventilation is of utmost importance for the health and well being of animals. However, none of the animals should be exposed to intense sunlight. The animal houses must preferably have some trees (say Neem) on its sides to provide coolness during hot weather. The trees can also be used to house the animals during noon hours. The height, length, width and depth of manger and water trough are the important considerations for the comfortable feed and water intake of animal. Provision of open paddock is very essential for exercise, obtaining mild sunlight and for certain farm operations.

3. Feeding

After weaning, the roughages should constitute major bulk of the animals' diet. Preference should be given to local and seasonal roughages. Feeding wheat straw in combination with greens has been quite popular. At some places, where it was not possible to economically grow/ purchase green fodder, it has been witnessed to keep even high producing animals without greens (but on dry fodder and concentrates) for years together. Such a practice would heavily cut down cost on feeding livestock. The concentrates are the important components of the animals' diet. Its regular feeding reduces age at maturity, inter calving period and maintain production and vigor of the animal. Farmers have always been seen concerned about high cost of concentrate feeds or their components in the market. It is therefore, necessary to arrive at some least-cost formulation for the concentrate feed. Feeding of minerals in accordance with its requirements is of paramount importance to save unnecessary expenses. In this regards, it is advisable to ascertain certain deficiencies (of Iodine, Zinc, Copper, etc.) in the fodder producing soils before going for using mineral mixture of a given Brand. The minerals are regularly required through out the life of the animal. However, their requirements are more during pregnancy and lactation. It is desirable to provide feed at certain scheduled intervals (3-5 times a day) to allow animals get enough time for rumination/ chewing the cud. Scheduling of feeding also helps animals remain away from indigestion troubles besides savings on labor and supervision cost. The availability of drinking water should be so arranged throughout to enable them take it at their will. It would help in reducing stress to a larger extent and would be beneficial in maintaining production.

4. General management

Animals are required to be dealt with politely. Any sort of sudden change in feed, schedule, environment, attendant, etc. is liable to damage their behavior. This can cause decline in the production. Animals like grooming. This is an act, which keeps them healthy, provides good look, increase blood circulation and fetches more price while selling. Constant keeping of animals at one or two places make them lethargic. This sometimes becomes the precipitating factor for certain hoof disorders as well. A mild exercise in comfortable weather improves body condition of the animals. It is difficult to obtain a germ free environment. However, it is always desirable to go for the better conditions, particularly for the cleaning of floors, smooth drainage, disposal of excreta, provision of clean and fresh water and feed for the animals in order to remain free from diseases. Occurrence of certain diseases, like Mastitis, Navel-ill, parasitic and fungal infestations, etc. can be drastically reduced by observing cleanliness.

Calves being the future generation of dairy enterprise, their protection and good growth can take the business far ahead of those who pay less attention on this aspect. Calves must be provided with colostrum as early as possible and frequently during first few days of their life. It helps them boost their resistance to various diseases and also helps in removing the meconium, (first voided faces) due to its laxative action. Calves can easily digest colostrum/ milk up to one-tenth of their body weight per day. Navel-ill and tetanus are the commonest sufferings of the new born due to poor care and cause great loss to the business. Proper care of navel region (aseptically cutting the chord at 10 cm away from the body and its dipping in tincture iodine, keeping the calf on dry and clean bedding) during first few days of life can save the calves from heavy mortality. Calf Pneumonia can be controlled by providing them warm environment and avoiding forced drenching of oils or medicines. Similarly, the calves must be protected from parasitic infestations to enable them grow faster. The weaning of calf is the next important factor in deciding economics of the enterprise. It is practiced either at birth, later at some stage or never. The economics of the enterprise also varies according to the stage of weaning the calves. It is advisable to take the calf away from the dam right at birth and feed it colostrum artificially through spoon/ bottle/ pail. Such a practice proves beneficial in cases of calf death, helps in recording actual milk yield of the animal and brings the cow in heat early.

Milking : The sole aim of a dairy enterprise is to obtain larger quantity of quality milk regularly. Having ascertained all other pre-requisites for this ultimate aim, without paying much attention on milking and its scheduling probably may put one in embarrassing situation. For the average cow/ buffalo it is advisable to go for milking twice a day and for high producing animals thrice a day at equal intervals in comfortable environment. Milking should be complete and fast. It may be done by full hand method or by milking machine (at large farms). In both the ways, animal should feel enjoying the act. Generally one milker is needed for 12-13 cow and milking machine is recommended for large farms. However, maintenance of machine is of paramount importance to keep in smooth working order.

Mastitis is the most common disorder of great economic concern. It is normally seen in high producing animals. The causes are many. However, the most common cause is the poor management. The teat has two openings: the upper one – that opens into the mammary gland and the lower one – that opens to the exterior. These are controlled by sphincter muscles for opening/ closing and for the discharge of milk. The outer opening is more important for the occurrence of mastitis due to poor management. It remains open for a certain period after milking, and may take variable time to seal depending upon the individual and breed type (from 20 to 45 minutes, Dave and Singh, 1999). Indian cattle are able to seal it promptly. Thus, this duration of time is most important when the teat tip may get infected due to poor floor hygiene. It is, therefore, recommended to dip the teat in mild iodine solution and keep the animal stand for some time by providing her concentrate or let her loose for grazing or exercise.

Heat detection and breeding : Supervisors must know that non-pregnant animals cows/ buffaloes keep cycling after every 19-22 days. Hence, it is possible for every one to watch them during this period more carefully. Generally, buffaloes come into heat during early morning hours. A slimy silky substance may be seen lying on the ground at the site and also sticking on the tail. Cow in heat can be easily recognized by mucus discharge, bellowing and mounting symptoms. The cow and buffalo uniformly show drop in milk yield, frequent urination, swollen vulva and reddened mucus membrane as cardinal symptoms. If possible, it is advisable to take help of teaser bull for confirming the status. Missing of the heat symptoms means extra expenditure on rearing of animals for next 21 days. Breeding is the most important decisive factor in long-term sustainability of the dairy enterprises. The time of breeding of the animal must be accurately judged for. Breeding should never be taken in casual manner nor it be handed over to the novice. Majority of animals conceive when they are bred 12 to 18 hours after the onset of heat. One may opt for natural service or can avail artificial insemination (AI) services. In any case the potentials of the bull must be known and desirable.

5. Specific management considerations

- *Distancing the calf*: Economically, the calf is a burden on farmer where housing, labor and other inputs are costly. In such cases it is advisable to distance the calf from main herd with some reliable herdsman on mutually agreeable terms and conditions.
- *Quality product*: The selling price of the milk is the sole determining factor on the economics of the dairy enterprise. The price is dependent on the quality, which includes colour, odour, density, particulate organic matter, fat per cent and microbial load. Normally, the milk does not have objectionable elements in it when it flows out of the mammary gland. However, majority of these elements get into milk due to poor hygienic practices followed during milking and storage. A vigilant farmer can easily pay attention to these. This aspect would keep the farmer always in heavy demand.
- *Record keeping*: The importance of record keeping for follow-up, monitoring, economics and decision-making cannot be ruled out. These should be maintained regularly for each and every event and for each animal. A good record should be clear, understandable by others and analyzable.
- *Fixing the targets to achieve*: One should always try to lay down certain targets for the dairy herd to perform. These may be-Body weight of calves at birth; at 1, 3, 6, 12, 18 and 24 months; age at weaning; age at first oestrus, conception, calving; service period, lactation period, dry

period; milk, fat and SNF yield, microbial load, resistance to certain disorders–abortion, retained placenta, prolapsed uteri, mastitis, hoof disease, etc., soft or hard milker, behaviour-suckling the calf, vices, etc. The goals for efficient reproductive health programme include an average calving interval between 12 to 13 months, breeding efficiency below 1.5 services/ conception, and less than 30% non returns after 60 days, repeat breeder of 8-10% or less, abortion 1-2%, retained placenta 5-10% and metritis 5-10% or less

- *Labor*: Cost on labor is becoming higher. Hence, it is advisable to go for some sort of automation where ever it is possible. For this some areas, like fodder harvesting, chaffing and delivery, milking, cleaning the floor and premise, etc. could be thought of. Contract labor system is also now becoming common at livestock farms.
- *Mixed farming*: Wherever possible, one should try to go for the dairy enterprise in combination with some agriculture farming. It would enable the farmer use agricultural byproducts for feeding the animals, land for grazing, and parting of some agricultural labor/ machinery in dairy business during emergencies. At the same time dairy business can also mutually benefit the agriculture enterprise through manures, consumption of superfluous milk or its byproducts.

Table 1 : Floor space requirements for cattle and buffaloes

Type of animal	Floor space req./ animal (m^2)		Max. animals/ pen	Height of shed (cm) at eaves	Manger/ water trough (cm)			Height of inner wall
	Covered area	Open paddock			Length	Width	Depth	
Young calves	1.0	2.0	30	175 cm in medium and heavy rainfall areas, and 220 cm in semi-arid and arid areas.	20-30	30	15	15
Old calves	2.0	4.0	30		40-50	40	15	20
Cow	3.5	7.0	50		60-75	60	40	50
Down Calves	12.0	12.0	1		60-75	60	40	50
Bulls	12.0	120.0	1		60-75	60	40	60
Buffalo	4.0	8.0	50		60-75	60	40	50

- *Herd Size and Culling*: This ever dynamic factor in dairy enterprise is of most concern to the farmer. If herd size is to be kept constant, it is advisable to spot out poor animals at their early age and prompt actions be taken for their culling. Wherever replacement is required, it should be done with superior stock.
- *Stress control*: It is very important to consider installing stress control measures from inclement weather, flies and insects (wallowing pond,

showering, misting, fanning, netting) to help in producing optimum milk. The cost incurred on these measures is generally recovered early by improvement in milk production.

It is seen from the above account that for running a successful dairy enterprise, one has various aspects to consider, where it is easy to cut down the cost and increase productivity. The sustainability of the enterprise is possible through judicious application of knowledge and skills one puts in an efficient business planning, execution and monitoring.

□□□

30

Farm Scale Management of Poultry

R.K. SHARMA

Domestication of livestock began with the process of civilization of mankind. Poultry farming is known to have existed in India since pre historic times. The Indian Red Jungle fowl is considered as the ancestor of modern chicken breeds. Chickens were first domesticated in S E Asia (India, Southern China, Burma, Malaya region) between 5000–3000 BC and then spread westwards and then Africa, Europe and America. Poultry in India was a leisurely and backyard village venture till 1960. The entire poultry scenario of Indian sub-continent changed only in last 4-5 decades and now India ranks 3rd in egg production (3.9%) after China (45%) & USA (8%) and 5th in broiler production (3.3%) after USA (26.5%), China (17%), Brazil (15.5%) & Mexico (4.3%) with about 45,000 and 1500 million eggs and broilers produced annually. But in terms of per capita availability it ranks amongst the lowest. Poultry production systems of the developing nations including India can be broadly categorized into two distinct classes: one-low external input traditional family poultry production system or free range or backyard poultry in unorganized sector in the villages and the other one is high input industrial or intensive poultry production system in urban and peri-urban areas. The former is highly unorganized and consists of thousands of small holders (5-15 birds) where birds scavenge and forage on house-hold wastes, shrubs, weeds, crop residues, insects and aquatic plants with no or little external inputs. The birds are looked after by the families specially women and children. The locally adapted indigenous chicken germplasm thriving well under stressful and low input production system in the countryside may appear to have low productivity in comparison to highly specialized industrial germplasms but it is highly efficient converter of available low quality feed and other resources which other wise are likely to go waste. Native birds are hardy and resistant to most of the diseases/ parasites. The industrial or intensive poultry is highly organized and uses state-of-the-art technology. Major constraints of village poultry are: non-availability of appropriate germplasm, high incidence of early mortality, lack of paravet facilities, and lack of knowledge regarding hygiene and shelters. Village poultry suffers from lack of viable technology, and suitable marketing and transport infrastructure. It also suffers from lack of R & D efforts to diversify poultry production by using other species like guinea fowl, ducks and quails etc. The challenge of fighting poverty and malnutrition can be effectively met to a large extent by strengthening traditional village poultry, which plays a major role in the rural economy of Afro-Asian countries as it accounts for about 80% of the world's poultry production and up to 90% in least developed nations. Unfortunately no attention has been paid in the past to promote and strengthen the rural poultry and as a result of this most of valuable poultry germplasm has either been lost or has been degraded.

What has been achieved?

Many developing nations including India have made spectacular achievements in the area of poultry production during last 4-5 decades but this has not improved the availability of poultry products and purchasing power of rural masses in the countryside of India. The poultry industry has today transformed itself into the fastest growing sectors of the Indian animal agricultural industry with broiler and egg segment growing @ 10-15 and 8-10% as against 5 and 2% growth in milk and food production. The annual egg production which was about 1780 million in 1960 rose to 45 thousand million in 2008. There is increase in per capita availability from a meager 10 eggs and 155 g meat in seventies to 45 eggs and 1.7 kg meat in India, per year against the recommended level of 180 eggs and 11 kg meat as compared to 120 eggs of world average and 250 eggs in developed countries. Moreover, there is a vast gap for the per capita availability of eggs in rural and urban India (15 vs. 100 eggs), which needs to be bridged to improve the nutritional status of rural masses. In rural areas the poultry products still continue to cost almost double the price of urban areas since most of the commercial poultry ventures are located in urban and peri-urban areas (Sharma, 2000). About 25% people living in urban areas consume about 75% of eggs and

100% broilers produced in India. The per capita availability of pulses has come down severely from 61 g in 1961 to less than 38 g which is causing malnutrition and protein deficiency.

Though, the number of native birds has increased marginally from 63 million in 1961 to 78 million in 2000, they are being indiscriminately replaced very fast by exotic birds but their contribution in providing supplemental income and nutritional security to poor and landless households is significant. The percentage of native poultry population has declined very rapidly from 94.2% in 1961 to 8% in 2000. The average egg production of native birds in India is less than half of native birds of other countries as it has reduced from 60-70 eggs in 1960 to 40-50 eggs in 2000 (Sharma, 2000; Sharma *et al.*, 2004). It is roughly estimated that 10 native hens can provide the same income as a woman earns from a day's work and can stay at home and take care of children and other household activities without affecting family income (Banerjee and Sharma, 1998). Poultry industry of India is providing employment to about 3 million people and contributing over 1% to the country's GNP. Poultry produces white meat and contributes about 25% of the total meat produced in the world. Private sector is playing a significant role by supplying about 95% of demand of day old chicks of commercial layers and broilers.

What can be achieved?

There exists tremendous scope to promote and strengthen both industrial and village poultry production system to support the further development of poultry. In order to create a demand of poultry products and help farmers in increasing their profit margin, there is strong need of well-developed poultry processing and marketing chain with a wide variety of poultry products with improved keeping quality and shelf life under Indian conditions. At present, only 2-3% of poultry products are sold as dressed and rest as live. Indian poultry industry, if it has to thrive and grow, needs to open new markets and create demands and increase profitability with new and diverse products through Public-Private partnership. This may be done by strengthening all fronts of poultry production, processing and marketing, as there are limitations for production of meat from other livestock. Organized development of poultry processing industry is essentially needed to provide easy access of quality products to a large variety of burgeoning consumers and for creating higher demand with better marketing, which in turn will support and act as growth engine for poultry production that will eventually ensure food security and provide livelihoods to millions. Government support and investments in poultry processing with appropriate marketing interventions are required for pushing poultry prospects further. Technological interventions in production, processing with appropriate marketing strategies are, therefore, an essential component of achieving sustained growth of poultry sector. Thus the challenge of fighting poverty and malnutrition can be effectively met to a large extent by promoting poultry through technological, managemental and market interventions, involving all stakeholders and strengthening linkages, which will increase its productivity and profitability by many folds.

Commercial or industrial poultry

Poultry farming can be profitably undertaken by unemployed persons on a full time basis as a means of gainful self employment or by employed/ semi employed persons on a part time basis to enhance the family income. Poultry farming has several advantages over other occupations as i) poultry requires less space and initial investment ii) broilers are ready for sale within 5-6 weeks and 6-7 crops can be taken per year) iii) poultry is most efficient converter of non-fibrous feed and mill-by-products into best quality animal protein iv) poultry requires no specialized training and utilizes spare time of family members v) poultry products are the best and cheapest source of the animal protein and vi) there is easy availability of inputs and markets for poultry. Specialized poultry stocks of very high production potential for egg, meat and dual type have been produced worldwide. Almost all the 10-12 brands of international repute are available in India through a network of franchisee hatcheries in the private sector. Some of the popular brands are Vencob, Ross, Lohman, Arbor Acres and Avian, broilers and Babcock B V 300, H & N, Deejay Dekalb layers. In addition, 5-6 brands of hybrid chicks are available in government sector. Day old broiler chicks can be purchased and may be reared either on deep litter (floor) of the brooder house or in the cages up to 5 to 6 weeks of age to be sold as live or dressed broilers/ table birds. Egg type chicks become layers at the age of 20 weeks and are capable of producing more than 300 eggs per year. Initially poultry farming can be started on small scale, which can later be expanded and diversified to make it an integrated unit. A small sized farm with a capacity of 1,000 commercial broilers per cycle can earn up to Rs. 5000 per month as net profit. Similarly, a layer farmer can earn up to Rs. 1, 00,000 per annum from 1000 layers.

Housing (deep litter)

Housing is a fundamental requirement which provides comfort and protection from natural predators to birds. Poultry houses should be located on an elevated and well-drained piece of land and soil type should be preferably sandy loam. Brooder house should be at least 300 feet away from grower/ layer houses. Orientation should be east-west to avoid direct sunlight, draft and rain. The space between houses should be at least 30 feet to allow enough ventilation. The width of house should not exceed 30 feet to allow proper house ventilation. The space between two rows of houses should be 60 feet. The length of the brooder house can be kept as per the requirement and its height (in middle) should be about 12-14 feet with an overhang of about 2-3 feet to prevent rain splashes. The height of the sidewalls should be about 6-8 feet in colder and 8-10 feet in warm areas. The masonry sidewalls of the brooder house may be ¾ closed (lower 6 feet bricked) and ¼ open (upper 2 feet wire meshed) in warm areas and fully closed with windows and ventilators in hilly areas. The house should have a cement concrete floor with a plinth of 2-3 feet to prevent seepage from outside. The shape of roof is kept shade type with about 4 feet (1/3) slope for every 12 feet roof. The roofing material should be chosen keeping in mind the cost, availability and suitability of materials. Asbestos-cement sheets are durable but costly. Light roofing material or thatches with locally available materials may also be used as they are cheap but roof should not leak in rainy season. Some trees should be planted to act as windbreakers.

Preparation of house (cleaning and disinfection)

After depopulation of house, remove old litter and other debris to a safe distance. Entire brooding premises should be thoroughly cleaned and sanitized. Remove all poultry house equipments to washing areas where they are washed, scrubbed, disinfected, repaired and painted for reuse. Repair the floorings, plug and seal all cracks and crevices, blow the wire mesh and all the concrete surfaces from all sides with the help of flame gun/ blowlamp to disinfect/ disinfest the house. Disconnect electricity of house before pressure cleaning twice, first with plain water and then add some disinfectant in the water, scrub the floor with 40% caustic soda. Paint the brooder house with lime and allow it to dry. Make the house airtight, bring in the equipments and fumigate with 3X concentration for 30 minutes. For this use 60 g of Potassium permanganate and 120 ml of Formalin for 100 cubic feet area. Keep the house empty for 1-2 weeks to break the cycle of pathogens. The floor, feeder and waterer space requirements are given in Table 1.

Table1. Floor, feeder and waterer space requirements on deep litter (per bird)

Age/ stage	Floor space	Feeder space	Waterer space
Under the brooder	8.00 sq.inches	1.0 linear inch	0.25 linear inch
1-2 weeks	0.25 sq.ft.1.5	linear inch	0.30 linear inch
2-4 weeks	0.50 sq.ft	2.0 linear inch	0.50 linear inch
5-6 weeks	1.00 sq.ft	3.0 linear inch	0.60 linear inch
7-8 weeks (layer chicks)	1.00 sq.ft	3.0 linear inch	0.60 linear inch

* Brooding period is up to 6 weeks in broilers and 8 weeks in layer chicks

Brooding management

Use any litter material, which is cheap, easily available, absorbs moisture and provides cushioning. Sawdust, rice husk, ground nut hulls and chopped grass may be used. It should be dry and free from moulds. Spread 2-inch thick layer of litter and spread 5 layers of English newspaper on it to prevent chicks eating litter materials and their subsequent death due to chocking. Place the brooder in the middle of the area and encircle the brooder with chick guard of about 1-foot height and have a gap of about 2 feet between brooders and the guards. Fix 2-4 bulbs of 40-100 watts inside the brooder depending upon the season in such a way that chicks can move freely in and out of brooder. Switch on the brooder 24 hours before the arrival of chicks and get brooder temperature of 90-95° F. Keep the temperature of brooder house below 15° F i.e. about 80 F. In order to obtain the brooding temperature in winters, close windows and opening in the house, use gunny bag curtains or polythene sheets, if necessary to check the excess ventilation but allow ample opening on roof side for ventilation to drive away ammonia, carbon dioxide and other obnoxious gases and also to keep litter dry. Place waterers (6 Nos. for 250 chicks) at the edge of brooder and fill up with medicated water (Vita. A, C and electrolytes). Scatter finely broken maize (500 g for 1000 chicks) over the newspaper for first two days. Hang a thermometer at the edge of brooder and about 2 inch above the floor. Also place chick feeders in between the waterers, half inside the brooders and half outside and fill them with broiler starter ration. Remove the chick guard after 3 weeks.

Try to obtain delivery of chicks during daytime in winter and during early morning hours or evening in summer. When the chicks arrive, open the boxes and stir the chicks to avoid pile-ups. Release the chicks quickly one by one under the brooder heat. While transferring chicks from boxes to brooder count them and dip their beaks in fresh medicated water for guidance of chicks to the source of water. Also make an examination for chick quality (appearance, weight, navels, eyes etc.) and transit mortality. Reject weak, unhealthy and chicks with pot bellies/ unhealed navels. Chicks are poikelotherm and without feathers hence cannot maintain their body temperature till they convert homoeiotherm and develop some feathers at about 4-5 weeks. Therefore, take special care of chicks in the first week as chicks have a tendency to huddle during first few days. Remove one layer of newspaper daily and chicks will be on litter on sixth day. Keep adjusting the height of brooder as the age and height of chicks' increase. The chick guard must be expanded from third day onwards to allow a greater area and keep adding sawdust to maintain litter depth of 2-3 inch. Purchase chicks only from a reputed hatchery so that you are sure of quality and genetic potential to obtain a profitable feed conversion ratio (FCR) of 1.7-1.8 i.e. chicks should have the genetic potential of achieving an average live weight of at least 1.6 - 1.7 kg by consuming 3 kg feed in six weeks. In case of layers, they should able to produce 1 dozen eggs per 1.7-1.8 kg feed. Now a days all the leading international brands of poultry stocks are available in India via their respective Indian collaborators through a network of associates/ franchise hatcheries scattered all over the country. Some government run hatcheries like Central Avian Research Institute, Izatnagar, UP also provide good quality stocks.

Temperature and light management

Chicks should be provided 24-hour light up to 3 days thereafter 23-hour light (including natural day light) and 1 hour dark regimen may be followed up to till brooding is over. Maintain the light intensity by providing 40-watt bulb per 200 square feet of floor space for the first 4 weeks and then 25-watt bulb per 200 square feet thereafter. Provide a brooding temperature of 90-95° F in the first week and decrease it at the rate of 5° F per week and there is no need to provide artificial heat after 4 weeks of age. Artificial heat may be given by using *Bukharis* or *Sigdies* or gas brooders or electric heaters.

Feeding and watering management

Feed may be provided in metal chick and grower feeders fitted with appropriate grills to prevent feed wastage. Height of feeder should be raised gradually with the help of feeder stands. Hanging type metal and pan type plastic feeders are available in the market. Automatic feeding and watering systems can be installed at a large-scale farm. Provide broiler starter mash for first four weeks and broiler finisher mash thereafter till they are marketed. Broiler starter and finisher mash should have 23 and 20 % protein and 2800 and 2900 K cal energy in addition to 1.2 % Ca, 0.5 % P, 6000 I U Vit A, 600 I U Vit D 20 I U Vit E and 1.4 g Choline. Ration for layer chicks should have 22 % protein and 2700 K cal energy. Feed can either be prepared at the farm or may be purchased from a reputed company. It is advisable to prepare feed at the farm, as it is cheap and fresh.

Growing management of layers (9-20 weeks)

Grower houses can be constructed in similar way as brooder house except that lower half (4-5 feet) of the side walls may be bricked and mesh may be put in upper half in warm areas. In colder areas lower two third part of the side walls may be bricked. Clean and disinfect the grower house as brooder house, add 3-4 inch layer of fresh litter material (rice husk or saw dust) and shift the grower chicks from brooder to grower house. Increase the floor, feeder and waterer spaces @ 1.5 square feet, 4 and 1 linear inch per bird respectively. Use grower feeders and waterers. Slowly replace the chick feed by grower feed which should have 16 % protein and 2600 K cal energy. There is no need to provide artificial light during growing period. Debeak the chicks at about 15-16 weeks to prevent cannibalism.

Layer management (21-72 weeks)

Layer houses can be constructed in similar way as brooder house except that lower one third part (2-3 feet) of the side walls may be bricked and mesh may be put in upper two third part. In colder areas lower half part of the side walls may be bricked. Prepare the layer house as brooder house, add 5-6 inch layer of fresh litter material (rice husk or saw dust) and shift the pullets from grower to layer house. It is advisable to shift pullets 2-3 weeks prior to start of laying so that they may adjust to new house and put nests to enable birds to select nest holes of their choice. Provide 1 nest hole for every 4-5 layers. Increase the floor, feeder and waterer spaces by 2.0 square feet, 6 and 1.5 linear inches respectively. Use layer feeders and waterers. Slowly replace the grower feed by layer feed which should have 18 % protein and 2700 K cal energy. Provide 16 hours of light (including day light) during laying period. Start giving 45 minutes of artificial light from point of lay and increase it @ 45 minute per week till it reaches 16 hours. Collect eggs 5-6 times a day to prevent breakage and heating of eggs.

Health management

The chicks must be vaccinated against all the prevalent diseases like Marek's, Ranikhet, Gumboro, Infectious bronchitis and Fowl pox as per given schedule. Keep a watch on coccidiosis after three weeks and start mixing coccidiostats in the feed or in drinking water. Consult a poultry veterinarian in case of any disease problem. Dispose the dead birds properly in an incinerator. Follow bio-security measures strictly which include restricted entry of visitors, minimum movement of personnel from one unit to another; plugging of leakage of water lines and their regular flushing; all-in; all-out system of rearing, proper disposal of old litter/ farm waste and dead birds and keeping premises free from rodents and wild birds.

Table 2. Vaccination schedule of broilers

S No.	Age	Name of the vaccine	Remarks
1.	1 Day	Marek's disease vaccine (F1)	0.2 ml S/C (hatchery)
2	5 day	Ranikhet disease vaccine	one drop in eye/nostril
3	12-14 day	Gumboro disease vaccine	one drop in eye
4	21-24 day	Infectious bronchitis vaccine	drinking water
5	28 day	Ranikhet disease vaccine (LaSota)	drinking water

Table 3. Vaccination schedule of layers

S No.	Age	Name of the vaccine	Remarks
1.	1 Day	Marek's disease vaccine (F1)	0.2 ml S/C (hatchery)
2	5 day	Ranikhet disease vaccine	one drop in eye/nostril
3	12-14 day	Gumboro disease vaccine	one drop in eye
4	21-24 day	Infectious bronchitis vaccine	drinking water
5	28 day	Ranikhet disease vaccine (LaSota)	drinking water
6	35-40 day	Gumboro disease vaccine	drinking water
7	6 week	Fowl pox vaccine	0.2 ml wing web
8	8-10 week	R2 B vaccine	0.5 ml I/M
9	14-15 week	Fowl pox vaccine	0.2 ml wing web

ANNEXURE-I : Suggested feed formulae for commercial broilers

Ingredients (%)	Broiler Starter (0-3 week)	Broiler Finisher (4-6 week)
Yellow Maize	55.000	58.000
D.O.R.B	5.680	10.680
Soyabean Meal	36.000	28.000
Mineral Mixture	3.000	3.000
Coccidiostat	0.050	0.050
Vitamin A+B_2+D_3+K	0.035	0.035
Vitamin B. Complex	0.035	0.035
Probiotic	0.050	0.050
Lysine	0.100	0.100
DL. Methionine	0.050	0.050
Total	**100.00**	**100.00**

Suggested feed formulae for commercial layers

Ingredients (%)	Chick mash (0-8 weeks)	Grower mash (9-20 weeks)	Layer mash (21-45 weeks)	Layer mash (46 -72 weeks)
Yellow Maize	50.000	51.500	56.000	57.000
D.O.R.B	12.700	27.220	12.725	12.725
Soyabean Meal	34.000	18.000	24.000	22.000
Mineral Mixture	3.000	3.000	3.000	3.000
Shell Grit	-	-	4.000	5.000
Vitamin A+B2+D3+K	0.020	0.020	0.030	0.030
Vitamin B. Complex	0.020	0.020	0.025	0.025
Probiotic	0.050	0.050	0.050	0.050
Lysine	0.100	0.100	0.100	0.100
DL.Methionine	0.050	0.050	0.050	0.050
Coccidiostat	0.050	0.030	-	-
Vitamin B-12	0.010	0.010	0.020	0.020
Total	100.00	100.00	100.00	100.00

ANNEXURE-II

Economics of broiler production (financial statement to produce 6,000 broilers (6-week cycle) of 2-monthly rotation of 1000 broilers (deep litter)

(A) Capital Investment

(a) *Building (Brooder house)*

i)	Floor Area 1,000 sq. ft. @ Rs. 100/ sq. ft.	Rs. 1,00,000
ii)	Poultry Equipment's @ Rs. 15 per bird	Rs. 15,000
		Rs. 1,15,000

(b) *Stock* : Cost of raising the first crop of 1050 broilers : (5% extra for mortality & culling)

i)	Cost of 1050 broilers chicks @ Rs. 14/ chick	Rs. 14,700
ii)	Cost of 1575 kg of feed @ Rs. 10.50 per kg (@ 1.5 kg av. cons. for various age groups / broiler)	Rs. 16540
iii)	Electricity, medicines, litter, laborer, etc. @ Rs. 5.00 per bird	Rs. 5250 **Rs. 36,49065,650**
	Total (a) + (b)	**= Rs. 1,51,490**

(B) *Recurring Expenditure (For six cycles to be taken in I year)*

i)	Cost of 6,300 broiler chicks @ Rs. 14 per chick	Rs. 88,200
ii)	Cost of feeding 6,300 chicks (After mortality & culling) @ 3.2 kg per bird (Total feed 202 qtls.) and feed cost @ Rs. 910.50 per kg.	Rs. 2,12,100
iii)	Cost of litter, medicines, electricity and laborer @ 15 % of feed cost	Rs. 13,230
	Total :	**Rs. 3,13,530**

(C) *Income*

i)	Sale of 6,000 broilers, each weighing on an average 1.6 kg @ Rs. 40.00 per kg	Rs. 3,84,000
ii)	Sale of 288 gunny bags @ Rs. 8.00 per bag	Rs. 2,304
iii)	Sale of 13 tones of manure @ Rs. 200 per tone	Rs. 2,600
	Total:	**Rs. 3,88,904**

(D) *Gross Profit*

Income (C) - Expenditure (B)	Rs. 75,374
Deductions	
i) Depreciation on : Building @ 10%	Rs. 10,000
Equipment @ 15%	Rs. 2,250
ii) Interest on 75% capital @ 12.5%	Rs. 14202
iii) Insurance @ Rs. 0.50 per chick	Rs. 3,150
Total:	**Rs. 29,602**

(E) *Net Profit* : Rs. 75,374 - Rs. 19,602 = Rs. 45,772

Net Profit : Amounts to Rs. 8 per broiler

ANNEXURE-III

Economics of small scale layer farming

[Financial statement for setting up of a 500 layer farm] (Deep litter system)

(A) Capital investment

(a) *Building*

i) Poultry shed consisting of brooder house and laying house - 1,000 sq. ft. @ Rs. 75 per sq. ft.	Rs. 75,000
ii) Equipment for 500 birds @ Rs. 12 per bird	Rs. 6,000

(b) *Stock* : Cost of rearing 500 chicks from one day old to the point of lay (20 weeks):

i) Cost of 525 day old sexed female chicks (5 per cent) extra to compensate for depletion and culling) @ Rs. 17 per chick	Rs. 8,925
ii) Cost of feeding 525 chicks up to 20 weeks @ 8 kg per chick and feed cost @ Rs.10 per kg (the average number of chicks fed for the full period would be 512)	Rs. 40,960
iii) Electricity, medicines, litter, laborer, etc. @ 15 % of feed cost	Rs. 6,144
Total (a + b)	**Rs. 1,37,029**

(B) *Operational recurring expenditure (during 12-month laying period)*	
i) Cost of feeding 500 layers for one year @ 40 kg per bird @ Rs.10 per kg	Rs. 2,00,000
ii) Cost of litter, electricity, medicines, etc. @ Rs. 10. per bird	Rs. 5,000
Total	**Rs. 2,05,000**

(C) *Income*	
i) Sale of 1,40,000 eggs @ Rs. 1.75 paise per egg (egg production at the rate of 280 eggs per bird per year)	Rs. 2,45,000
ii) Sale of 17.5 tons of poultry manure (40 layers produce one ton of manure; 100 chicks produce one ton) @ Rs. 200 per ton.	Rs. 3,500
iii) Sale of 320 gunny bags @ Rs. 10 per bag	Rs. 3,200
iv) Sale of 500 spent hens @Rs. 50 per bird	Rs. 25,000
Total (Rs.)	**Rs. 2,76,700**

(D) *Gross profit* : (Income- Recurring expenditure)		
(Rs. 2,76,700- Rs. 2,05,000)	=	Rs. 71,700

(E) Net profit		
Gross profit	=	Rs. 71,700
Less (a) Depreciation on Building @ 5 %	=	Rs. 3,750
Equipment @ 15 %	=	Rs. 900
(b) Interest on 75 % of capital @ 12.5 % (12.5 % on Rs. 1,02,772)	=	Rs. 12,846
(c) Insurance @ 50 paise / bird	=	Rs. 250
Total	=	**Rs. 17,746**
Net Profit	=	**Rs. 53,954**

NET PROFIT : About Rs. 0.38 per egg

References

Banerjee, A.K. and Sharma, R.K. 1998. Breeding possibilities in adverse climatic conditions with special reference to Indian conditions. Proc. of X European Poultry Conference, Jerusalem, June 1998, Vol. 2 ; 505-508.

Sharma, R. K. 2000. Final Project Report. U. P. Council of Agricultural Research, Lucknow, INDIA, Sponsored Project.

Sharma, R. K., Tripathi, S. C. and Neupane, D. 2004. Conservation of poultry genetic resources of Indian sub-continent with special reference to India. Abs. XXII World's Poultry Congress, 8-13 June, 2004, Istanbul (Turkey): 198.

□□□

period [illegible] and SMR [illegible] disorders-abortion, retained placenta, [illegible] disease etc., soft or hard milker, behaviour-suckling the calf, vices, etc. The goals of efficient reproductive health programme include an average calving interval between 12 to 13 months, breeding efficiency below 1.5 services/conception, and less than 30% non returns after 60 days, repeat breeder [illegible] or less, abortion 1-2%, retained placenta 5-10% and metritis 5-10% or less.

[illegible] farming [illegible] to go for [illegible] the [illegible] of [illegible] also now becoming common at livestock farms.

- *Mixed farming*: Wherever possible, one should try to go for the dairy enterprise in combination with some agriculture farming. It would enable the farmer use agricultural byproducts for feeding the animals, land for grazing, and parting of some agricultural labor/ machinery in dairy business during emergencies. At the same time dairy business can also mutually benefit the agriculture enterprise through manures, consumption of superfluous milk or its byproducts.

Table 1: Floor space requirements for cattle and buffaloes

Type of animal	Floor space req./animal (m²) Covered area	Open paddock	Max. animals/pen	Height of shed (cm) at eaves	Manger/water trough (cm) Length	Width	Depth	Height of inner wall (cm)
Young calves	[illegible]	2.0	30	175 cm in medium and heavy rainfall areas and 220 cm in dry areas	20-30	30	15	15
Old calves	2.0	4.0	30		40-50	40	15	20
Cows	3.5	7.0	50		60-75	60	40	50
Down-calver	12	[illegible]	1		[illegible]	[illegible]	[illegible]	[illegible]
Bulls	12	120	1		[illegible]	[illegible]	[illegible]	60
Buffalo	4.0	8.0	50		[illegible]	[illegible]	[illegible]	[illegible]

- *Herd size and culling*: This ever dynamic factor in dairy enterprise is of most concern to the farmer. If herd size is to be kept constant, it is advisable to spot out poor animals at their early age and prompt actions be taken for their culling. Wherever replacement is required, it should be done with superior stock.

- *Stress control*: It is very important to consider installing stress control measures from inclement weather, flies and insects (wallowing, pond

31

Income Augmentation through Rearing Small Ruminants

D.V. SINGH

The large livestock population of India includes 178 m cattle, 98 m buffaloes, 65 m sheep, 125 m goats, 13.52 m pigs, 0.40 m rabbits and 2.56 m other species like equines, camelids, etc., besides 585 m poultry for uses, like - food (milk, meat, egg), work (draft), pleasure, protection (guarding, caring, skin, fur, leather), companion (dog, cat), etc.

The livestock based agribusiness has always drawn deep attention of rural masses, scientists and planners, especially in India. This is simply because 70% rural mass is either landless or marginal farmers, whose livelihood depends on livestock farming and agriculture. The ever increasing human population coupled with ever decreasing cultivable land (particularly for large ruminants) forces us to look for better options for securing high quality animal protein rich food obtainable in comparatively shorter lock-in periods. Husbandry of pigs, rabbits, goats and sheep are positive steps in this direction. All these animals have the power to convert non human food materials in to high quality utilizable proteins at faster rates. Moreover, their reproduction rate is also enormous. This all direct us to look further in to more details about farming options of these livestock depending up on the marketability, individual choice and capital needs.

1. Piggery

The rate of gain in body weight in pigs is excellent. Piggery is gaining popularity due to increasing interest of all classes towards pork as it is tasty and cheap. In the past, the occupation was limited to certain sections of the society. As the initial investment on piggery is comparatively meager, it is now being adopted by all sections of unemployed and poor youths in rural and urban areas as an easy means of self employment and poverty alleviation. Pig has the potential to convert non utilizable hotel and kitchen wastes and market residues in to high quality animal protein. It has a remarkable potential to convert 4-5 kg feed in to 1 kg body weight at faster rate. It attains 80 kg body weight within 6 months ages (100 times of birth weight) and becomes ready for sale. The sows have high reproducing rate. Their conception rate is very high (up to 90%), gestation period is short (112-115 days) and they deliver 8-10 piglets two times in a year. This helps them to multiply faster. Pigs can easily survive in all kinds of agro-climatic conditions. Pork can be easily preserved for long and thus it has great potential for export.

Selection of foundation stock is an important activity in any livestock farming. The breed of choice should be made based on climate, capital investment, land/ labor availability and kind of business (short/ long term). The common breeds found in India are Yorkshire, Berkshire, Landrace, Duroc and Hampshire. Large White Yorkshire and Middle White Yorkshire can be reared throughout India while Berkshire, Landrace and Duroc are popular in North-East and Southern regions.

The cost of feeding alone is about 70% of total cost. Hence, judicious planning and alternative sources of feed must be taken in to consideration while formulating the ration. Cheaper sources of feed include feeding of various items like kitchen waste, bakery waste, grocery waste, potato (75% with grains), sweet potato (30-50% with grains), molasses/ sugar cane juice *maili* (20 to 40% of ration of growers and adults), rice polish (50% of ration), wheat or rice bran (60% of ration), berseem/ maize (2-3kg/ day) can reduce the cost on feeding heavily.

Piggery can be profitable if entrepreneur considers selling the produce in right value at right time. Middlemen can cause a lot of inconvenience in direct selling activities. Some of the government and private organizations involved in marketing of pigs are CDF, Aligarh, Nanded and Goregaon, M.S., Perumber and Saidapet, T.N., Alakbeer, Hyderabad, Deemapur Mandi, Assam. There is high demand of pork in Kerala and North-eastern States, particularly Assam.

Economics: Considering 1 male and 10 female pigs were purchased at one year age, female delivering 8 piglets two times in 15 months and average body weight of pig at selling is 80 kg.

Table 1. Economics of piggery

Particulars	Fixed cost (Rs.) Unit	Rate	Total cost	Particulars	Recurring cost (Rs.) Unit	Rate	Total cost
Shed (40 ft²/ adult, 10-15 ft²/ young)	1300 ft²	100	130,000	Conc. feed - Adult	2.0 kg x11 x 550d	8	96,800
Utensils	11	500	5,500	-Young (160no.)	300 kg	8	348,000
Pigs (10+1) having 100 kg weight	11x100	60	66,000	Green fodder - Adult	3 x 11 x 550 d	1.0	18,150
Total			201,500	-Young(160no.)	400 kg	1.0	640,400
Interest on capital	1.5 yr	15%	30,225	Health cover	11	100	1,100
Depreciation on shed	1.5 yr	15%	19,500	Electricity	18 m	200	3,600
-equipment	1.5 yr	15%	825	Labor	18 m	1000	18,000
Insurance cover		6%	3,960				
	Total (Rs.)		54,510		Total (Rs.)		585,650
Returns							
Sale of young (160 no. @80kg BW; @ Rs.60/ kg							768,200
Manure (Adult@ 4 kg; young @ 2 kg; *18m) =200 ton, @Rs.75/ ton)							15,000
Value of parent stock (@250 kg BW*11*Rs.60/ kg)							110,000
Total Returns							948,000
Net Returns (in 18 months)							307,840

2. Rabbitary

Rabbits are raised mainly for meat, fur, skin, wool and research and development of technologies. About 50% of total fur skins utilized world over come from rabbits. They adapt to a wide range of environmental conditions. None can question their potential as meat, fur and wool producing animal, besides laboratory, show and fancy uses, based on feeds that are not fit for human consumption. That is why there is a magnetic attraction about rabbitary for profit and pleasure. Rabbit, a docile and pleasing look animal, produces pearly white meat from grain free/ high forage-low cereal diets. They are suited to both small (backyard self sufficiency) and large scale commercial production in rural and urban areas. It can utilize forage protein more efficiently comparatively to other livestock. They have faster growth rate and better feed conversion efficiency with feed : grain ration as 1:2.5, comparable with pigs and poultry, but with an advantage of consuming roughages. The doe with 16 kits/ year require <33% feed energy than a cow (with 0.9 calves) and < 50% than ewe (with 1.2 lambs) for each kilogram of market weight. They can be bred anytime of the year (even on the day of parturition), have short gestation (30 days) and can add 20-40 young/ year. Their biological productivity is 10-25 times of their weight in offspring (cow-0.6 times; ewe 0.8 times) per year. Young rabbits attain 2 kg weight in 12-14 weeks and their meat is pearly white, highly delicious, low in fat and calories but high in protein value and consumed without religious taboo. Thus, rabbit farming is comparatively cheaper to other livestock.

Table 2. Economics of rabbitary

Particulars	Angora rabbit		Broiler rabbit	
A. Capital Expenditure				
Sheds & cages	15′x 40′ @ Rs.50/ ft²	30000	12′x30′ @ Rs.30/ ft²	10800
Wire cages	Rs.60 x 100	6000	Rs. 50 x 100	5000
Daily use articles: Scissors, buckets, brushes, feeder, waterers, nest boxes		4000		3200
Rabbits	10 F, 2 M; @ Rs. 800	9600	10 F, 2 M; @ Rs. 300	3600
Total (A)	Say	50000		22600
B. Recurring Expenditure				
Feed	12x120x365x650Q (12 adults)	3500	Adults (12 months)	3500
	134x50x365x650Q (134 young)	15000	Young (2 months)	3000
Hay	12x50x365x100 Q (12 adults)	200	Lumpsum	1000
	134x30x365x100Q (134 young)	1500		
Miscellaneous - Water, electricity, medicines, etc. @Rs.10/ each (150 rabbits)		1500	Approx.	500
Total (B)	Say	21000		8000
C. Depreciation	@ 10% of fixed amount	4000		2600
D. Interest	@ 14% on Rs. 70000	9800	@ 14% on Rs. 40000	5600
E. Cost of Production	Say (B+C+D)	34800	Say	1600 0
Income during first year				
Wool/ Meat/ Skin	12x700g; 134x300g:49kg @ Rs.1000/ kg	50000	100 kg, @ Rs.100/kg	10000
			(1 kg/ young) @ Rs. 50 x 100	5000
Manure	150 x 40	6000	150 x 40	6000
Surplus rabbits	40x250	10000	40 x 25%	10000
Total income		61000		31000
Profit during first year		26000		15000

Angora rabbit produces one of the finest wool. German angora weighing 4 kg yields about 800 g wool/ annually (i.e.200 g/ kg live weight) without the need of scouring, which is 4 to 6 times higher than a Merino sheep (3 kg greasy wool or 1.5 kg clean wool/ 40 kg body weight). Production of Angora wool world over is estimated to be around 3000 MT (China - 2000 MT). India produces 80-100 MT Angora wool and also imports large quantity to meet the need of spinning industry. The breeds like, German Angora, British Angora, Russian Angora and particularly the Crossbred Angora are commonly reared by Indian farmers. The breeds for biological research purpose are New Zealand

White, California, Dutch and Black Brown. Whereas, the breeds like, Polish, Palminao, Havana, Beveran, New Zealand Red, English Spot White, Dutch, etc. are reared for fancy purpose. Rex and Satin breeds are raised for fur skin.

Rabbit raising can be suitable done under temperate climatic (10°-25°C) with 65-70% RH through out the year. Broiler rabbits can tolerate 30°-35°C. Selection of trios (1 male and 2 females) from the same genetic line is desirable as breeding stock. The doe at selection time should be about 6-7 months old, 2.5-3.0 kg weight and should have 8 functional teats. Rabbits are housed in two systems - hutch system and shed/ cage system. Maximum production from rabbits can be obtained by paying good attention towards hygiene, cleanliness, ventilation, light, no dampness and protection from predators.

It is seen from the above that meat rabbits are not as much rewarding as Angora rabbits. Moreover, the demand of rabbit meat and skin is not assured. Therefore, Indian farmers are not coming forward for broiler rabbit production.

3. Goatry

The socio-economic importance of goats in India is evident by the sharp increase in their numbers and contributions during the last 30 years. Goats contribute milk, meat, fibre, skins, manure and transportation to the subsistence of small holders and landless poor. They play an important role in capital storage, income and employment generation and house hold nutrition. It is easier to increase the population of goats as capital investment is relatively low, land requirements are small, reproductive rates are higher both due to shorter breeding interval and high prolificacy and they can be managed by spare family labor and do not require any serious facilities and management skills. There is much less risk in goat farming in drought prone areas. They are much resistant to adverse effects of toxic plants. They have higher dry matter and fibre digestibility and can subsist on poor vegetation. Thus, they act as insurance against disaster under pastoral and agriculture subsistence system. Goats have religious and ritualistic importance throughout India. Goats can be milked as often as required (moving refrigerator). Goats are more economical than sheep (2.5 times) under natural grazing conditions. They contributed up to 55% of the total cash income of a farm family in hot and arid region. Goats make important contribution to the economy of the country. National Commission on Agriculture (NCA, 1976) recommended reduction in the number of goats from then existing 67 to 47 million. The population has since then doubled, and increasing @ 3.6% per annum. As per FAO (2005) statistics yearly production of chevon 4.1 MT (37% of India), milk 2.35 MT (3% of India), skins 100 million, pashmina 0.03 MT, manure 0.85 MT.

Goat produces lean and juicy meat that excellently meets the kitchen needs of all Indian dishes. All non-vegetarians prefer chevon to beef and mutton and thus it is costliest. Goat milk has smallest fat globule size and thus it is easily digestible and found suitable for infants, old, sick and old. Goat hair, due to higher tensile strength are preferred for rope making, and mohair from Angora goats is used for making fancy bag. Chegu and Changthangi goats of North Himalayan regions produce finest (8-14 um) under coat, Pashmina, used for the manufacture of superfine fabric. The goat leather, particularly of Black Bengal, is used in the manufacture of fancy and high valued products and has high export potential. Goat bones are utilized for bone meal manufacturing.

Goats can be reared under extensive (migratory, transhumance, free range, pasture and range grazing) for larger flocks under low resource use. Under semi-intensive system the animals of several owners are combined for grazing for 4-6 hours and the supplanted with kitchen waste, concentrate mixture, crop residues, green and dry fodder, leaves, etc. as per the availability as well as intensive systems.

Economics (95 does+5 bucks)

A few case studies indicated that starting with a loan for raising 100 goats, the owners could repay all the debt in 5 years. The input–output ratio was 1:2.3.

4. Sheep rearing

The sheep contributes greatly to the agrarian Indian economy, especially in arid/semi-arid and mountainous areas. Rajasthan is the largest sheep rearing state. Sheep provide mutton and wool beside manure, which is an important source of organic fertilizer especially in southern states. Of the total wool produced in the country, roughly 10.5% falls above 56s (suitable for worsted sector and hosiery), 22.3% between 44s and 56s, 34% between 36s and 43s, and remaining below 36s in quality. Country requires 15m kg of 64s quality wool for the manufacture of fine apparels, 10m kg of 56s quality for medium quality apparels and 20m kg lower quality wool for carpet manufacture. Most of the breeds of sheep are very well adapted to harsh climate, long migration, tropical diseases, poor nutrition and shortage and poor quality drinking water. Few newly evolved have arisen from crossing with exotic fine wool breeds. Indian breeds have been classified according to their suitability for meat or wool (carpet/apparel) or their combination. The apparel wool breeds are Hissardale, Nilgiris, Kashmir Merino, Bharat Merino, Nali/ Chokla synthetics/ Nilgiris Synthetics and Avivastra. Whereas, coarse carpet wool breeds are Malpura, Sonadi, Muzaffarnagri, Jalauni, Deccani,

Bellary, Coimbatore, Chottanagpuri, Balangiri, Bhakarwal and Shahbadi. The exotic breeds are Merino, Rambouillet and Polworth (for fine wool), Suffolk, Dorset and Southdown (for mutton), Corriedale (Dual) and Karakul (for pelt). The breeding strategy will involve decision on objectives and approaches aiming for genetic improvement depending upon socio-economic and ecological considerations, existing genetic resources and their productivity.

Table 3. Economics of 100 sheep farming (100 ewes + 5 rams)

A. Initial capital investment		**Amount (Rs.)**
Shed	Covered area 130 m^2, @Rs.300/ m^2	51,000
	Open area 200 m^2, @Rs. 60/ m^2	
Animals	100 ewes (1-1.5 yr) @ Rs.1400/-, 4 rams @ Rs.2000/-	148,000
Equipment	Buckets, ropes, spades, scissors, etc.	6,000
Total (A)		205,000
Financial Plan: Bank Loan – 165000; Margin money – 40000		
B. Expenses		
Fixed	Interest on loan @10%	16,500
	Depreciation on animals @17%/ annum	25,160
	" shed @17%/ annum	8,670
	" equipment & machinery @20%/ annum	1,200
	Total (Rs.)	**51,530**
Variable	Dry Fodder for 3 m, @2kg/d/animal, @ Rs.150/Q	28,200
	Conc. – 140 ewes, @ Rs. 6.25/ kg	
	@ 250g during advanced preg. (95 x .25 x 30)	7,688
	@ 600g during 150 d lactation (95 x .6 x 150)	
	4 Rams for 3 m @500g (5 x .5 x 90)	
	Labour-1 @ Rs.2100/ m	25,200
	Miscellaneous – medicine, etc.	5,000
	Total (Rs.)	**66,088**
	Total (A+B)	**117,618**
C. Income	Sale of wool, @2kg/ adult, 0.6 kg/ lamb; @Rs.50/ kg	14,600
	Sale of 140 lamb, @Rs.1200/-	168,000
	Sale of 60 cart load manure @ Rs.250/ cart	15,000
	Total (Rs.)	**197,600**
Net income/ annum		79,982
Net income/ month		6,665
Annual rate of return on initial capital investment (%)		39.0

An adult sheep requires about 2 l water/ day during winter and 4 l/ day during summer. Throughout the country, they depend on meager range vegetation, natural pasture and crop residues. Tree leaves form integral part of feed for sheep during scarcity. Other sheep feeding practices include extensive natural range management, cultivated pasture, silvipasture, grazing with supplementation as well as intensive feeding. Sheep breed throughout the year. Indigenous sheep should be bred at 1.5 to 2.0 years of age either naturally or artificially. Synchronisation of estrus provides better results in sheep. They don't require elaborate housing facilities but minimum provisions will definitely increase productivity, especially during inclement weather conditions. Adult sheep should be provided 1.0 m² space. Shed of 18x6 m can accommodate about 120 sheep. Long tailed sheep require docking and all sheep require dipping at least two times a year.

□□□

32

Feed and Fodder Management for Commercial Livestock Production

S.K. SINGH AND JYOTI PALOD

The economical production of livestock products depends largely upon three main factors–the efficiency of an animal, its nutrition and management. The efficiency of livestock is the result of inheritance and improved breeding. Its productivity, however depends upon the adequate inputs in terms of feed and fodder with scientific management and care. Where livestock are kept under required desired conditions, more than 70% of the total expense goes for feed, while remainder represents on housing, labour depreciation, interest and other miscellaneous items. These items are practically the same regardless of the total livestock produce. Thus, scientific and judicious combinations of various available feed ingredients to produce least cost balanced ration for meeting the nutritional requirements of different livestock species for maintenance, growth and production at different stages of their life have a greater significance.

Feeding dairy animals

Feeding of calf up to 3 months of age: Feeding young calves during first 3 months is most important part of calf raising program though it is a relatively small proportion of total feed cost on a dairy farm. Due to the absence of functional rumen, nutrient requirement of young calf are very much similar to non-ruminants.

Proper care and feeding of calf during initial critical period of life is essential. Underfeeding of calf during growing period causes general debility, stunted growth, unthriftness in later life, delayed sexual maturity, reproductive problems, late calving and lowered milk production after calving. Similarly due to overfeeding growing calf become very fat which is responsible for reduced breeding and milk capacities in future.

Calf feeding systems

Natural system in which the calf sucks milk from mother's udder. Milk being an important constituent of human diet, calf is allowed to suck initially for a few seconds only to induce let down of milk and only a little quantity of milk is left in the udder after milking due to which calf is unable to grow at a proper rate. On the contrary if calf is allowed to suck the mother's udder for a long time, it over-consume the milk which causes digestive problems.

Artificial system of calf feeding is scientific and is followed in all well organized dairy farms. The details of this system are as under.

One of the most important point to be given due attention in calf rearing is that calf should receive colostrum (the first postpartum mammary secretion) as its first feed just after birth (preferably within an hour). Colostrum is an excellent source of essential nutrients and it contains antibodies which are nature's way of providing protection to the young calf against many diseases. It contains high level of immunoglobulins like lgM, lgG and lgA which are synthesized by the dam.

Colostrum has laxative action also, therefore it helps in removal of first faeces. If mother does not give colostrum, colostrum from other cows or a substitute of equal nutritive value should be provided for first few days. Colostrum substitute can be prepared as under.

Whip an egg in 275 ml of warm water, add half teaspoonful of castor oil, 10,000 IU of vitamin A, 525 ml of warm whole milk and 80 mg of aureomycin, stir well and give it at 40°C. This is sufficient for one meal. The calf should be fed such meal three times a day.

Limited whole milk feeding should be done after 4 days of colostrum feeding. Milk is a highly nutritious complete feed for calf and best growth of calf is obtained by providing adequate milk. As milk is deficient in iron, copper, manganese and zinc therefore these micro-minerals should be supplemented along with some salt.

Milk replacer feeding

To make the calf feeding economical and increase the availability of milk for human consumption milk replacer feeding is used. Milk replacer is a highly nutritious, palatable, properly mixed concentrate powder which resembles the composition of milk and is suitable for calf feeding (Table 1). Milk replacer feeding is started at 10 days of age. It is introduced gradually so as to avoid reduction in growth rate. It should contain minimum 22% protein, 10%fat, 95%TDN, 0.7% calcium, 0.6% phosphorus and other nutrients. It should contain such ingredients which are readily soluble so that it can be dissolved quickly at the time of feeding. Milk replacer of relatively low quality can be used after 3 weeks of age. Milk replacer is prepared by mixing 7 parts of water to 1 part of powder by weight so that a mixture of about 12% total solids similar to whole milk is formed. After mixing with water it is fed similar to whole milk.

Table 1. Composition of a milk replacer developed at NDRI, Karnal is as under

Feed component	Quantity (kg)
Wheat flour	10
Fish meal	12
Linseed meal	40
Milk	13
Coconut oil	7
Linseed oil	3
Butyric acid	0.3
Citric acid	1.5
Molasses	10
Mineral mixture	3
Aurofac (antibiotic)	0.3
Rovimix (vitamin mix)	15

Calf starter feeding

Calf starter is a palatable and highly nutritious concentrate mixture (Table 3) which should contain high quality protein and low fibre and silica content. It should be introduced in calf feeding at 2 weeks of age and gradually increased so as to replace milk entirely at 8 weeks of age. Ingredients of calf starter must be fresh, highly palatable and free from toxic materials. It should contain minimum 23-26% crude protein, 4% crude fat and maximum 7% crude fibre & 2.5% acid insoluble ash (B.I.S.). Initially calf will eat only a small quantity of calf starter but later on free choice availability will increase its consumption. At this stage dairyman has to fix its amount to keep the feed cost economical and to prevent the growing calf from excessive fattening. When the calf becomes 3 months old, an economical concentrate mixture containing less protein and other essential nutrients can be used. Feeding schedule based on calf starter is given in Table 2.

Table 2. Feeding schedule based on calf starter

Age (days)	Calf starter (g)	Colostrum (litre)	Milk (litre)	Skim milk (litre)
0 – 5	-	1/ 10th of body wt.	-	-
6-30	-	-	1/10th of body wt.	-
31-60	125	-	1/15th of body wt.	1/25th of body wt.
61-90	250	-	1/25th of body wt.	1/15th of body wt.

Table 3. Ingredient Composition of calf–starters

Ingredient	Proportion (%)
Maize	42
Groundnut-cake	28
Fish meal	7
Wheat bran	20
Mineral mixture	3
Total	100

Hay, silage, pasture or roughage feeding

One of the objectives in economical calf raising is to utilize nutrients from the cheaper source as soon as possible. Therefore, it is desirable to begin offering free choice high quality hay at about 2 weeks of age to the calf. The fodder for hay making should be grown on land with adequate fertility and harvested at an early stage of maturity to ensure high nutrient content. Leguminous hay is the best for calf feeding. Silage is less desirable than hay. Grass silage may reduce growth rate. Pasture can be utilized in place of hay. When high quality forage is provided to the calf in sufficient amount, it will provide essential nutrients as are provided by hay.

Initially calf will consume very little amount of roughage but later on it will start consuming roughage and at about 3 months of age rumen will be developed for microbial digestion. At this stage calf will be in position to grow on concentrate mixture and roughage.

Feeding of calves from 3-6 months of age: At this stage calves are fully kept on calf starter, succulent green fodder and good quality hay.

Ration schedule from 3-6 months

a. Calf starter 1 - 1.5 kg with 10 kg non-leguminous green.

b. Calf starter 0.75 - 1.0 kg with 2.5 kg leguminous green and 2 kg dry fodder.

With the advancement of age weight of calf increases therefore quantity of feed and fodder offered should be increased accordingly.

Feeding of calves from 6-12 months of age : At this stage rumen is fully developed and calves could meet most of their requirement from succulent roughage. By 6 months 70% of their requirement are met from roughages.

Ration schedule from 6–12 months

a. Concentrate 1 kg with 15-20 kg non leguminious green.

b. Concentrate 1 kg with 7-10 kg leguminious green and free choice dry fodder.

If roughage available is only straw the entire requirement are to be met from concentrate.

Feeding of calves from 12 months to age of maturity : At this stage requirement are met accordingly to availability of feeds and fodders along with body weight of the animal. For an average sized animal following schedule may be followed.

Ration schedule from 12 months to age of maternity -

a. Concentrate 1.5 kg with 20-25 kg non leguminous green.

b. Concentrate 1 kg with 5-10 kg leguminous green and 3 kg dry fodder.

Feeding of Adults : While considering the feeding of adult dairy cattle and buffaloes, proper consideration should first be made for the purpose for which the animal has to be fed. Cattle generally eat 2.0 -2.5 kg dry matter for every 100 kg of live weight daily. Buffaloes and CB animals are slightly heavy eaters and their dry matter consumption varies from 2.5 to 3.0 kg daily per 100 kg body weight.

Conventional/ orthodox method for total dry matter: Under Indian conditions, while ration computation is made, amount of grazing is negligible as it is extremely poor. Since the bulk is essential for rumen, the dry matter allowance should be divided as under.

Total dry matter	2/3 as roughage	2/3 dry roughage or up to 3/4 if sufficient legume is available
	1/3 as concentrate	1/3 green roughage or 1/4 when green fodder is legume.

Although the method described is founded on scientific lines however it is neither economical nor practicable under Indian conditions. There is non-availability of greens during a large part of year in many parts of the country. Therefore thumb rule is used to guide feeding of animals as per the availability of ingredients in a simple way. The approach here is based on practical experience rather than scientific basis.

When nutrient requirement are to be fulfilled by straw and concentrate only: In most parts of India, green fodder is seldom available. In urban areas and most parts straw is considered to be the only basic roughage.

Under such circumstances, the object should be to compound concentrate mixture which will provide at least 20% protein (14-16% DCP) and 68-72% TDN. Variety of feed ingredients should be added in compounding concentrate mixture so as to make it palatable, slightly laxative and balanced with minerals and vitamins. Variety of feed ingredients also have other advantages e.g. the imbalance of a. a. or minerals of one feed ingredient can be corrected by the other.

Now the question remains as to how to formulate the concentrate mixture that will provide 14-16% DCP and a minimum of 68% TDN with out taking into consideration computation. For this approximately following parts of different feed ingredients should be used for preparation of conc. mixture.

Oil cakes - 25-35 parts

Millets/ Cereals - 25-35 parts

Cereal by products - 10-25 parts

Pulses/ Chuni - 5-20 parts

The amount of concentrate mixture and straw that will provide optimum maintenance requirement for an adult cattle is given in table 4.

Table 4. Straw and concentrate requirement of an adult cattle

Feed component	For zebu cattle	For CB/ purebred Indian cows/ buffalo
Straw (kg)	4	4-6 kg
Concentrate (kg)	1 - 1.25	2.00 kg

It should be fortified with 1% mineral mixture, 1-2% salt and 20-30 gm. of Vit. AD_3/100 kg containing 50,000 I.U. Vit. A and 5, 000 I.U. Vit. D_3/g

When non leguminous green fodder is available: By feeding non leguminious fodder alone the nutrient requirements for maintenance of adult animals can be met.

When leguminous green fodder is available: By feeding about 5.5 kg straw and 9.5 kg of leguminous green fodder the nutrient requirement of adult animals can be met.

Gestation ration : In pregnant animals, further allowance from the fifth month onward upto last quarter of pregnancy must be made for proper growth of the foetus and to keep the mother fit for optimum milk production on calving. For this, in addition to maintenance ration a further amount of 1.25 and 1.75 kg concentrate mixture is recommended for Zebu and CB cow/ buffaloes respectively. It is desirable to go for liberal feeding of pregnant animals particularly CB cows/buffaloes from 6 weeks before calving so as to secure full development of mammary glands for optimum milk production and to increase body reserves. For this 2.0 -3.0 kg of concentrate for Zebu and between 4.0 -5.0 kg for CB/ purebred Indian cattle/buffaloes over and above maintenance requirement are recommended.

The quantity of conc. mixture should be fixed on the basis of feeding value of forages and milk yield of cows and can be calculated as 0.1% of expected milk yield i.e. if a cow is expected to yield 2000 kg/lactation should be given 2 kg concentrate mix daily over and above the maintenance requirements. This practice of giving extra amount of concentrate mixture is known as steaming up. Animal eat-less feed until involution of uterus takes place so firstly dam utilize body reserve for milk production and involution takes place in 45-60 days. During last few days (about 10 days) before calving the amount of extra concentrate mixture should be reduced to 1.5 kg and 1-1.5 kg little warm bran mash should be given to the animal. Wheat bran contains adequate amount of crude fibre and phosphorus. Due to crude fibre it is laxative in nature and prevents constipation otherwise due to constipation prolapse may occur. Similarly phosphorous in wheat bran prevents its deficiency due to which there is prevention of metabolic diseases like postparturient haemoglobinuria. Before parturition animal should be provided cooked dalia and gur.

After partutition the cow should be given fresh warm water and warm mash consisting of 1 kg wheat bran, 1.5 kg crushed cereals, 0.5 kg gaggery and 25 g each of common salt, mineral mixture, ginger, ajwain and methi seeds. It will help in evacuation of placenta. Few days after calving concentrate mixture feeding should be started. The amount of concentrate mixture should be increased gradually by which after a week or two the animal can start consuming its full quota of concentrate requirement.

Production ration

It is the additional allowance of ration for milk production over and above the maintenance requirement. For zebu cattle 1 kg additional amount of conc.

mixture is required for every 2.5 kg of milk over and above the maintenance requirement while same amount of conc. is required for every 2.0 kg of milk production for C.B. cattle/buffaloes. The requirement of dairy cattle can be met easily by using thumb rule method but for buffalo where milk is extremely rich in energy due to high fat percent, thumb rule method does not work as efficiently as for cattle. The energy requirement for buffalo is not satisfied by thumb rule method. For high yielders with high fat percent in milk, this type of situation is very common. To overcome this sort of critical situation generally observed in buffaloes at least 10 kg extra amount of green fodders like paragrass, maize etc. should be supplied to meet the demands of energy requirement. An attempt to use the quantum of straw to fulfill the energy requirement is likely to fail as buffaloes will not consume 9 kg of straw for reasons of unpalatability.

Feeding of high yielders : It is a common experience that nutrient requirements, particularly energy requirements, of high yielding cows above 15 kg milk/day and buffaloes above 10 kg milk/day are difficult to meet through normal concentrate mixture and fodder because of their limited rumen volume. Once the rumen is filled, animals refuse to consume more feed whatever may be the quality till the previous material passes into the lower digestive tract. For this category of animals, feeding of high amount of concentrate mixture is also not advised as it affects the fat content of milk and may also cause acidosis. Therefore, a concentrate mixture of high energy density comprising either whole oilseeds such as cotton seed, sunflower seed or expeller pressed oil cakes are recommended. Sometimes rice polish having 15-18% oil content is used but problem of rancidity is encountered as the oil gets rancid during the storage of rice polish as a result of endogenous lipase activity. Therefore either fresh or parboiled rice polish may be used in animal ration. High yielding cows may be given the same quantity of concentrate mixture even after declining the milk yield during the last phase of lactation to replenish their body reserves. After drying, such animals may be provided good quality fodder ad lib and at the last quarter of their gestation one kg conc. mixture should be provided for the normal development of foetus.

Feeding systems for dairy animals

A. *Separate feeding system (conventional system)* : In this system roughage is group fed or placed before the animal whole day where as concentrate is offered individually on the basis of body weight, milk yield, milk fat percent, pregnancy and growth. In this system overfeeding of concentrate may lead to inefficient use of concentrate, reduced fat content of milk and health problems.

B. *Complete feeding system* : This system for dairy animals is now a days used to simplify the process of feeding by saving labour, reduced cost of storage (on building) and by providing balanced diet. It also improves voluntary feed intake, digestibility of nutrients and efficiency of utilization of digested nutrients for production, all these ultimately increase production of animal and thereby provide more profit. There are different methods by which complete feeds are prepared. These methods are based on the ingredients and their processing used for complete feed preparation.

   i. Chaffed mixture in which chaffed roughages are evenly mixed with ground concentrate mixture.

   ii. Silage mixture in which chaffed silage is evenly mixed with concentrate at the time of feeding.

   iii. Mash mixture in which ground roughage and ground concentrate mixtures are mixed evenly.

   iv. Pellets/ Blocks in which roughage and concentrate are ground and mixed along with a binder then converted to pellets/ blocks.

Feeding of sheep

Sheep posses a unique ability to survive on natural grasses, shrubs and farm waste products like residues of the field. With their small muzzle and split upper lip, they can nibble tiny blades of vegetation which can not be eaten by large animals. Maintaining sheep on concentrates is not the natural method. They consume 2.5 to 3.0 kg dry matter per head per day when fed good quality roughages in the stall.

Feeding of lambs : Just after birth lamb should be provided colostrum from ewe. Cow colostrum can be used if colostrum from ewe is not available. Colostrum feeding should be followed by milk feeding. If milk is not available milk replacer feeding should be practiced.

Creep feeding : It is the practice of providing the nursing lambs the opportunity to eat supplemental feed. It is the first step in getting lambs to market at an early age.

Lambs begin to nibble creep feed at about 1 week of age however upto about 3-4 weeks of age the ewe's milk provides the majority of nutrients. Although lambs upto 3-4 weeks of age do not consume sufficient amount of creep feed, the small amounts consumed help in establishing rumen function and habit of eating. Upto 12 weeks of age lambs suckling the mother should be supplemented with creep mixture.

Composition of creep mixture

Maize flour	67%
GNC	10%
Wheat bran	10%
Fish meal	10%
Mineral mixture	2%
Common salt	1%

To 100 kg of above feed, addition of 150 g TM-5 and Vitablend (Vit. AB_2 D_3) gm should be done.

Feeding of growing and finishing lambs : Feeding of growing and finishing lambs depends upon the availability of fodder.

i) When good quality fodders or their hays are available then following quantities of concentrate may be fed

Body wt. (kg)	Concentrate mixture (g/day)
10-15	50
16-25	100
26-35	150

ii) When good quality fodders are not available and animals are kept on mature grasses like straws, stovers etc, then following quantities of concentrate may be fed

Body weight (kg)	Concentrate mixture (g/day)
10-15	300
16-25	400
26-35	600

Feeding of ewe flock : Feeding of ewe flock is closely related with its breeding cycle. Accordingly it can be divided into 3 periods.

i. *Feeding during resting period* : The nutrient requirements during this period are low and ewes can be kept on a low level of feeding. During this period if the ewes maintain their body weight or gain 0.02–0.03 kg per day then there is no need of supplementation. On the contrary, if the ewes loose their body weight, grazing should be supplemented with additional conc. mixture.

Flushing : It is the practice of increasing the feed intake of ewe prior to breeding so as to increase the weight, ovum production and number of twins. It is usually done either by allowing the ewes to graze excellent pasture or by feeding 200-300 gm. of grains/head/ day.

ii. *Feeding during gestation period* : The gestation period of ewes ranges from 144-152 days. First half of gestation period is less critical nutritionally but ewes should be provided 50 percent more nutrients than the maintenance needs during last 6 weeks of pregnancy.

iii. *Feeding during suckling period* : After lambling ewes should be properly fed. To avoid udder troubles, reduced amount of grain mixture should be fed for first 10 days after lambing. During this period leguminous hay should be provided ad lib. The amount of conc. should be gradually increased upto 250 -300 gm. After 3 months only maintenance ration should be given.

Feeding of goats

Goats have peculiar feeding habits. They are all together different from other species of domestic livestock in choosing and consuming a diet. They discriminate plant fractions or feed particles which appear identical, by consuming one and leaving the other. They prefer fresh, clean and previously untouched diet. They can distinguish between bitter, sweet, salty and sour tastes and have high tolerance for bitter tastes than cattle.

For rearing of goats on scientific lines their feeding schedule should be as under:

Feeding of Kids : Kids must receive colostrum from the doe within one hour after birth and should continue for 3 days. Weaning should be done just after the birth of the kids in a large herd. Kids may be fed colostrum or dam's milk by bottle or pan. After colostrum feeding, whole milk or milk replacer feeding should be done. Milk replacer feeding may be continued till the kids are 4 months old. At 3-4 weeks of age good quality legume hay and starter feed should be started so as to provide early rumen development.

Feeding after 4 months : From 4 months to breeding, kids may be fed good quality roughages to provide enough nutrients for normal growth. With low quality roughage supplementation of 350-400 g concentrate mixture is needed.

Finisher ration : Goats to be used for meat purpose should be provided 30-40% of the dry matter from roughage source and the balance amount from conc. mixture containing 12-14% protein and 60-65% TDN. This will provide lean meat.

Feeding of pregnant goats : Feeding of pregnant goats during last 6-8 weeks of gestation should be given due consideration. During this period if quality fodder is not available 400 gm concentrate should be given every day. With quality fodder only 250 gm concentrate will be sufficient. Mineral mixture lick should be available free choice so as to take care of calcium and phosphorous requirements of dam and foetus. A few days prior to kidding amount of conc. mixture should be reduced to half and bran should be added in the diet.

Feeding of lactating goats : Nutrient requirements of lactating goats are higher. They should be provided 400 gm conc. mixture with poor quality roughages. If high quality roughages like lucern, berseem are available then their maintenance requirement is satisfied. Additionally they should be provided 350 gm conc. mixture per litre of milk production.

Feeding of pigs

Pigs are non-ruminants, highly prolific among meat producing livestock, exceed all other type of farm animals in the efficiency of converting feed into high quality pork. In India, in recent years a large of piggeries have come up where pigs are commercially reared. A skillfully formulated feeding programme is essential for a profitable piggery unit. Feeding of pigs to be followed should be as under.

Feeding of piglets : Just after birth piglets are fed colostrum up to 4 days. Thereafter sow milk is fed and piglets are gradually shifted to creep feed. The feeding depends on body weight.

Creep feed : A ration to be fed up to 20 kg live weight when the litter with sow. The composition of creep feed is given in table 5.

Table 5. Composition of creep feed

Ingredients	Quantity (kg)
Ground Yellow Maize	40
Skimmed milk	10
Ground Nut Cake	10
Til cake	10
Wheat bran	10
Molasses	10
Fish meal	6
Brewer's yeast	2
Mineral mixture	2
Vitamin mixture AB_2, D_3	0.01

Pig growth meal : A ration to be fed from weaning up to 35 kg live weight. Compostion of pig feed is given in table 6.

Table 6. Composition of pig growth meal

Maize	30.00 kg
Ground Nut Cake	20.00 kg
Wheat bran	40.00 kg
Fish meal	7.50 kg
Mineral mixture	2.50 kg
Vitamin mixture $AB_{2,}D_3$	10.00 gm.

Pig finishing/ breeding meal : A ration to be fed to pigs over 35 kg live weight is given in table 7 and 8.

Table 7. Composition of pig finishing / breeding meal

Ingredients	Quantity (kg)
Maize	40
Wheat bran	30
GNC	12
Til cake	10
Fish meal	5.5
Mineral mixture	2.50
Vitamin mixture $AB_{2,}D$	0.01.

Table 8. Gestation ration (cereal based)

Ingredients	Quantity (kg)
Maize	50
GNC	20
Molasses	5
Wheat bran	18
Fish meal	5
Mineral mixture	2
Vitamin mixture $AB_{2,}D$	0.01

Furrowing ration : It is a good practice to feed highly and with bulky feeds from 4-5 days before and after furrowing by substituting wheat bran, oats, ground legume hay or dehydrated lucerne meal. At furrowing about one third of the ration may be made up of these bulky feeds. Adding bulk to the ration at furrowing help to prevent constipation and reduce problem of mastitis-mertitis agalactia. There should be a good supply of fresh water at furrowing time.

Lactation ration : The feed requirement for lactating sow is more. Lactating sow require 3.5 kg of meal with 0.2 kg meal per piglet in the litter. Crude protein content of feed should be 16%. A full feed of high energy low fibre diet is necessary for maximum milk production during lactation.

□□□

Harnessing Advanced Technologies

33

Precision in Commercial Agriculture

RAJEEW KUMAR, VINEETA AND RAMPHOOL PUNIYA

Commercial agriculture is the need of hour due to increasing crop competition, profits from crop quality as well as quantity, costs and productivity improvement, more sophisticated machineries and equipment per farmer and sustainable growth. Environmental concern for land and environment conservation, pollution reduction and natural disaster damage control also pressurised to scientists and farmers evolve to new farming technology. Precision agriculture is the alternative, which facilitates the commercial agriculture by catering above mentioned things. Précising farming is an emerging concept in commercial agriculture. It is a micro management system to arrive at improved agricultural and land management decisions that result from using information delivered by geospatial technologies. Precision farming implies a management strategy to increase productivity and economic returns with a reduced impact on the environment. It is based on the application of information technology to a description of variability in the field, variable-rate operations and the decision making system.

Precision agriculture

Precision agriculture has been defined as the application of technologies and principles to manage spatial and temporal variability associated with all aspects of agricultural production. Precision agriculture has the capacity to enhance production and protect the environment while conserving soil and water resources. Precision conservation is a technique, where use of precision technologies and procedures, across spatial and temporal variability, to achieve conservation objectives. By definition precision agriculture is a suit, where variability among the agriculture system is catered by using different technologies related to automated crop management, weather independent system (RS) and decision support system (MIS ,GIS, Crop Model) are used for precise estimation and application of resources considering individual as a unit (a plant, an animal etc). It aims towards achieving socio-economic and ecological sustainability. Precision agriculture has been touted as "agriculture of the future" by which production profitability will be increased, agrichemical use will be reduced, nutrient use efficiencies will be increased and off-field movement of soil and agrichemicals will be reduced.

Precision agriculture facilitates conservation and production better than whole-field, conventional management practices. However, only a few studies have been conducted to determine whether these management strategies meet this high expectation. Most field-level studies have been either computer model-driven comparisons, or statements of reduced loadings with precision applications. Further, these studies typically focus on only one or two management practices/inputs when comparing site-specific management to uniform management. Results from these studies have been mixed. Often the likelihood that a precision agriculture approach improved production and/ or environmental parameters was dependent on the degree and type of variability initially found in the experimental area. Some studies have found that the decision rules developed for uniform management were insensitive to management recommendations when developing a site-specific plan. For some aspects of management [e.g., nitrogen], seasonal climate influences have had more impact on production than spatial field variability, and thus temporal information may dictate the optimal management.

Dimensions of precision agriculture

The management of agricultural production is undergoing a change, both in philosophy and technology. Until recently, agricultural managers have generally made decisions regarding fields based on average conditions within those fields, with data that was often sparse and qualitative in nature. Soil fertility was determined by compositing soil cores into a single sample that was intended to best describe conditions across a field. Field scouting for crop conditions or pest infestations was done at a few locations within the field, and observations often have been more qualitative than quantitative. For the most part, whole fields have been considered to be the basic agricultural production units, and have been managed for the mean condition or, in the case of pest management, managed intensively to overcome variability within that field.

Components of precision agriculture

Precision agriculture has three components: i) capture of data at an appropriate scale, ii) interpretation and analysis of that data, and iii) implementation of a management response at an appropriate scale and time. Each particular manageable factor has its own scale of variability. Area-wise management of insects and weather forecasting for crop management decisions are examples of variables that are managed at a scale larger than the individual field. Other factors like soil fertility and pest distributions can vary significantly at the subfield level and over the growing season. Therefore, it is natural and important to perceive precision agriculture in terms of finer spatial or temporal units of decision making:

- Capture of data at an appropriate scale,
- Data Management, interpretation, analysis and
- Output implementation of a management response at an appropriate scale and time

Enabling technologies

A fascinating aspect of precision agriculture is that a single technology is not being undertaken to improve a single practice. Instead, across the crop-production sector of the United States, precision agriculture is emerging as the convergence of several technologies with application to several management practices. However, every technology is not necessarily required or applicable for every practice on all crops, and development and enhancement of several of the potentially relevant basic technologies are being driven by forces outside of the agricultural sector. Thus, it is difficult to develop a generally accepted view of the dimensions of precision agriculture. Every area of information technology microelectronics, sensors, computers and telecommunications is in an evolutionary process of continuous improvement. As these introductions take place, some products will become economically feasible for agricultural applications. Research and development of many technologies used in precision agriculture have occurred outside the agricultural community. Precision agriculture technologies such as the global positioning system (GPS), geographic information systems (GIS) and remote sensing have their core constituencies outside agriculture. Crop and soil sensors operating on farm machinery, variable-rate fertilizer applicators, and yield mapping systems are technologies that have been developed within the agriculture sector by private industry.

1. Geo-referenced information

Geo-referencing refers to relationships among data based on their geographic locations. This spatial emphasis implies a new way of looking at agricultural information and site variability. Although spatial variability has always been recognized, data comparisons have often been made without specific information on site location, yielding qualitative results. Comparisons of data detailed from specific locations which are obtained from specifications by using various precision agriculture methods will be one of the important new techniques that can improve farm management. The value of a database for precision agriculture practices increases when the data layers are spatially referenced to each other. Co-registration of data will become critically important as management units get smaller and as more precise field data (location precision from sub-meter to a few centimeters) become available. It is expected that data referenced to physical location will allow different types of information to be compared and quantitatively analyzed at multiple locations. For example, physical properties of soil core samples collected from a field could be compared with other spatially explicit data available for decision making, such as characteristics of the mapped soil unit and topography, yield monitor data, and irrigation, nutrient, or pesticide applications recorded during variable-rate applications.

2. Global positioning system

The Global Positioning System (GPS) is a system of satellites, emitting electronic signals that can be received by mobile field instruments sensitive to the transmitting frequency. Positioning is achieved through the use of simultaneously received satellite transmissions from four or more satellites above the horizon. With a constellation of 24 satellites, any location on earth can have four or more satellites in view for 24 hours each day. By referencing the satellite's exact location and the time the signal takes to travel between the transmitter and the receiver, the location of the receiver can be determined by triangulation. Use of the GPS receiver allows latitude and longitude coordinate information to be associated with data obtained from a specific site on the field. The GPS can also be used to provide navigational guidance, enabling a producer to revisit a spot in the field and check the efficacy of management decisions. The GPS is an essential field component for most mapping-based precision agriculture and other measurements of field characteristics that would be used to determine product application maps. Even for operations with real-time sensing and control of inputs, GPS positioning will be valuable. If the sensed parameters and application rates are recorded and geo-referenced, these data can be included in the management database. Adoption of GPS and other spatial referencing technologies will have a widespread impact on data collection and analysis.

The positions provided by GPS receivers currently are not sufficiently accurate for dynamic real-time precision agriculture uses. Various errors, including those introduced by the U.S. Department of Defense (DOD) for security purposes (selective availability), contribute to the inaccuracy (National Research Council, 1995c). The present system under selective availability has an accuracy of about 100 meters. However, technical solutions are available to improve the positioning accuracy. A technique known as differential correction is widely used to remove the effects of the error sources. Position error is determined by using one or more fixed base stations to compare the calculated position with the station's known location. By combining the error values with the GPS signals, position accuracy can be improved to about 2 meters or less. The augmented positioning is known as differentially corrected GPS (DGPS). These corrections can be made either by software in a post processing operation or by hardware for real-time positioning. Most precision agriculture operations require the availability of real-time positioning, necessitating the transmission of the differential correction signals to the GPS receivers in the field. Differential correction procedures are cumbersome, prone to signal loss, or expensive depending on the method used to generate and transmit the differential correction signal.

3. Geographic information systems and mapping software

Digital geographic data that can be stored, analyzed, integrated, and displayed in different representations, form the core of precision agriculture. The software packages used to handle such data, Geographic Information Systems (GIS), are available with a wide range of capabilities and costs, but all are able to graphically display geo-referenced data. Although a single data layer (*i.e.*, yield data) can be mapped with the use of less-sophisticated software, more complex relationships (*i.e.*, temporal patterns or multivariate comparisons) are best performed with full-function GIS packages. The data layers derived from combinations of raw data can generate information about spatial variability among factors in crop production.

4. Yield mapping systems

Yield mapping systems record the relative spatial distribution of yield while the crop is being harvested. These systems collect geo-referenced data on crop yield and characteristics such as moisture content. The resulting maps can dramatically illustrate the areas of yield variability from either natural processes or agricultural practices. Because yield is a primary factor in most management decisions, precise yield maps are desired to confirm spatial treatment decisions.

5. Variable-rate technologies

Precision agriculture was pioneered by domestic U.S. industry, beginning with the conception and implementation of Variable-Rate Technology (VRT). VRT applicators spatially vary the application rates of agricultural inputs such as seed, fertilizer, and crop protection chemicals. VRT systems include specialized controllers that vary specific material flow rates, even multiple product rates simultaneously, in response to a desired change in the local application rate (on-the-go). VRT systems can be designed in different ways depending on the products to be applied and the source of the information utilized to specify local rates.

Present commercial VRT systems are either:

- Map-based, requiring a GPS/DGPS geo-referenced location system and a command unit that stores an application plan of the desired application rate for each location within the field, or
- Sensor-based, which does not require a geo-referenced location system, but includes a dynamic command unit that specifies application through real-time analysis of soil and/or crop sensor measurements, for each location within the field as it is encountered.

6. Ground-based sensors

Basic research is needed to investigate soil and crop processes applicable to development of ground-based sensing systems. Sensors offer the opportunity to automate collection of soil, crop and pest data at a level of intensity not economically feasible with manual sampling and laboratory methods. Fields are highly heterogeneous. Increased sampling will result in accurate characterization of within-field variability. Improvements to VRT and crop modeling are expected to advance rapidly with a higher spatial density of measured soil and crop parameters. Sensors are needed that are fast, efficient, and can assess factors important to crop production. Sensors have been developed or are under way to measure soil and crop conditions including soil organic matter, soil moisture content, electrical conductivity, soil nutrient level, and crop and weed reflectance.

7. Remote sensing

Remote sensing—the acquisition of information from remote locations such as an airplane or satellite—is a potentially important source of data for precision agriculture. In the long term, remote sensing could provide numerous forms of information, both spatially and temporally. However, improvements are needed in the analytical products and delivery systems if remote sensing is to meet its promise for precision agriculture.

A sequence of remotely sensed images over time can provide information about crop growth and the spatial variation within fields. Detailed spatially distributed multitemporal information, in visual form, is not readily obtainable from conventional crop management systems or from site-specific crop management methods. Remotely sensed images (*i.e.*, color infrared aerial photographs or multispectral images acquired from satellites or airplanes) show spatial and spectral variation resulting from soil and crop characteristics. These images show the state or condition of fields when the images were acquired. One of the most useful aspects of remote sensing is its ability to generate images showing the spatial variation in fields caused by natural and cultural factors.

8. Decision support systems

Decision support systems (DSS) are used in agriculture for tactical, strategic, and policy-level decision support. Because producers are continually faced with making tactical decisions, such tools are becoming increasingly useful on the farm. However, few DSS are in general use by agricultural producers today, in part due to difficulty in use and limited information provided—from their point of view. They have been used to aid in decisions that are complicated by large amounts of information and data. Data collected by a consultant, obtained through a weather forecasting service, or acquired through a sensing operation are analyzed and linked with appropriate decision rules that identify actions to assist in producer decision making.

How precision farming can be helpful in improving crop productivity

Agriculture, the dynamic system governed by several biotic and abiotic factors, needs to be sustained. There is a need to transform low-yielding food production systems in to high yielding ones through the convergences of agrotech (mainly production related), biotech (productivity related) with space technology (RS and GIS). Precision farming is a truly comprehensive approach which covers all phases of production from planning to post harvest. Precision farming basically depends on measurement and understanding of variability. The PF can help to increase the crop productivity by following ways

- Variation in soil nutrient supply may be most significant within a large field, or among many small fields within a landscape. Both spatial and temporal variations are catered by precision farming with aims at improving the input - output characteristics of the soil and crop system as they vary in space and time.
- Information and management technology are combined into a production systems to cater the variability, that can increase production efficiency by allow more efficient input use through efficient application.

- In precision farming, instruments are able to complete in-field operations like cultivation, seed sowing, application of fertilizers and pesticides and harvest timely, which further enhance the crop productivity.
- Precision farming provides sufficient understanding of the processes involved to apply inputs in such a way as to be able to achieve a particular goal. The goal, however, might not necessarily mean maximum yield but may be to optimize financial advantage while operating within environmental constraints.
- Precision farming involves the use of spatial asset allocation and management to distribute available time and money where it is most needed and will provide the best return.
- Precision agriculture provides tools for tailoring production inputs to specific plots within a field, thus potentially reducing input costs, increasing yields, and reducing environmental impacts by better matching inputs applied to crop needs.
- Precision agriculture technology is paving the way for agricultural producers by allowing for precise management of inputs. The appropriate management processes and information needs vary among different environments, but also among different decisions to be made.

Prospects of precision farming in india

This system is not for developing countries like India, where the farmers are poor, farming is mostly subsistent and the land holding size is small. Although precision farming research in India is meager but the results obtained from different components of precision farming have been exciting. Followings are the pioneer workers in area of precision agriculture in India.

Tata Kisan Kendra (TTK) : The concept of precision farming being implemented by the TKKs has the potential to catapult rural India from the bullock-cart age into the new era of satellites and IT. TCL's extension services, brought to farmers through the TKKs, use of remote-sensing technology to analyze soil, inform about crop health, pest attacks and coverage of various crops predicting the final output. This helps farmers adapt quickly to changing conditions. The result: healthier crops, higher yields and enhanced incomes for farmers. Some of the research institutes such as Space Applications Centre, ISRO, M.S. Swaminathan Research Foundation, Chennai, Indian Agricultural Research Institute, New Delhi, and Project Directorate of Cropping Systems Research, Modipuram, had started working in this direction and in soon it will help the Indian farmers to harvest the fruits of frontier technologies without compromising the land quality.

Despite the many obstacles listed earlier, business opportunities for precision farming technologies including GIS, GPS, RS and yield monitor systems are immense in many developing countries. The scope for funding new hardware, software and consulting industries related to precision agriculture is gradually widening. Punjab and Haryana states, where farm mechanization is more common than others, may be the first to adopt precision farming on a large scale. Recently, the governments of certain Asian countries initiated special efforts to promote precision farming. Precision farming is useful in many crops like rice, wheat, sugar beet, onion, potato and cotton among the field crops and apple, grape, tea, coffee and oil palm among horticultural crops are perhaps the most relevant. Nutrient stress management is another area where precision farming can help Indian farmers. Most cultivated soils in India are acidic and spatial variation in pH is high. Detecting nutrient stresses using remote sensing and combining data in a GIS can help in site-specific applications of fertilizers and soil amendments such as lime, manure, compost, gypsum and sulphur. This in turn would increase fertilizer use efficiency and reduce nutrient losses. In semi arid and arid tropics, precision technologies can help growers in scheduling irrigation more profitably by varying the timing, amounts and placement of water. For example, drip irrigation, coupled with information from remotely sensed stress conditions (*e.g.*, canopy temperature) can increase the effective use of applied water from 60 to 95 per cent there by, reducing runoff from 23 to 1 per cent and deep percolation from 18 to 4 per cent. Pests and diseases cause huge losses to Indian crops. If remote sensing can help in detecting small problem areas caused by pathogens, timing of applications of fungicides can be optimized.

□□□

34

Advances in Production of Quality Planting Material

VIRENDRA SINGH

The procurement of good quality seed and vegetative material needs greater care for the production of quality planting stock. The planting stock is used to raise the plantation of different trees at different sites under social forestry/ reforestation programmes. One of the important elements of the social forestry is the adoption of agro-forestry practices that provide fuel, fodder, small timber, fruits, etc. alongwith agricultural produce (food grains). The planting stock determines the future potential of the programme as better the planting stock, greater is the future yield. Therefore, it is vital initially to collect the good quality seed and the vegetative material for the production of quality planting material required for healthy plantations.

Collection of seeds

In India, due consideration to the genetic gains of the seed source is not given and seed collection is generally made from local stands. With few exceptions, there is no or very limited number of organised seed supply establishments in the country. Thus, the natural or artificial stands in India are the main source of tree seeds (Khanna, 1984). Scientifically, the stands are neither classified nor managed for seed production purposes. The seed is usually collected by labour at the schedule rates fixed by the Forest Department of the State concerned. The schedule rate, in general, is so low that the labourers hardly think about quality of the collected seeds. Such seed is not only collected from malformed trees but also from the ground because their main criteria are to collect as much quantity of seed as they can to fetch more money. Such seeds are usually of low quality. The growth rate of the seedlings obtained from such seeds is inferior and hence do not produce quality planting stock. The selection of seed source is the most important aspect of quality seedling production in nursery (Mc Millan, 1979; Dirr and Heuser, 1987). In the past, we were not much concerned about the collection of quality seed, but these days, there have been an increasing demand for the information of original seed source and the kinds of the trees or stands from which seed has been collected/ is to be collected. This has necessitated the cautious collection of the seed, especially the selection of areas to produce seed of improved genetic quality. Some progress has been made to identify seed production areas, plus trees selection and establishment of seed orchards for some economically important species in different states of India. Following are the guidelines that may be followed to select the trees for seed collection (Stein *et al.*, 1974).

- The seed should be collected only from healthy, vigorous trees of reasonably good form that are making average or better growth.
- The seed should not be collected from isolated trees of naturally cross-pollinated species because these are likely to be self-pollinated, and the proportion of viable seeds from such trees is generally low and the seedlings originating from such seeds are weak.
- The collection of seed should be made from nearly mature trees as the seed from mature trees have high viability.
- The collection of seed should be avoided from areas containing numerous poorly formed abnormal, disease and insect infested trees.
- The seed collection from unproven plantation of unknown origin should be avoided.

Generally, four types of seed stands are used for seed collection (Anon., 1974)

A. Unclassified stands

B. Classified stands

C. Seed production areas

D. Seed orchards

The seed obtained from unclassified stands, classified stands, seed production areas and seed orchards represents the increased genetic control of seed quality.

A. Unclassified stands

The naturally occurring stands are used to collect the seed. Such stands are not rated on the desirability of their trees as parents, but are used because seed collection is easy and convenient. The collection of seed from such stands may include open-grown trees as they produce abundant seed frequently. The bulk of the seed crop within stands is usually produced on the most vigorous and dominant trees receiving light both from above and sides. However, the proportion of viable seeds from such trees is less and the resultant seedlings are weak.

B. Classified stands

The way by which the genetic quality of existing supplies of tree seed can be maintained is to confine the seed collections to the best stands (Rudolf, 1948). This requires a survey of available stands for seed collection in each collection zone and classifying them according to quality. A seed collection zone is an area with defined boundaries and altitudinal limits in which soil and climate are sufficiently uniform to indicate high probability of reproducing a single ecotype (Anon., 1974). Different criteria are to be employed for different uses as timber production, fodder production, soil stabilisation, shelterbelts or environmental planting. The seed production is largely concentrated on the most vigorous and best formed trees.

C. Seed production areas

Seed production areas also called as seed stands are good natural or planted stands from which the phenotypically inferior trees are removed. The establishment of seed production area enables to procure improved seed in a short period of time for meeting the immediate seed requirements. The seed production areas is "a crop of vigorously growing, middle aged to mature trees of good quality, properly thinned and left out to contain trees of good vigour and well developed crowns with clear boles and managed exclusively for seed collection" (Khanna,1984). Management of seed production area is also easy.

The establishment and management of a seed production area involve the following essential steps.

1. *Selection of stand*: A number of seed production areas may be required within each planting zone for continuous supply of planting stock adapted to a locality or site. The selected stands should be very well

adapted for the site and they should be away from the poor stands (Matthews, 1962). Small admixture of other species is acceptable if they do not cross pollinate with or seriously hinder the growth and development of the desired species. The stand should have a minimum area of 4 hectares with a topography that facilitates frequent access for the cultural treatment and seed collection. The stands should be well stocked to permit the selection of an adequate number of quality seed crop trees. The trees should have been developed under competition as it is not feasible to adopt the rigid stocking standard for the stand. The dominant trees should have at least 30 cm average diameter (Cole, 1963) or they should be old enough (middle aged) to produce a good quantity of seed, to be adapted to the site and should be superior over trees of average stands. The number of dominant trees may vary from species to species but should vary between 50-75 trees per hectare after thinning (Zobel and Talbert, 1984). Seed production areas may also be established in plantations provided they are of a known seed source and suitable for the area.

2. *Selection of crop tree*: The first step is the selection of best trees and to mark them as seed crop trees. Seed crop trees should be healthy, dominant and fast growing. They should also possess the following characteristics: -
   - The stem should be straight without any defect like knots, fluting etc.
   - The crown should be compact with large foliage area
   - The branches should be small with wide branch angle
   - The trees should be free from insect-pests and disease
   - The trees of the stand should have been in a state of flowering and fruiting

3. *Isolation zone*: The seed production area should be surrounded by an isolation strip to minimize the contamination of the selected trees by pollen from poor stands or undesirable trees. The isolation or dilution zone of at least 150 m wide should be established around the area. However, wider isolation zone may be required for those species in which pollen dispersal is greatly affected by the wind. As a general rule, the average spacing between seed crop tress should be equal to one-half the average height of the dominant and co-dominant trees of the stand. The removal of the competing tress may be completed in one or more felling at an interval of 5-10 years to check the excessive wind throw of the seed crops. Seed production areas for different tree species in India are in table 1.

Table 1. Seed production area (SPA) in India

State	Species	Area (ha)
Chattisgarh	*Tectona grandis*	301.30
	Emblica officinalis	10.00
Haryana	*Eucalyptus tereticornis*	2.00
Jammu & Kashmir	*Pinus wallichiana, Cedrus deodara, Robinia pseudoacia*	238.60
Jharkhand	*Dalbergia sissoo, Gmelina arboea, Tectona grandis*	249.90
Madhya Pradesh	Different species (*Tectona, Tamarind* etc.)	1366.00
Karnataka	*Tectona grandis*	316.60
	Santalum album	7.00
	Acacia catechu	10.00
Kerala	*Tectona grandis*	1100.00
	Casuarina equisetifolia	4.50
	Santalum album	22.60
	Dalbergia latifolia	5.30
Orrissa	*Casuarina equisetifolia Tectona grandis*	500.5
Uttarakhand	*Shorea robusta*	489.80
Tamilnadu	*Eucalyptus tereticornis*	8.70
	Tectona grandis	92.8
	Eucalyptus grandis	26.4

Source: Anon. (2003)

4. *Protection*: The protection of seed production area from wildfire, pests and diseases, undergrowth etc. is essential to get the uninterrupted supply of seed from these areas. The controlled fire or burning may be used as this provides the benefit of protection against certain insects and diseases. The undergrowth is removed from the stand to protect the trees from nutrient uptake competition. The area is also protected against insect pests and diseases by the application of suitable insecticides and fungicides. The protection against animal damage can be ensured by erecting a suitable fencing around the seed production area. Manure and fertilizers are also applied to encourage the flowering, fruiting and seed production.

D. Seed orchards

The main function of a seed orchard is to produce seed of improved genetic quality on regular and sustained basis. The establishment of artificially planted forest tree crop from harvested seed, a silvicultural method, had seldom been practised before the start of eighteenth century. The idea of establishment of plantation for the production of seed appeared in literature as early as 1787. The seed orchard of tree species has been set up for different purposes as to improve the quality and quantity of forest produce. The genetic improvement in seed orchard is usually greater than seed production areas and depends on the intensity of selection and the heritability of the traits on which selections are based. The first known seed orchard of *Larix eurolepis* was established in Scotland, Great Britain in 1931. The practice of establishing seed orchards for the production of quality seed in other parts of the world was started after the establishment of seed orchard in the Great Britain. The seed orchard establishment is relatively a recent concept in India. The seed orchard programme for teak was initially started in two states (Venkatesh, 1974). A seed orchard is "a plantation of genetically superior trees, isolated to reduce pollination from genetically inferior outside sources and intensively managed to produce frequent, abundant, easily harvested seed crops" (Zobel *et al.*, 1958). Organisation for Economic Co-operation and Development (O.E.C.D., 1974) has defined seed orchard as "a plantation of selected clones or progenies which is isolated from outside source, managed to produce frequent, abundant and easily harvested crops of seed". It is established by setting out clones (as grafts or cuttings) or seedling progeny of trees selected for desired characteristics. The usefulness of seed orchard has been stressed as a tool for supplying quality forestry (genetically improved) seed. Most of the orchards established before 1973 were of clonal material but there is a growing trend towards using seedling material, particularly for those species which flower at an early stage despite the fact that most of the breeders favour clonal seed orchards.

Objectives of seed orchard establishment: The seed orchards are managed with the following aims and objects.

- The seed orchards provide seed on a regular and sustained basis
- The genetically improved seed of forest tree species is produced
- The seed of known origin is made available from the seed orchards
- The seed harvesting is easy and the mechanized harvesting can also be employed in seed orchards
- The germination capacity of the seed produced in seed orchards is better than those produced under normal forest condition. Thus the seeds from seed orchards are superior in quality
- The seed orchards can also be used as breeding orchards. The controlled crossing between selected parent clones can be carried out to further improve the species.

Types of seed orchard: Seed orchards are of two types:

a) Clonal seed orchard

Clonal seed orchards are characterized by growing orchard plants from vegetatively propagated material usually taken from relatively mature plants. It is an orchard which has been raised by grafting clones in the form of scion or bud of plus trees on the stock of two or three years old seedlings raised at proper spacing in advance or by planting rooted cuttings of plus trees at proper spacing . Orchards of most tree species have been established with grafts although rooted cuttings have been used for a few species (*Libbey et al.*, 1972). The grafting may be done in the glass house or in the nursery beds. Sometimes, grafting is also done on plants established in plantations. The best period of the year for grafting depends upon the method of grafting, the species and the environmental factors at the place of grafting on orchard site or in the open nursery. A number of grafting methods have been described by Garner (1950) and many of them have been tried by different workers in the field of tree improvement (Nienstaedt, *et al.* 1958). The most commonly used methods are cleft, side and bark grafts. When both scion and stock are in favourable condition and grafting is properly done, successful unions occur in about 75 per cent of the grafts. There might be instances of some failures later on due to incompatibility between stock and scion, insect damage, diseases, inadequate root development after field planting or other damaging agents. Some of the advantages of clonal seed orchards (Masters and Beineke, 1973) are:

- Good genotypes are preserved.
- Maximum genetic profit gains are obtained.
- The genetic material and combining ability are known.
- Early flowering of the grafts take place.
- It is easier to manage for increased growth and flowering control
- Pressures to rogue for early seed production are less
- Problems with early mating are less.
- Full- sib progeny tests will eventually be available to verify and early roguing
- Improved seed is available before progeny testing is complete

b) Seedling seed orchard

It is a seed orchard, which had been raised from the seedlings obtained from seeds of plus trees. The names of plus tree species for different states of India are given in table 2. Such orchards take longer time to produce seed but have the advantage of having a broader genetic base combined with progeny testing. There are very few seedling seed orchards, which have been established

Table 2. Selected plus trees

Tree species	Number of trees	State
Acacia catechu	37	Maharashtra, Rajasthan, U.P., Uttarakhand
Acacia nilotica	48	Rajasthan
Albizzia lebbek	04 (candidate)	U.P.
Azadirachta indica	93	Tamilnadu, U.P., Uttarakhand, Rajasthan
Bombax ceiba	57	Arunachal Pradesh, Kerala, Uttarakhand, U.P.
Casuarina equisetifolia	91	Different states of India
Dalbergia latifolia	92	M.P., Karnataka, Kerala, Tamilnadu
Gmelina arborea	98	Arunachal Pradesh, Tripura
Pinus roxburghii	04	Uttarakhand
Prosopis cineraria	77	Rajasthan
Santalum album	79	Karnataka, Tamilnadu, Kerala, Andra Pradesh
Tectona grandis	1000	Maharashtra, Karnataka, Andra pradesh, Arunachal Pradesh
Shorea robusta	50	Himachal Pradesh, U.P., Uttarakhand

Source: Tewari (1994, 1995)

Seed orchard establishment

Location, establishment and management of seed orchard involve the goal of maximising the production of seed of high genetic and physiological quality as soon as possible. The minimum area of the seed orchard should not be less than 4 hectares. A list of seed orchards in different states of India is given in table 3. The establishment of seed orchard involves the following steps.

1. *Site selection*: The selection of site is one of the important criteria in the establishment of a seed orchard. The following points should be taken into consideration at the time of site selection for a seed orchard (Faulkner, 1964; Krugman, 1966; Limstorm, 1965; Matyas, 1969; Zavadil, 1969; Sarvas, 1970; Zasada, 1971; Beckers, 1972
   - The climatic conditions should be favourable for tree growth, flowering and continuous production of fruits and seeds
   - The location of a seed orchard should be within its natural zone but at short distance south or at slightly lower elevation than the location of the parent trees
   - The air drainage should be sufficient at the site to avoid the frost damage to flowering and fruiting
   - Protection of the site from strong winds, animal damage, insects etc. is necessary

Table 3. Seed Orchards in India

State	Species	Area (ha)
Chattisgarh	*Tectona grandis*	98.00
	Emblica officinalis	20.00
Haryana	*Dalbergia sissoo, Tectona grandis, Ficus benghalensis, Eucalyptus species, Acacia nilotica, Melia azederach etc.*	43.00
Jammu & Kashmir	*Robinia pseudo acacia,Cedrus deodara, Pinus roxburghii, P. wallichina, Abies pindrow*	178.55
Jharkhand	*Acacia catechu, Cassia siamea, Schleichara oleosa, Dalbergia sissoo*	60.00
Karnataka	*Azadirachta indica*	1.15
	Tamarind indica	170.00
	Casuarina equisetifolia	3.00
	Tectona grandis	44.20
Kerala	*Tectona grandis*	50.65
Maharashtra	*Tectona grandis*	6.50
Orrissa	*Tectona grandis,Casuarina equisetifolia*	12.37
Tamilnadu	*Santa album*	0.80
	Tamarind indica	6.00
Tripura	*Tectona grandis, Gmelina arborea* (Gamar)	1.70
Uttar Pradesh	*Tectona grandis Gmelina arborea*	2.62
	Dalbargia sissoo	15.80
	Acacia nilotica	6.00
Uttarakhand	*Cedrus deodara, Pinus roxburghii, P. wallichila, Fraxinus micrantha, Juglans regia, Abies pindrow*	214.00

Source: Anon.(2003)

- The soil of the site should be of intermediate fertility unless it is desired to grow the trees under high or very low soil fertility
- The site should be situated at a place where input sources as labourer, machine etc. could be managed easily
- The orchard site should be well drained and as level as possible
- The area should be accessible for heavy machines and equipments
- The site should be selected in the proximity of a permanent water source because the water is essential during the establishment of a seed orchard and during drought period

2. *Site preparation*: The basic site preparation work (trees and shrub clearance and drainage) should be conducted at least one year in advance of planting. To facilitate operation of mechanised equipment and to minimise the maintenance of equipment, it is essential to prepare smooth

ground. If ditch or flood irrigation is to be used, the ditches or ridges should be made during ground preparation.

3. *Isolation*: It is difficult to obtain complete isolation because wind borne pollen travel great distances. Thus, a seed orchard should be established in a large open area or in a stand consisting of unrelated species. If the seed orchard is established near stands of species with which orchard plants can cross-pollinate, the orchard should be surrounded by an isolation zone of unrelated species. The width of isolation zone may differ from one species to another. The minimum width of isolation zone has been recommended to be about 120 m (Squillace, 1967) or 150 m (Zobel *et al*., 1958).
4. *Planting*: Planting can be done in squares, rectangle or quincunxes. The planting of seedlings, grafts or cuttings to establish the seed orchard should be done at a sufficiently wide spacing, which can permit rapid growth and development of large crowns. The spacing, however, should not be much wider which can promote self-pollination (Fowler, 1965). The spacing should be uniform to increase easy operation of the machinery between trees in more than one direction i.e. cross cultivation. The initial planting distances in seedling seed orchards are normally much less than in clonal seed orchards. The distance varies from 0.6 m to 6.0 m in both the directions with the majority at 1.5 m to 2.0 m. Initial spacing in early stages of clonal establishment of seed orchard may vary from 2.5 m x 2.5 m to 5.0 m x 5.0 m and 6.0 m x 6.0 m for fast growing species (Schultz, 1969).

   At present the trend is to use wider initial spacing since it is now generally accepted as being more economic in the long run to accept lower early seed yields than to systemically thin expensively raised grafts before they are fully productive. It is essential to remove half of the trees before final spacing if planting had been done at closer spacing (Kellison,1969). Several clones or families are roughed out if they do not have combining ability or if their progeny lacks superiority. Such roughing results in irregular spacing. The final average distance between trees should be 6.0 m to 9.0 m (Zobel, *et al*., 1958). The recommended number of clones or families varies from 10 to 50 per orchard. The clones or families are so arranged in the seed orchard as to promote cross-pollination. More than 50 number of clones or families can be planted initially which enhances the possibility of raising the average seedling quality through the roughing of phenotypically inferior or poorly performing clones or families. The pot - raised stock generally have curled root systems and may create serious problems later on (Zobel, *et al*., 1958).Therefore, attempts should be made to loosen the roots before planting. The rooted cuttings in nursery are not generally used for clonal grafted orchards and root stock in many countries are established and grown for many species on the orchard sites before grafting (Larsen, 1956; Kieding *et al.*, 1964). The seed orchards of second, third and subsequent generations can be established with elite trees (best parents) after progeny evaluation at each level.

5. *Design*: Different types of designs can be employed to establish the clonal or seedling seed orchard. The term clone or ramet is used for clonal seed orchard and the term progeny or family is used for seedling seed orchard. There are a number of designs, which can very well be used.

6. *Ground Cover*: Weeds and other vegetation crops should be kept under strict control within a half meter radius of each plant until it is established. The ground may be kept bare by disking or treating with a suitable weed killer. Mulch around each tree may be beneficial during the first two years after planting.

7. *Fertilization*: Fertilizers containing nitrogen, phosphorus or potassium are commonly used. It is difficult to formulate rules and provide prescriptions for general guidance on their use. Effective fertilizer prescriptions depend upon the knowledge of the nutrient requirements of the plant and the amount of available mineral nutrients in the soil. Faulkner (1966) has emphasized the need for complete grass control if the seed trees are to benefit from the fertilisers. Good responses of fertilization on flowering and seed yield have been demonstrated by different workers on pine species (Brinkman, 1962; Schmidtling,1971) and in Douglas fir (Steinbrenner *et al.*, 1960 ; Harry *et al.*, 1968).The fertilizer application has shown an increase in seed weight and germination energy in *Pinus elliottii* (Mergen and Voigt, 1960).

8. *Irrigation*: Irrigation is essential to maintain the soil moisture at an optimum level through out the growing season to get maximum tree growth and seed yields. The optimum level of soil moisture regime can hardly be achieved by rainfall as its distribution is unequal and erratic. Irrigation can also be used to reduce frost damage in seed orchards. The duration and frequency of irrigation may vary from species to species. It is also depends on the climate, soil and other factors.

9. *Thinning*: The initial spacing between orchard trees should be sufficiently wide to support the full development of the crowns to obtain cone or fruit crops for several years before thinning is required. Narrow spacing can be used for those species which take short time to start flowering. The tree crown in an orchard should always be kept isolated to admit

Table 4: Nursery technology of some important tree species of Uttarakhand

S. No.	Scientific/ Local name	Fruit/ Seed collection time	Seed storage capacity	Pre- treatment	Germi- nation (%)	Sowin g time	Depth of so- wing (mm)	Spacing (cm)	Age in nursery (Month)
1.	*Bauhinia variegata* (Kachnar)	May-June	1 yr	Soaking in water for 24 hrs.	90-95	May-June	5-10	20 x 15	4-5
2.	*Celtis australis* (Kharik)	Oct-Nov	1-2 yrs	Soaking in hot water for 12-24 hrs.	60-70	Feb-March	20	15 x 15	10-12
3.	*Grewia optiva* (Bhimal)	Oct-Dec	1-2 yrs	Soaking in hot water for 12 hrs.	65-70	March-April	20	20 x 10	10-12
4.	*Acacia catechu* (khair)	Jan-March	1-2 yrs	Soaking in water for 24 hrs.	60-70	March-April	15	8 x 8 or poly bags	4-5
5.	*Robinia pseudoacacia* (Robinia)	June-Sept.	2-3 yrs	Acid scarification for 10 minutes	75-80	March-April	20	20 x 10	10-12
6.	*Melia azederach* (Bakain)	Jan-Feb	2-3 yrs	-	75-85	Feb-Mar	20	15 x 15	11-12
7.	*Terminalia chebula* (Harar)	Jan-March	1 yr	24 hours soaked seeds are to be put in the cow-dung	60	Jun-July	20-25	15 x 15	12
8.	*Terminalia arjuna* (Arjun)	April-May	1 yr	Soaking in water for 96 hrs	70-80	April-May	20-25	20 x 10	3-4
9.	*Terminalia bellirica* (bahera)	Nov-Feb	1 yr	Acid scarifica-tion for 12 minutes	50-60	March-May	20-25	20 x 20	3-4
10.	*Albizzia lebbek* (Kala siris)	Jan-March	3-4 yrs	Soaking in water for 24 hrs	60-70	March-	10-15	15 x 15	4-5
11.	*Fraxinus micrantha* (Angu)	Sep-Oct	1 yr	Not required	50-70	March-April	20	15 x 15	24
12.	*Alnus nepalensis* (Uttis)	Dec-Jan	Less than one yr	Stratific-ation	60-70	March-April	Surface	15 x 10	05
13.	*Toona ciliata* (Toon)	April-May	Less than six month	Not required	60-70	May-June	Surface	15 x 15	12
14.	*Adina cordifolia* (Haldu)	Feb-April	1 yr	Not required	50	May-June	Surface	20 x 10	12
15.	*Acer oblongum* (Maple)	Jan-April	Less than one yr	Not required	50	March-April	10	20 x 20	3
16.	*Quercus leucotrichophora* (Banj)	Dec-Feb	Short period	Not required	70-80	Dec-Feb	20	20 x 25	18-24
17.	*Quercus floribunda* (Moru)	Aug-Sep	Short period	Not required	70-80	Aug-Sep	20	20 x 15	24
18.	*Quercus semecarpifolia* (Kharsu)	June-Aug	Short period	Not required	65-70	June-Aug	20	20 x 15	12-13
19.	*Cedrus deodara* (Deodar)	Oct-Nov	Less than one year	Stratifi-cation	80	March-April	10	15 x 25 or Poly bags	24-28
20.	*Pinus roxburghii* (Chir)	Feb-March	1-2 yr	Not required	80-90	March-April	15	Poly bags	4-6

Contd...

21.	*Abies pindrow* (Raga fir)	Oct-Nov	Less than one year	Stratification	30-50	March-April	15-20	20 x 10	36-48
22.	*Picea smithiana* (Parsaral)	Oct-Nov	More than one year	Stratification	40-80	March-April	10	15 x 15	28-30
23.	*Jujlans regia* (Akhrot)	Sep-Oct	1 yr	Stratification	70-80	November	50-60	20 x 20	24-36
24.	*Betula alnoides* (Saud)	May	1 yr	Not required	60-65	June-July	Surface	10 x 10	7-8
25.	*Ouginia oojeinensis* (Sandan)	May-June	1 yr	Soaking in water	75	May-June	10	15 x 15	3-5
26.	*Sapindus mukorossi* (Ritha)	Dec-Feb	1 yr	Soaking in water	90-95	March-April	10-20	15 x 15	13-15
27.	*Bombax ceiba* (Semal)	March-May	1-2 yr	Not required	60-75	May	Surface	20 x 20	12-13
28.	*Rhododendron arboreum* (Burans)	Jan-Feb	1 yr	Stratification	50-65	March-April	Surface	20 x 20	12-24
29.	*Dalbergia sissoo* (Shisham)	Dec-Feb	1 yr	Not required	80-90	March	15	10 x 10	4-5
30.	*Aesculus indica* (Pangar)	Oct-Nov	Short period	Stratification	80-85	Oct-Nov	40-60	15 x 15	8-9

Source: Troup, 1921; Khullar et al. (1991); Luna, 1996

sunshine and to permit the easy movement of cone or seed harvesting equipments. A systematic approach of spacing calculation in orchards of proven good genotypes can be practised on the basis of sun exposure on the crowns. Asakawa and Fujita (1966) obtained a substantial increase in seed production after thinning. Orchards in which thinning have been neglected may show a marked decrease in seed production which can last for several years even after thinning are made.

10. *Crown pruning*: An important aim of pruning is to retard height growth to make seed harvesting simple, cheaper and easier. Pruning has been tried to shape the crown or limit height growth to facilitate cone and fruit collection. However, severe pruning in Norway spruce (*Picea abies*) and Scots pine (*Pinus sylvestris*) have been found to reduce cone production.
11. *Protection*: Seed orchards are essentially mono culture and as such are highly vulnerable to damaging agents. Orchards should be established on sites suited to the species, but well isolated from stands of the same or closely related species. The high investment in seed orchards calls for a intensive control measures to protect them from insets, diseases, birds, frosts, ice and snow, fire and destructive animals especially those that eat seeds. Each orchard is subjected to its own particular damaging agents.

References

Anderson, E. 1963. Seed stands and seed orchards in the breeding of conifers. In World Consult. Forest Genet. Tree Improvement. F.A.O./FORGEN 8/1. 16p.

Anon., 1974. Seeds of woody plants in the United States. Agriculture Handbook No.450. Forest Service, U.S. Department of Agriculture, Washington, D. C.

Anon., 2003. Forestry Statistics India- 2003. Indian Council of Forestry Research & Education, Deharadun. 118 pp.

Asakawa, S. and Fujita, K. 1966. Studies on the management of seed stands. I. The establishment of experimental seed stands of *Pinus densiflora* and *Larix leptolepsis* and the results obtained for three years (1962-1964). Bull. For. Exp. Sta., Meguro. Tokyo, **184**: 81-134.

Beckers, B. 1972. Factors Influencing Seed Production In Douglas Fir. Bull. Soc. r. for. Bela: **79(4)**:253-261.

Brinkman, K.A. 1962. Fertilizers increase seed production of shortleaf pine in Missouri. Tree Plant. Notes. **53**:18-19.

Cole, D. E. 1963. Management of pine seed production areas. Proc. South. Conference Forest Tree Improvement. 7: 44-49.

Dirr, M.A. and Heuser, C.W. Jr. 1987. The Reference Manual Of Woody Plant Propagation. Ga. Varsity Press, Athens.

Faulkner, R. 1964. An outline of the British breeding programme. In : Forest Tree Breeding In The World (ed. R. Toda) p.1-8. Tokyo: Yamato-ya Ltd. Tyuuoo-ku.

Faulkner, R., 1966. A review of flower induction experiments and trails 1948-1963. Extr. From Pap. For. Res. For. Comm., London 1964/65, 207-219.

Fowler, D.P. 1965. Natural self fertilization in three Jack pines and its implications in seed orchard management. *For. Sci.*,11 : 55- 58.

Garner, R.J. 1950. The Grafter's Handbook. Faber and Faber, London.

Harry, J., Smith, J.H. G., Walters, J., and Kozak, A. 1968. Influences of fertilizers on cone production and growth of Douglas fir, western hemlock and western red cedar on The University of British Columbia Research Forest Bull. Fac. For. Univ. British Columbia 5.

Kellison, R. C. 1969. Establishment and management of clonal seed orchards of pine. In Second World Consult. Forest Tree Breeding, FO-FTB- 69-11/7, 7p.

Khanna, L.S. 1984. Principles and practice of silviculture. 3rd Edition. Khanna Bandhu, Dehradun.

Khullar, P., Thapliyal, R.C., Beniwal, B.S., Vakshasya, R.K. and Sharma, Ashok. 1991. Forest seed. ICFRE, Dehradun.

Kieding, H., Jones, N. and Webb, D. 1964. A programme of tree breeding for Nigeria. Commonw. For. Rev., **43(4)**: 319-326.

Krugman, S.L.1966. Freezing spring temperatures damage knob cone pine. U.S. Department of Agriculture, Forest Service, Research Note PSW-35,6p.

Larsen, C.S. 1956. Genetic in silviculture (Translated by M. L. Andersson). Edinburgh, Oliver and Boyd.

Libby, W.J., Kiang, Y.T. and Linhart, Y.B. 1972. Control pollination seeds from cuttings of coast red wood. Silvae Gent. *21*: 17-20.

Limstrom, G. A.1965. Interim forest tree improvement guides for the central states. U. S. For. Serv. Res. Pap. Cent. St. Exp. Sta. CS -12. London.

Luna, R.K. 1996. Plantation trees. International Book Distributors, Dehradun.

Masters, C. J. and Beineke, C. J. 1973. Clonal vs. half-sib seedling orchards for black Walnut. Proc. 20th North-East Tree Improvement Conf. Durham, 52-61,New Hampshire.

Matthews J. D. 1962. Seed selection and tree breeding in Britain. For. Comm., 8th British Commonw. For. Conf. East Africa. 5p.

Matyas, V. 1969. Weather influence on beech flowering. 2nd F.A.O./I.U.F.R.O. World Consult For. Tree Bree., Washington. F.O. - FTB 69- 11/12/.

Mc Millan-Browse, P.D.A. 1979. Hardy woody plants from seed. Growers Books, London.

Mergen, F. and Voigt, G. K. 1960. Effects of fertilizer application on the generations of Slashpine. Proc. Soil Sci. Soc. Am., **24(5)**: 407-409.

Nienstaedt., H., Cech, F. C., Mergen, F., Wang, C. W. and Zak, B. 1958. Vegetative propagation in forest genetics research and practice. J. For. **56:** 826-839.

OECD. 1974. Scheme for the control of forest reproductive materials moving in international trade. Organization for Economic Co-operation and Development, 2 rue Andre Pascal 75016 Paris, France.

Rudolf, P.O. 1948. How about our seed supply? *J. For.* **46**: 741-743.

Sarvas, R. 1970. Establishment and registeration of seed orchards. Folia For. *89*: 1-24.

Schmidtling, R.C. 1971. Cultivating and fertilizing stimulate precocious flowering in Loblolly pine. Silva. Genet., **20(5)**: 220-221.

Schultz, R.P. 1969. What is being learned from intensive management of slash pine. Forest Farmer, **28(6)**: 8-9.

Squillace, A.E. 1967. Effectiveness of 400 -foot isolation around a slash pine seed orchard. J. For., **65**: 823-824.

Stein, W.I.; Slabaugh, P.E. and Plummer, A.P. 1974. Harvesting, processing and storage of fruits and seeds. In Seeds of woody plants in the United States. U. S. Department of Agriculture, Forest Service, Agriculture Handbook 450. Washington, D. C.

Steinbrenner, E.C., Duffield, J.W. and Campbell, R.K. 1960. Increased cone production of young Douglas fir following nitrogen and phosphorus fertilization. J. For., **58(2)**: 105-110.

Tewari, D.N. 1994. Biodiversity and forest genetic resources. International Book Distributors, Rajpur Road, Dehradun.

Tewari, D.N. 1995. Agroforestry for increasing productivity, sustainability & poverty alleviation. International Book Distributors Dehradun. 799pp.

Troup, R.S. 1921. The Silvicure of Indian Trees. Volume I,II,III. Clarendron Press, Oxford.

Venkatesh, C. S. 1974. Forest tree improvement work in India: History, organization and present activities. In Forest Tree Breeding In The World (ed. R.Toda). Tokyo. p. 137-145, Yamato-ya Ltd. Tyuuoo-ku,.

Zasada, J.C. 1971. Frost damage to white spruce cones in interior Alaska. U.S. For. Serv. Res. Note Pacific Northwest Forest Range. Exp. Stn., PNW – 149.

Zavadil, Z. 1969. Evaluation of Douglas fir (*Pseudotsuga menziessii*) graft tree losses in the First Seed Orchards in Czechoslovakia. 2nd F.A.O./I.U.F.R.O. World Consult. For. Tree Breed., Washington. FO-FTB-69-11/14.

Zobel, B. J. And Talbert, J.T. 1984. Applied forest tree improvement. John Wiley & Sons, New York.

Zobel, B.J., Barber, J., Brown, C. L. and Perry, T.O. 1958. Seed orchards- their concept and management. J. For., 56(11): 815-825.

□□□

35

Nanotechnology in Commercial Agriculture

RAJEEW KUMAR

Commerce is a division of trade or production which deals with the exchange of goods and services from producer to final consumer. It comprises the trading of something having economic value such as goods, services, information or money between two or more entities. Commercial agriculture is the production of crop for sale, crops intended for widespread distribution *(e.g. supermarkets)*, and any non-food crops such as cotton and tobacco. Commercial agriculture includes livestock production and livestock grazing. It does not include crops grown for household consumption (e.g backyard garden or from a vegetable garden or fruit tree's).Success of commercial agriculture depends on Genomics, Technology and Market. To cover all these, technological triads namely biotechnology, information technology and Nanotechnology is the hot spot in the world scenario.

Nanotechnology

Nanotechnology is emerging technological platform for the next wave of development and transformation of agricultural systems. Nanotechnology is attracting large-scale investment from global food corporations, is backed by academic science, and has captured financial and ideological support from many governments around the world. As a result, nanotechnology is rapidly moving from the laboratory to the use. For example, a new range of 'smart' agricultural inputs and products are being developed, such as nano-seed varieties with in-built pesticides that will release under certain environmental conditions; nano-encapsulation techniques may make it possible to alter the nutritional composition, flavour and other attributes of food to match consumers' personal tastes and physiological requirements; and 'smart' food packaging able to detect the presence of pathogens. Other applications of nanotechnology across the agri-food system are emerging from a growing alliance between the corporate food sector and scientific communities. Nanotechnology represents the most powerful set of techniques et developed to take apart and reconstitute nature at the atomic level. In terms of economic relations, nanotechnology provides new opportunities for the extension and further integration of corporate ownership and control within and between sectors of the agri-food system.

The realization that there are small things in the world that are not visible to the naked eye extends back into human history. The development of the natural sciences created an interest in the microworld, in order to enable a better understanding of the world and the processes therein. Therefore, the development of new microscopic imaging methods represents certain milestones in the natural sciences. The microworld was approached by extending the range available for the direct visualization of objects through the enhancement of microscopic resolution. Microtechnology has changed our lives dramatically. The most striking impact is apparent in computer technology, which is essential for today's industry, and also for our individual life styles. Apart from microelectronics, microtechnology influences many other areas. The size of typical structures that is accessible is in the sub-micrometer range, which is at the limits of optical resolution and barely visible with a light microscope. This is about 1/1000 smaller than structures resolvable by the naked eye, but still 1000 times larger than an atom. Today's developments are addressing the size range below these dimensions. Because a typical structure size is in the nanometer range, the methods and techniques are defined as nanotechnology.

The prefix "nano," derived from the Greek "nano" signifying "dwarf," is becoming increasingly common in scientific literature. "Nano" is now a popular label for much of modern science, and many "nano" words have recently appeared in dictionaries, including: nanometer, nanoscale, nanoscience, nanotechnology, nanostructure, nanotube, nanowire, and nanorobot. Although the idea of nanotechnology producing nanoscale objects and carrying out nanoscale manipulations, has been around for quite some time, the birth of the concept is usually linked to a speech by Richard Feyman at the December 1959 meeting of the American Physical Society where he asked, "What would happen if we could arrange the atoms one by one the way we want them?"

Applications of nano-technology in commercial agriculture

In the agricultural sector, nanotech research and development is likely to facilitate and frame the next stage of Agricultural development like *genetically modified crops, animal production inputs, chemical pesticides and precision farming techniques.* While nano-chemical pesticides are already in use, other applications are still in their early stages, and it may be many years before they are commercialized. These applications are largely intended to address some of the limitations and challenges facing large-scale, chemical and capital intensive farming systems. This includes the fine-tuning and more precise micro-management of soils, the more efficient and targeted use of inputs; new toxin formulations for pest control, new crop and animal traits; and the diversification and differentiation of farming practices and products within the context of large-scale and highly uniform systems of production.

1. Genetically modified crops

Through the convergence of nano and bio techniques, it may be possible to improve the precision of genetic engineering and breeding programs, thereby ensuring greater control in delivering new character traits to plant and crop varieties. Researchers are attempting to use nanoparticles, nanofibres and nanocapsules to introduce foreign DNA and chemicals into cells. For example, silica nanoparticles have been used to deliver DNA and chemicals into plant and animal cells and tissues. Researchers in this field have also already succeeded in "drilling" holes through the membrane of rice cells to enable the insertion of a nitrogen atom, to stimulate rearrangement of the rice DNA. This technique has been successful in altering the colour of rice, and researchers aim to use this technique to extend the growing season for rice, enabling year round production. There is, however, little evidence of any commercial applications of such nano-genetic engineering research at this stage. The perhaps more distant prospect of not merely re-engineering existing plants, but of creating novel plant varieties from scratch using synthetic biology would enable significantly greater control over crop traits.

2. Chemical inputs

Techniques at the nano-scale are also being applied in an attempt to enable the targeted delivery or increased toxicity of pesticide applications. This includes the insertion of nano-scale active ingredients into pesticides. The specific properties of these nano-scale materials, such as their ability to dissolve in water or their increased stability, are designed to maximize the effectiveness of these pesticides. Leading agri-chemical companies including BASF, Bayer Crop Science, Monsanto and Syngenta are engaged in nanotech research in these areas. Pesticides may also be encapsulated via nano-encapsulation techniques. These encapsulation techniques enable greater control over the circumstances in which encapsulated pesticides will be released. For example, pesticides could be released quickly or slowly–depending on need–and under specific conditions, such as moisture and heat levels. Syngenta have obtained a patent for a 'gutbuster' microcapsule containing pesticides that will break open in alkaline environments, including the stomach of certain insects. Such nano-encapsulation techniques not only provide in-built pesticides for crops–in some ways similar to genetically modified *Bt* insecticidal crops–but also in-built switches to control the release and subsequent availability of pesticides. One of the rationales for these nano-particle pesticide applications lies in their improved capacity for absorption into plants compared to larger particles. As such, they may not be washed off as readily, thereby increasing their effectiveness, but also posing a new order of risks to consumers of these products. Farm workers and rural residents are also being exposed to these nano-pesticides, in the absence of any required safety testing or regulation of nano-scale formulations of already approved chemical pesticides. The size and dissolvability of nanoparticle pesticides may also mean they contaminate soils, waterways and foodchains across a wider geographical area, while nano-encapsulated pesticides may release their toxins in other environments or in the stomachs of other living organisms. Nano-pesticide research and development is concentrated within a small number of large agri-chemical companies that already dominate the agri-chemical and seed market, and these corporate actors are likely to further extend their control of these markets, and therefore over farmers. Proponents argue that pesticidal applications of nanotechnology promise to reduce pesticide use, due to their more precise and targeted nature. As such, nanotechnology is frequently portrayed as introducing environmental benefits.

3. Slow release fertilizer (nano-encapsulation)

Slow-release fertilizers are excellent alternatives to soluble fertilizers. In this nutrients are released at a slower rate throughout the season, plants are able to take up most of the nutrients without waste by leaching. Slow-release fertilizers is more convenient, since less frequent application is required and may be more expensive than soluble types, but their benefits outweigh their disadvantages

4. Nano sensor

Another development being looked at is a network of nanosensors. The sensors recognize when a plant needs nutrients or water, before you could see any sign that the plant is deficient. The dispensers then release fertilizer, nutrients, or water as needed, optimizing the growth of each plant in the field one by one. Nanosensors-or nano-scale, wireless sensors-represent the intersection of nanotechnologies and information technologies. Alongside geographical positioning systems and other information technologies, nanosensors could be scattered across farmers' fields to enable the 'real time' monitoring of crops and soils, and the early detection of potential problems, such as pest attacks and declining soil nutrient levels. Nanosensors have the capacity to extend the logic of precision farming in new and novel ways-to both identify and rectify agronomic problems in a very short time frame. The US Department of Agriculture, for example, is reported to be developing a "Smart Field System" that "automatically detects, locates, reports and applies water, fertilisers and pesticides going beyond sensing to automatic application". Nanosensors may thereby introduce greater efficiencies within and thereby facilitate the expansion of large scale farming operations.

Table 1. Research findings in different crops

Nano particles	Country	Response
Buckyball	Japan	Effect not clear
TiO_2 Nano mixture	Korea	Patent on a liquid after which time it dissolves in the groundwater and becomes indistinguishable from naturally occurring iron.
Seeding iron	Russian	Improvement in the germination of the seed
Soil binder	USA	Patent of soil set
Soil clean-up	China	Groundwater and soil cleanup

Nano research area

- Nanoscale phenomena and processes: Plant/animal cells; genomics/ proteomics; biosafety; crop/ animal production processes; natural resources cycles
- Nanomaterials: nano fertilizers and nanocides for efficient use of inputs, soil erosion control; packaging
- Nanodevices and systems: Biosensors for precision agriculture; diagnostics; pathogen/contaminant detection
- Smart delivery systems for genes/drugs/vaccines

- Nanofiltration: Nanobrushes for soil & water purification
- Smart systems integration
- Environment, social, ethical, health implications
- Education

Conclusion

Nanotechnology could facilitate a second green revolution in Indian agriculture if the following are emphasized:

- Strategic approach to nanotechnology research and development across the agricultural value chain
- Environmentally and socially responsible development of the technology
- Anticipatory design of effective regulatory mechanisms and strong governance systems designed with involvement of all the stakeholders
- Ultimate acceptance by the stakeholders

References

Roco, M.C.; Williams, S.; Alivisatos, P. (Eds). 1999. Nanotechnology Research Directions: International Working Group on Nanoscience, Engineering, and Technology Workshop Report. http://www.wtec.org/loyola/nano/ IWGN.Research. Directions/.

Stearns, R.C., Paulauskis, J.D., Godleski, J.J. 2001. Endocytosis of ultrafine particles by A 549 cells. *Am. J. Respir. Cell Mol. Biol.* 24 : 108-115.

Tsien, A., Diaz-Sanchez, D., Ma, J. and Saxon, A. 1997. The organic component of diesel exhaust particles and phenanthrene, a major polyaromatic hydrocarbon constituent, enhances IgE production by IgE-secreting EBV-transformed human B cells in vitro. *Toxical. Appl. Pharm.* 142 : 256-263.

□□□

36

Crop Production Using GIS and GPS Tools for Sustainable Agriculture

A.K. KANOJIA

The collection of data about the spatial distribution of significant properties of the earth's surface has long been an important part of the activities of organized societies. From the earliest civilizations to modern times, spatial data have been collected by navigators, geographers, surveyors and rendered into pictorial form by the cartographers. As scientific study of the earth advanced, new material needed to be mapped. The development in the assessment and understanding of natural resources: geology, geomorphology, soil science, ecology that began in the 19th century and have continued to this day, provided new themes to be mapped. Resources management requires to analyse phenomena, processes and their interaction throughout space and time. The resource information system for agricultural management has to be considered as multidimensional *i.e.* attribute dimension, spatial dimension and temporal dimension Singh and Singh 1998. Geographic Information System (GIS) offers such capabilities as they integrate multi-sector, multilevel and multi-period databases. New information systems have to be used by the scientific community to have overall idea and understanding of the existing pattern of the phenomena as they are associated with and to provide real time information for executing the necessary strategies by the administrators, the planners, the researchers etc. both effectively and timely. Since our earth is not a small piece of land of similar type of climate, soil, flora and fauna, water body etc. but it is full of diversities in every sphere of life and it is clear from these facts also that it is having total surface area of 509,700,000 sq. kms., The area of land is 148,400,000 sq. kms., while water bodies occupy 361,400,000 sq. kms. The variations in surface features, ocean depth, temperature, atmosphere etc. vary from one region to other over the earth. Consequently varied agro ecological regions, cropping patterns, plant pests, natural enemies are distributed on our earth. The diversities are also found in colours in our country and it is clear from the facts that it is 12 times of UK, 8 times of Japan, having 16% of world population, it is 7th largest and 2nd most populous country in the world with geographical area of 32,87,263 sq. km, while its north-south extent is almost 3214 kms and east-west extent: 2933 kms., its coast line is around 7516 kms (Kalpana, 2000) with different kinds of soils, vegetation, socio-ethnic groups, farming systems, patterns and modes of indigenous wisdom of local communities. Therefore, it is imperative to understand intricacies and complexities involved in the development of a phenomena over earth and that is why the use of such system like GIS helps decision making effectively as it is a powerful integrating technology which needs to be utilized by the community of natural resource scientists and the concerned planning people.

Scare water resources must be used intelligently

Valuable water resources are found in many forms

Spatial Technologies: GIS and GPS

The rapid development of spatial technologies in recent years has made available new tools and capabilities in crop production for management of spatial data. In particular, the evolution of geographic information systems (GIS), the global positioning system (GPS), and remote sensing (RS) technologies has enabled the collection and analysis of field data in ways that were not possible before the advent of the computer.

Geographic Information Systems (GIS)

GIS applications enable the storage, management, and analysis of large quantities of spatially distributed data. These data are associated with their respective geographic features. For example, water quality data would be associated with a sampling site, represented by a point. Data on crop yields might be associated with fields or experimental plots, represented on a map by polygons. The ability to interface GIS with relational databases enables integration of large data sets and many variables to support management decisions (Arvanitis *et al.*, 2000).

There are several definitions to define GIS: "A system which uses a spatial database to provide answers to queries of a geographical nature." (Good Child, 1985). "A powerful set of tools for collecting, storing, retrieving at will, transferring and displaying spatial data from the real world." (Burrough, 1987). "An organized collection of computer hardware, software, geographical data and personnel designed to efficiently capture, store, update, manipulate, analyze and display all forms of geographically referenced information".

Thus, GIS is a combination of hardware, software, database and personnel and it needs to have geographical area over and for which pattern and trends of the selected phenomena has to be shown and queries are to be made. Once geographic information of any kind is translated into the digital form in a GIS, it becomes easy to copy, edit, analyze, manipulate and transmit it.

A complete GIS must answer:

What exists at a particular location?

Where can a specific feature be found?

What has changed over time?

What spatial pattern exists?

What if...?

Activities involved in building a GIS includes (a) getting spatial data into GIS by digitization, (b) making spatial data usable, (c) getting attribute data into GIS, (d) managing the data-base, (e) performing geographic analysis and (f) presenting the results of the analysis.

The data for a GIS may be from manually surveyed data, remote sensing data and secondary data. A Global Positioning System (GPS), which is a handholding device, may also be used for capturing the spatial data. It is used to provide accurate position of the captured data.

GIS acts as a decision making tool and therefore its application is found extensively in research organizations, government concerns, business - for a wide range of analysis including environmental monitoring, land use planning, locational analysis, water resources management, agronomical practices, tax appraisal, utility and infrastructure planning, demographic analysis, pest distribution, pest modelling, pest forecasting, habitat studies, archaeological analysis etc.

Basic functions of GIS

Functions	Sub-functions
Data acquisition & preprocessing	Digitizing, editing, topology building, projection transformation, format conversion, attribute assignment
Database management and retrieval	Data archival, hierarchical modeling, network modeling, relational modeling, attribute query
Spatial measurement and analysis	Measurement operations, buffering, overlay operations, connectivity operations
Graphic output and visualization	Topographic map, statistical maps, 3D bird's eye view

Why GIS?

- Digital World
- Accessibility of data
- Compatibility
- Accuracy
- Spatial Analysis
- Multiple Themes

It is an essential and user friendly tool for precision agricultural or site-specific management and may use viable inputs from GPS and remotely sensed data.

Requirements of a GIS

Basic requirements for GIS work are the following:

- *Digitizer*: to convert map or survey-recorded geographic features like rivers, ponds, wells, district/village/site boundary.
- *Work station*: to act as a desktop device like a PC in which GIS software are installed
- *GIS software*: to undertake and accomplish GIS task like popular (ARC-GIS) from ESRI.
- *Plotter/printer*: to take printout in the way you want.
- *Internel Map Server (IMS)* : to publish useful GIS information over web in the same way as seen on GIS on desktop and user may query the thing from anywhere in the world.
- *Data*: to act as viable input both spatial and attribute data of the study.
- *Man power:* to act as a controller of GIS and data and presenter of the themes.

There has always been advancement in spatial technologies too and it helps working the same task with less number of devices as these days digitizers are becoming out of fashion since desktop digitizing is quite easy and becoming popular. After scanning any image on-monitor digitizing is done in spite of pasting the image/map over the table and tracing the concerned features manually.

Inputs to GIS

- Manually surveyed data
- Aerial photos
- Remotely sensed satellites data
- Census data
- Published data
- Geocordinates of site using Global Positioning System

The lobal Positioning System (GPS)

GPS technology has provided an indispensable tool for management of agricultural and natural resources. GPS is a satellite- and ground-based radio navigation and locational system that enables the user to determine very accurate locations on the surface of the Earth. Although GPS is a complex and sophisticated technology, user interfaces have evolved to become very accessible to the non-technical user. Simple and inexpensive GPS units are available with accuracies of 5 to 3 meters, and more sophisticated precision agriculture systems can obtain centimeter level accuracies.

GPS Receiver

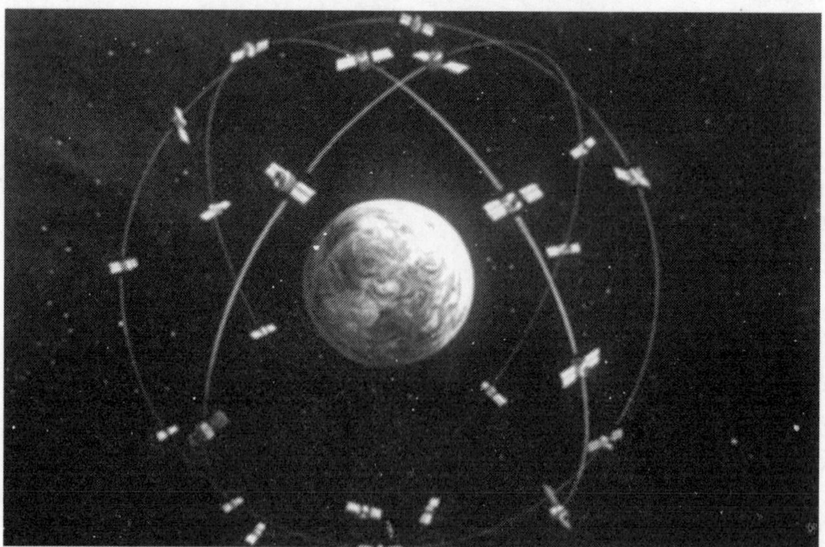

Concept of the GPS satellite constellation
Photo courtesy : U.S. Department of Space

The Global Positioning System (GPS) is actually a constellation of 27 Earth-orbiting Satellites (24 in operation and three extras in case one fails) , controlled by the US Department of Defense and paid for by the US taxpayer but provides free commercial service to everyone. A GPS receiver's job is to locate four or more of these satellites, figure out the distance to each, and use this information to deduce its own location.

A GPS satellite
Photo courtesy : U.S. Army

A GPS receiver calculates the distance to GPS satellites by timing a signal's journey from satellite to receiver. At a particular time (let's say midnight), the satellite begins transmitting a long, digital pattern called a pseudo-random code. The receiver begins running the same digital pattern also exactly at midnight. When the satellite's signal reaches the receiver, its transmission of the pattern will lag a bit behind the receiver's playing of the pattern.

A standard GPS receiver will not only place you on a map at any particular location, but will also trace your path across a map as you move. If you leave your receiver on, it can stay in constant communication with GPS satellites to see how your location is changing. With this information and its built-in clock, the receiver can give you several pieces of valuable information: (1) how far you've traveled (odometer), (2) how long you've been traveling, (3) your current speed (speedometer), (4) your average speed and (5) A "bread crumb" trail showing you exactly where you have traveled on the map. The estimated time of arrival at your destination if you maintain your current speed.

Costs of GIS

60% of the costs attributed to data, 30% to personnel and 10% to system hardware and software. It is easier to get hardware & software donated but no one can give you your data.

A GIS can manage different data types occupying the same geographic space. For example, a biological control agent and its prey may be distributed in different abundances across a variety of plant types in an experimental plot. Although predator, prey, and plants occupy the same geographic region, they can be mapped as distinct and separate features. Once patterns and correlations are elucidated, management practices can be modified to optimize yield and production costs, and minimize environmental impacts caused by excessive applications of fertilizers and pesticides. Site-specific applications of fertilizers, pesticides and others can be implemented by dividing a field into smaller management zones that are more homogeneous in properties of interest than the field as a whole (Zhang *et al.*, 2002). Since precision farming techniques are employed to increase yield, reduce production costs, and minimize negative impacts to the environment (Zhang *et al.*, 1999). Using GIS analytical capabilities, variable parameters that can affect agricultural production can be evaluated. These parameters include yield variability, physical parameters of the field, soil chemical and physical properties, crop variability (e.g., density, height, nutrient stress, water stress, chlorophyll content), anomalous factors (e.g., weed, insect, and disease infestation, wind damage), and variations in management practices (e.g., tillage practices, crop seeding rate, fertilizer and pesticide application, irrigation patterns and frequency) (Zhang *et al.*, 2002). The power of a GIS lies in its ability to analyze relationships between features and their associated data (Samson, 1995). This analytical ability results in the generation of new information, as patterns and spatial relationships are revealed.

A	→	Soil
B	→	Temperature
C	→	Rainfall
D	→	Humidity
E	→	Season
F	→	Crop
G	→	Pest data/irrigation/pesticide
H	→	Location

Data in GIS

1. *Spatial data* : describes locations of entity (where) i.e. map of Points, lines and polygons
2. *Attribute data* : characteristics of spatial answers (what)
3. *Image data* : gives feel of entity using camera or video (how it looks)
4. *Identifiers* : link mechanism for spatial attribute an image data, very important must be unique, used to control data management process

Two major data models in GIS: Raster and Vector

1. Raster

- Area divided into grid of cells, each cell contains a value
- Cells numbered, one set of cells and values forms a layer
- Model tells "what occurs everywhere"

2. Vector

- Data represented by points, lines, areas/Polygons
- Model tells "where everything occurs"

GPS/GIS Applications

a) *Guidance*: It acts as a guiding force and tool for

- *Point Guidance*: Point to point information is provided for attaining precise application of viable inputs in the field.
- *Swath Guidance*: Whole band of land is taken like scanning and the homogenous or heterogeneous patches are identified in the feild.

b) *Control*: Its application helps to control variable rate application of critical inputs, variable depth tillage, variable irrigation etc.

- *Mapping*: Its application helps taking basic and applied decision for the field/region under study
  - Soil properties
  - Fertilizer Applications
  - Chemical Prescriptions
  - Tillage Maps
  - Irrigation potential and actual usage
  - Yield Mapping
  - Pest Mapping
  - Topographic Maps
  - Planting Maps

GIS acts as an integrating technology: The various themes may be generated and viewed integrating many layers of concerned information over one map to arrive at a concrete decision for the region of study. This is called a Visual correlation of the selected themes.

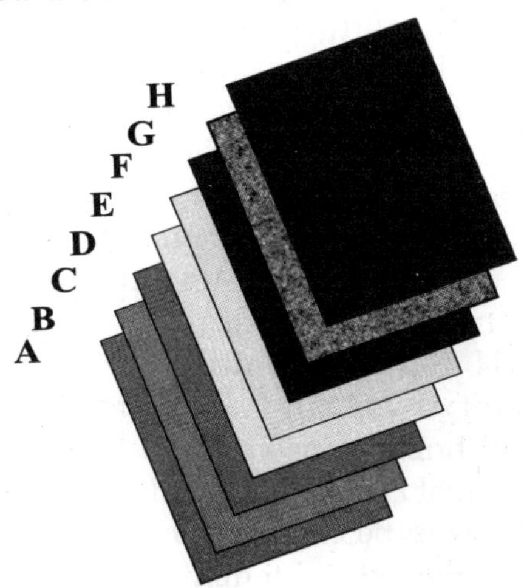

Sending the results to the end/potential users through GIS

The following diagram throws light on how results may be sent to potential users over web so that the generated information may be used timely and efficiently by the concerned ones anywhere in the world and advisory may be issued accordingly:

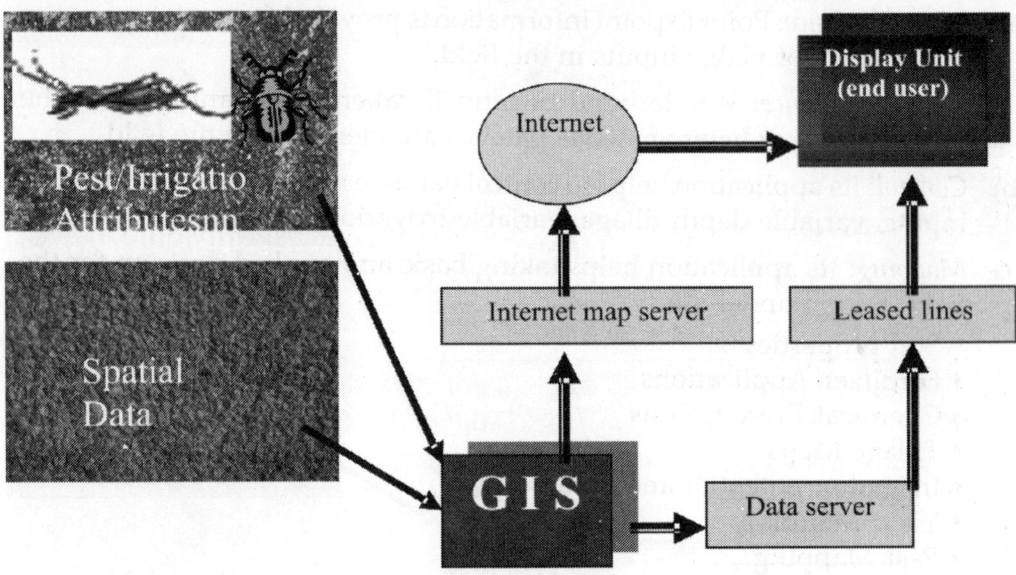

Water management in crop production

Importance of Water management in Crop Production

Photosynthesis, respiration, absorption, translocation, utilization of minerals/nutrients, cell division etc

Efficient and Economic utilization: Proper appreciation of the relationship of Soil-Crops-Climate-Water is essential

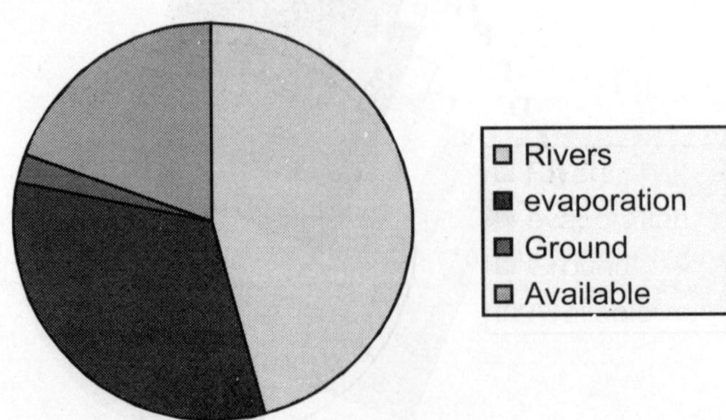

Statistics may be generated for water resources for the region under study in the following way as it has been generated for the country to manage it according to priorities:

Water resources are valuable and indispensable for any society especially for agriculture and India is not an exception, and it may be quite visible through following facts.

Geographical Area of the country: 328m hectare

Average annual rainfall: 112 cm

Total annual precipitation: 37,00,000 million cu m

80% received from SW Monsoon

80,000 million cu m seeps into the ground

17,00,000 million cu m flows into the rivers

12,00,000 million cu m evaporates back into the atmosphere

This kind of data generation helps in knowing available resources and planning them according to priorities and site specific requirement.

How much to irrigate

Basic and valuable inputs must be considered for generating spatial and non spatial data. Quantification of geo-referenced data may be taken as inputs for developing auotomated model for decision making using GIS and GPS

- *Measurement of irrigation water*: Application of precise quantities at rest or motion
- *Depth of irrigation*: The quantity of water needed for net irrigation on different soils types per metre depth of soil profile at 50 percent soil-moisture availability

Soil Type	**Irrigation depth (cm)**
Sandy	2.5-5
Sandy loam	45-80
Loam	70-100
Clay loam	80-120
Heavy clay	100-140

Root-system of different crops: annual/perennials

Shallow-rooted : Nursery-beds/Leafy vegetables/Garlic/Onion

Medium-rooted : Rice/Maize/Wheat/Papaya/Groundnut

Deep-rooted: Cotton/Sorghum/Sugarcane

Very deep-rooted: Fruit trees

Apply light and frequent irrigations for increased uptake of nutrients and lesser leaching.

Soil-moisture measurement

This kind of data may be generated by taking geo-referenced soil samples using GPS at field and may be analyzed using experience or at laboratory.

Testing by following the established procedures: The generated data may become decision making parameter and may be used in GIS for irrigation application on the basis of the location specific requirement, as knowledge of amount of moisture present in the soil is essential for deciding the time and the quantity of water to be applied.

Systems and methods of irrigation

Since there are many ways/systems to irrigate the field such as surface irrigation system, Sprinkler irrigation system, Subsoil irrigation system, Drip irrigation system and by quantifying their supplying potential and the requirement of the location, planning may be done on the basis of these parameters and visual modeling may be applied for arriving at the suitable application.

Irrigation practices for important crops

Irrigation practices for important crops must be taken into consideration for designing the expected model for water use in the study area. Interval between two irrigations depends on the rate of soil-moisture depletion. Soil-moisture depletion may be known by applying the following formula in your model development exercise for your following area:

Moisture deficit can be calculated by the following formula:

$$D = \frac{FC - Mc \times Bd \times DR}{100}$$

Deficit (D), Field Capacity (FC), Bulk density of soil (Bd), Depth of root-zone (DR)

Useful inputs

Useful quantified inputs for developing application are the data of :

Availability of soil water for crop growth

Suitability of soil for irrigation

Plant and water absorption

Quality of irrigation water

Practices for crop production under conditions of soil or water salinity

Assessment of availability of water, soil and crop factors

The proper assessment of availability of water, soil and crop factors needs the following inputs:

Water availability : Quantity per day, Rotation or turn period, System and methods of irrigation, Water-quality class;

Soil: Texture, Structure, Depth up to the water-table, Infiltration and permeability, Depth of soil, Slope of land;

Crops: Type of root system, Life span, Critical period, Acreage, Market value, Yield response in relation to water supply.

Factors affecting water resources

Climatic: Rainfall-intensity, duration, distribution from Snow as well as Evapo-transpiration.

Physiographic: Drainage area, Shape, Slope, Stream density form, Land use, surface infiltration, soil type, Storage capacity of channel.

Geological: Lithologic composition, texture, thickness of rock formation, Faults and folds, aquifers, permeability, porosity.

Assessment of surface water resources

Monthly Runoff= monthly Rainfall – Monthly Evaporation losses

Monthly Evaporation losses = 0.481 x Mean Monthly Temp

Watershed characterization and runoff assessment using GIS

Following inputs are required to build a GIS model for assessing runoff and watershed characterization:

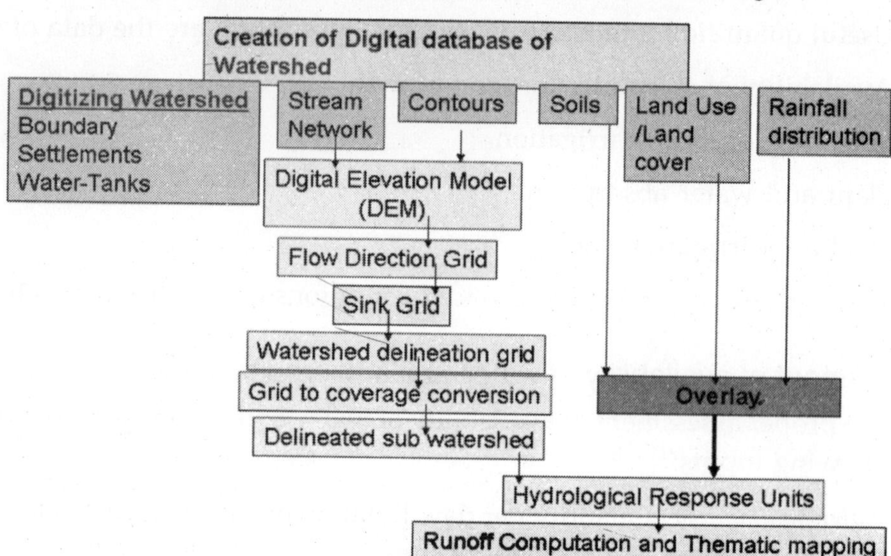

Creation of Digital database of Watershed needs digitizing of watershed including its boundary, settlements, water- tanks. Stream Network, Contours, Soils, Land Use /Land cover, Rainfall distribution.

By using the above mentioned inputs from the watershed area Digital Elevation Model (DEM) may be generated using contours and then Flow Direction Grid, Sink Grid, Watershed delineation grid, Grid to coverage conversion, Delineated sub watershed can be generated. Finally hydrological response units are obtained before runoff computation and thematic mapping by overlaying all the gathered information. This has been depicted in the given schematic diagram where flow of information and step by step procedure is given.

Conclusion

Geo-spatial technologies are becoming important tools in managing natural resources especially agricultural resources. Readily available visual interpretation in geo-referenced mapping through such tools provides valuable inputs for arriving at the safety and mitigating measures to safeguard the standing crop of economical strength and vigor. Crucial information on agriculturally important parameters along with real time weather data can definitely help timely management of concerned crop. The crop database created using spatial technologies can be used by agronomists, plant protectionists, planners and researchers for further analysis and planning. There is a high potential to use these technologies in agricultural sector in the country and elsewhere.

References

Arvanitis, L., Ramachandran, B., Brackett, D., Rasoul, H., & Du, X. 2000. Multiresource inventories incorporating GIS, GPS and database management systems: A conceptual model. Computers and Electronics in Agriculture, 28, 89-100

Kalpana, R. 2000. Handbook of General Studies. Spectrum India Pub., New Delhi, pp. 241.

Samson, S. A. 1995. Determining need for a Geographic Information System (GIS). Journal of Extension [On-line], 33(5).

Singh, R.L. and Singh, R.P.B. 1998. Elements of Practical Geography, Kalyani Publ., New Delhi, pp.1-7, 186-200.

Zhang, N., Runquist, E., Schrock, M., Havlin, J, Kluitenburg, G., & Redulla, C. 1999. Making GIS a versatile analytical tool for research in precision farming. Computers and Electronics in Agriculture, 22, 221-231.

Zhang, N., Wang, M., & Wang, N. 2002. Precision agriculture — A worldwide overview. *Computers and Electronics in Agriculture*, 36 : 113-132.

□□□

Promoting Agro-Industry

37

Agro-Industrial Development in India

V.P.S. ARORA

Pre independence era

Agro-industry' is an omnibus expression. It could cover a variety of industrial, manufacturing and processing activities based on agricultural raw materials and also activities and services that go as inputs to agriculture. However, traditionally, Indian agriculture drew most of its inputs locally from the village and the farm. Modern inputs, like chemical fertilizers, pesticides, tractors and other agricultural implements, pump sets, diesel, gas and power, had little import in the overall agricultural inputs. Agro-industries were essentially perceived as first level post-harvest processing of farm produce. Agro, village, cottage and rural industries meant the same set of economic activities. The expressions were interchangeable. Even when agro-industries have been assigned a special place in the successive Five Year Plans, a good deal of confusion continues to persist regarding their coverage. It is not very clear whether agro-industries are to denote only the activities directly related to agriculture or the total farm output and related activities.

In a typical Indian village, economic activities were and, to a considerable extent, even today are generally associated with a particular 'caste' and community, which has practiced a particular profession for generations. The nature of activity undertaken was, in general, hereditary and occupational mobility within the village nearly non-existent. Generally, each village had at least one household each of blacksmiths, carpenters, weavers, potters and those engaged in oilseed crushing. Agricultural and non-agricultural activities had direct and strong organic links within the village. Landowners needed professionals to take care of their plough and other agricultural implements. A typical Indian village in this sense was an organic whole, in which, while each activity had its own distinct identity, there was a considerable degree of inter-dependence.

The view, that village industries and crafts are an important part of rural life and should be vigorously protected to ensure sustenance to a self- reliant village, is essentially an outcome of the traditionalist philosophy, articulated forcefully by Gandhi. It was in fact, a reactive approach- a defence mechanism against the onslaught of the British industry. The vision of agro-industries/ rural industries during the pre-independence era, had some serious limitations, the most significant ones being that it did not project itself to the likely impact of: (i) the spread of literacy and technical education, (ii) availability of alternative technologies, (iii) the growth of mass media, (iv) the changing aspirations of the people and the youth in particular, and (v) easy availability of power and electricity. The truth is that agro and village industries were seen in terms of production in a traditional village and not the village of the future in a free and modern India. Agro industries, as traditionally understood, could also not accommodate regional specialization to exploit comparative and locational advantages. It also remained un-appreciated that techniques of production are not always independent of the socio-economic system.

Post independence era

The pre-independence view that the cottage and small industry faced direct competition from the big industries stood revised. The big was not a substitute for the small and vice versa. Both were to grow; in fact, in a mutually supportive manner. With a view to reducing the areas of direct confrontation, the Government of India adopted a number of specific measures such as a favourable treatment to Khadi, cottage and village industries. During the period, 1952 to 1954, the All India Khadi and Village Industries Board and a Board each to promote silk, coir, and handicraft, handloom and small scale industries were instituted. These Boards were required to recommend general policies and prepare action plans for promoting activities in their respective areas. There was, however, no special category of industries called agro-industries. Machine was no more a synonym of colonial exploitation.

The First Five Year Plan made a distinction between village industries, small industries and crafts. Agricultural development and village industries were looked upon as Siamese twins that could hardly be separated from each other. The First Plan also visualized that an increase in agricultural production would raise farmers' incomes and expands opportunities for processing raw materials in the villages. The Planning Commission decided to draw up Village Industry Programmes covering the industries like village oil industry; soap making with neem oil; paddy husking; palm gur industry; gur and khandsari; leather industry; woolen blankets; high grade hand-made paper; bee keeping; and cottage match industry. Other industries assigned special priority were: Khadi, coir, sericulture, fisheries, forests, dairying and horticulture.

In addition to stressing the role of heavy industry, the Second Plan, based on Mahalanobis Model, also assigned a special place to rural, cottage and small industries. It envisaged that the expanding demand for consumer goods sector would be met from outside the large units. This would reduce pressure on the capital and the limited savings of the economy and the strategy would fit in well with the need to expand employment opportunities.

The cottage and small scale industries were visualized as a panacea for most economic ills. It was probably the emotional appeal and romance of the heroic memories of the national struggle for independence that mesmerized the planners so that very few questioned the raison d'etre of small scale and village industries in the long-term scenario of national development. This is best illustrated by a Planning Commission note of August 1961, which had been prepared by Jaya Prakash Narayan and outlined the Gandhian philosophy of rural industrialization that stressed on rural industrialization. He suggested that the aim and total long term effect of rural industrialization should be to convert the present lopsided purely agricultural communities into balanced agro-industrial communities. The Planning Commission identified consequently 40 rural areas for intensive development of small industries.

The Rural Industries Programmes were to cover all kinds of small industries and processing industries based on agriculture. It was recognized that: "With the increase in the production of cereals, pulses and a number of cash crops like sugarcane and oilseeds visualized in the Third Plan, there will be considerable scope for the expansion of processing industries in rural areas. With a view to providing fuller employment and strengthening and diversifying the rural economy, it will be desirable to develop these industries to the maximum extent in the decentralized and small scale sector and on a cooperative basis".

Different varieties of the decentralized sector (cottage, rural, small or agro) continued to enjoy a special place in the successive Five Year Plans. The inherent strength and weakness of the policies towards small and village industries are now better appreciated. The official patronage to small scale, rural or cottage industries does not appear to have made a noticeable impact on the problem of rural-urban migration. The establishment of cottage and village craft emporia and training of artisans and design workshops has also not mitigated the situation. The Khadi and handloom sectors do seem to continue providing an additional source of income in villages. It is, however, doubtful if these industries, left to themselves, have the inherent strength to face competition by the modern mill sector.

Post green revolution era

The period of 'sixties' witnessed the beginning of the green revolution in some parts of India. In the Punjab, Haryana and western Uttar Pradesh, agricultural output per hectare rose markedly due to the enhanced canal and well irrigation, widespread adoption of new and improved seed varieties, enlarged inputs of chemical fertilizers and use of pesticides. While managerial practices are important, it is an undisputed fact that the green revolution was a direct consequence of high levels of agro inputs per unit of land. The enlarged inputs were not obtained from the farm itself or from traditional sources. The switchover to electricity, diesel and pump sets was almost dramatic; the high-yielding seeds were brought in from research centres; and tractors and agricultural implements, supplied by national and international sources. The green revolution brought Indian agriculture in close contact with industry. The nature of agro-industry relationships extended themselves to supply of industrial inputs instead of agriculture playing the raw material supply function only. The demand for a variety of industrial inputs had to be satisfied, if agricultural development was to be optimized. With a view to reducing problems of procurement of industrial inputs for agriculture, the State Governments were advised by the Centre to set up Agro-industrial Corporations.

The Agro-industrial Corporations were to promote agro-processing and generate additional employment opportunities in rural India. These objectives did not, however, find a priority in their actual working, for they chose the easier course of promoting sales of tractors, agricultural machinery, fertilizers, pesticides, etc. None could have been happier than the large enterprises manufacturing tractors and other agricultural machinery, when they found these state-sponsored corporations were most willing, even anxious, to undertake marketing for the private sector. In some states, beginning with the early 'seventies', subsidiary corporations were also set up to provide cold storage and crop processing facilities.

The official efforts at promoting village, rural and agro-industries were grossly inadequate compared to the magnitude of the task involved. They were only too thinly spread both in relative and absolute terms. The 'eighties' witnessed a strong plea for promotion of agro-industries in India. The orientation and the context of the assertion, however, have been vastly different from the arguments of Gandhi, Karve, Mahalanobis and Jaya Prakash Narayan. Agro-industries of the 'eighties' were essentially understood in terms of food processing industries. The arguments, briefly put, were:

- India' huge population with surging middle upper class offers a large market for modern durables and agro-based products, especially semi-processed and convenience foods.
- There also exists a huge unexploited potential in the international market, where India has competitive edge over many other supplier countries.
- Growth of food processing industries triggers expanding demand for farm produce, vegetables, fruits and other greens that improve agricultural incomes.
- The industry gives consumers in having access to vegetables, fruits and other farm products throughout the year. This would mean better returns and incentives to Indian farmers.
- Establishment of modern plants with sophisticated technology would help reduce crop wastage due to seasonal gluts and the perishable nature of farm products.
- Urban centres witnessing a substantial change in the intensity of woman employment and families with both husband and wife doing formal jobs create potential for consumption of convenience and semi-processed foods.

The post-1985 period has witnessed a substantial change in the official policy towards technology import and role of private foreign capital, in general, and in the food industries, in particular. The change is evident from the fact that, of the 111 foreign collaborations in food related industries approved during 1951-1985, more than 60 per cent of the agreements were for technical know-how and only 40 per cent for financial collaboration. The distribution of the collaborations in 35 years in term of the nature of processes shows that nearly 40 per cent of the collaborations were for "plant and machinery". As a single product area, "marine" was the main sector, which enjoyed a dominant position. In contrast to the pre-1986 period, the number of collaborations in the area of food- related industry approved during 1986-93 stood at 293.

The 'eighties' also witnessed a keen interest in investments in the area of food processing and soft drinks. The TNCs visibility in this area is indeed a marked one. For instance, Pepsi entered into collaboration with Punjab Agro

Industries Corporation and the Tatas to establish processing facilities for tomato juice and paste along with soft drink concentrates. Though a failure, General Foods of US also entered India during this period. Kellogg came in production of breakfast foods. Nestle, known for its interest in coffee, has started marketing "Maggie" convenience foods, ketchup, chocolates, etc; Hindustan Lever, the first entrant to the hydrogenated edible oil industry in India, handed over the Dalda production and marketing to its sister company, Lipton. The Levers, however, have acquired control over another large manufacturer of soap and oil products, TOMCO. They have also taken over Kissan, a company known nation-wide for jams and squashes, and are reported to have acquired rice-milling facilities. Brooke Bond, an associate of the Levers, has entered marketing of masalas. Among the new entrants to the edible oil industry are ITC and Britannia. Parle, the market leader in the soft drinks segment, which had fought tooth and nail against the entry of Pepsi, was obliged to abandon its fight with TNCs and join hands with Coca-Cola. It appears that the withdrawal of restrictions on the use of foreign brand names has speeded up the process of domination of the Indian consumer goods market by transnational corporations.

The entry of the U.S. based TNCs has coincided with the Indian policy to give high priority to private foreign direct investments and revision of the licensing policies to permit entry of large Indian companies and TNCs into the food processing industry. The establishment of the new Ministry of Food Processing Industries (MFPI) at the Centre was an indication of the Government's thinking. The MFPI is obliged to create increased job opportunities in the rural areas with specific reference to women, and unemployed youth by development of primary produce through a network of processing units. The MFPI is expected to promote modern technology and marketing techniques in aid of the farmers. Modern technology could help raise the average productivity in food processing, but to expect modern food processing industries to create a substantial rural job opportunities may not be realistic. Modern automatic plants of large capacities with "not touched by hand" operations would add little to create new employment, on the one hand, and may, in practice, replace the traditional technologies and practices that were labour-intensive on the other. The employment potential of such processing units cannot, therefore, be large. Also, the employment may only be of seasonal nature. This being the scenario, it is but logical to fear that the objective of making a meaningful impact on the level of employment of women and rural youth is hard to achieve.

Food processing industry in the modern sector is likely to have its main characteristic in "brand" domination. The poor hygiene, in which the traditional food processing is generally undertaken, does not help sustain high confidence in the processed food produced and packaged by the cottage and small units. Bulk food processing is amenable to standardization and offers certain economies of scale. It is because of such well known advantages that brand names carry substantial premium. Emergence of near monopoly situation in food processing is not unknown. And once a brand name gets accepted widely, it opens up possibilities for launching new products with a massive thrust. Indian consumers are well aware how `Maggie' noodles were followed up by Maggie sauce and Maggie soups. It seems that, instead of offering more employment, the modern food processing plants can cause the closure of many cottage and small establishments.

Location of a large food processing unit would bring in a substantial increase in the demand for the required farm produce. This would also provide better market prices to the farmers. Given the fact that there would be thousands of small farmers who would be suppliers of farm produce to one buyer/processor, the market situation would invariably be of monopsony, the stronger one seeking to exploit the situation to its maximum advantage. The past experience in two important agro-based industries, viz sugar and tobacco, shows that the processing units tend to thrive at the cost of the farmers. The sugar industry of U P and Bihar is a case in point. On the other hand, whenever processing is undertaken in a cooperative framework, there is a faster growth both for the farmer and the industry. The sugar cooperatives of Maharashtra, Gujarat and Karnataka have won a name for themselves. The same holds true of the edible oil (groundnut) cooperatives of Gujarat and the Amul experiment of dairy development on cooperative foundations. Thus, location of large private sector agro-processing units may be accompanied by perpetual conflicts, supply or processing uncertainties, labour problems, farmers' agitations and political interventions. On the other hand, if there were large farms specializing in a crop to meet requirements of the agro-processing unit, there could be a more organized and stable relationship between the farms and the factory.

An important policy assumption is that agro-industries have a large potential for exports from India. One wonders, if at the present stage of India's level of food, vegetables and other farm production there is a scope for generating export surpluses. Can India be a net exporter of food in the foreseeable future? The impact of exports of essential vegetables was well demonstrated by the abnormal price rise for onions during the mid-'seventies. The argument here is not against strengthening the food processing industry; it is against the export argument for setting up the industry.

Post economic reforms era

Indian food-processing sector is undergoing rapid transformation since mid nineties. As the country moves on the path of development, agriculture sector evolves from traditional subsistence level farming to commercial agriculture producing high value and processed products. Strong macro-economic fundamentals (sustained economic growth rate of about 8% during last three-four years) and the changing socio-economic scene are driving what was once a traditional, small-scale processed food production system into a modern industry aimed at catering to the evolving tastes and needs of discerning consumers. There is high potential demand in the domestic as well as export market. On the supply side, with vast arable land and agro-climatic diversity, India has the potential to be food basket of the world. With the interaction of demand and supply forces the food processing sector has the potential to be the driver of economic growth and enhance rural incomes. Food processing industry is widely recognized as a **sun rise** industry, with a vision to make India the food basket of the world.

India is a one of the leading producer of agricultural and livestock commodities-grains, fruits, vegetables, milk etc. however, the share of these products going for processing is very small, Also, India's share in world trade of processed food in 1.6% and value addition in 20% (Ministry of Food Processing, Annual Report 2005-06)

Table 1. Size of processing relative to production

Commodity	Production (m.t)	World position	% Processed
Milk	91	I	35
Fruits & Vegetable	150	II	2
Livestock	485	I	21
Grains	204	III	-
Fish	6.3	III	-

Source: Ministry of Food Processions Industries, Annual Report 2005-06

Key issues in food processing in India are the low level of processing, consumer perception regarding processed food, and unorganized nature of processing sector. Lower level of industry level processing does not mean that Indians do not consume processed foods. Instead, most of the processing is done at the household level-jams, pickles, jellies, and other traditional items are usually made at home and preserved through traditional means such as storing with salt, sun drying etc. Another important aspect of the processing sector relates to traditional habits and perception that "**processed food in not fresh**". However, this notion is changing fast because of exposure to a variety of processed foods in the organized food retailing outlets. Currently most of the processing is in the unorganised sector. Technically, organised sector means the units registered under Section 2m of Factories Act, 1948, whereas the rest in categories as organized sector. Primary food processing is a major industry with several thousand rice-mills/ hullers, flour mills, pulse mills, and oilseeds mills. There are several thousands of bakeries, traditional food units and fruit, vegetable and spice processing units in unorganized sector. In the organised sector, there are over 820 flour mills, 418 fish processing units, 5198 fruit and vegetable processing units, and 171 meat processing units.

Given the Indian reality, a number of factors should be taken note of in formulating a policy mix for the agro-industries.

- Promotion of agro-industries has to be seen in the perspective of inelastic nature of land available for cultivation. Future developments in agriculture will require more cropping intensity and better water management, adoption of new and improved varieties of seeds and crops, and use of appropriate quantities of manures and pesticides.
- Indian agriculture continues to have low productivity. This may necessitate larger and newer varieties of industrial development.
- Indian agriculture suffers from substantial pre and post harvest wastage. Systematic efforts should, therefore, be made to minimize wastage and perishing of farm output and to adopt practices and technologies that cut down processing losses.
- There can be a large national and international market for agro-based health products, products based on herbs, plant roots, fruit, leaves and vegetables provided these are processed, packaged and marketed in accordance with internationally accepted standards. There is also a growing awareness of the use of environment friendly organic insecticides and manure from farm waste material. The emergence of 'neem' preparations as safe and effective insecticide is a case in point. A good deal of progress has already been made in the area of bio-gas technology for a healthier, more convenient and cheaper source of energy in rural households than any known to them so far.
- The additional and associated advantage that the food processing industry offers is the possibility of affecting addition of appropriate vitamins and minerals in which the average Indian diet is considered deficient.
- Cooperatives are to be encouraged to take up agro-processing in a big way to realize the economies of scale. They offer, in addition, the best means of harmonizing the interests of the producer, the processor and the consumer. Problems of pest control, canal or underground water management and preservation of seed purity against pollen transfers

would also require special coordination. More importantly, introduction of harmless bio-fertilizers and bio-pesticides is easier in a cooperative setting. The farmers should be encouraged to form cooperatives to establish and run agro-industrial complexes for total processing of organic produce of specified types of crops.

- While more research is necessary to discover new uses for today's 'waste' materials, there are already many proven commercially viable processes, which could be taken up for implementation in these agro industrial complexes. Byproduct systems for paddy, sugarcane and groundnut are well known.
- A modernized industry in a rural setting would also find employment for the educated unemployed, semiskilled and the unskilled rural workers. It would also help realize the objective of "back to villages" while breaking simultaneously the stranglehold of an endemic occupational structure.
- Marketing is an area that requires a very special attention. Large corporations, Indian or foreign, have an edge in marketing. Without unduly hurting the interests of the consumers, one should be able to utilize the existing infrastructure to distribute the products of cooperatives. One could also enlist the support of non-governmental consumer organizations to educate consumers about the inherent qualities of the products offered by the cooperatives. Foreign brand names, whether original or in hybrid form, might be disallowed in agro-industrial products to give the local processors an equal chance of competing. Preferential treatment in purchases by public bodies might also be extended to cooperatives; provided the products meet quality standards and prices are within the acceptable limits. For example, dressing cotton and bandage cloth of a cotton complex and intravenous dextrose fluids of a corn complex could be preferred in Government Hospitals. One could even think of transferring certain public and private sector industrial units to the cooperatives.

Advancement of agro-industries, fully integrated with the developmental needs of rural population, calls for bold and possibly unconventional measures. At the administrative level, it requires urgently the creation of a district level authority vested with wide-ranging powers and having majority representation of cooperative bodies. This authority would be the nodal agency for providing various support services such as information on crop varieties, weather, products, markets and coordination of technology acquisition.

❑❑❑

38

Industrial Utilization of Sugar Crops

R.S. VERMA

Sugarcane and sugarbeet are two main sugar crops of the world. Out of world's total white crystal sugar production, about 70% comes from sugarcane and 30% from sugarbeet. More than 90 countries in the world cultivate sugarcane, while 35 countries produce sugar from sugarbeet. About 12 countries produce sugar both from sugarcane and sugarbeet. At present, India produces sugar only from sugarcane.

Worldwide sugarcane occupies an area of 20.4 million hectares with a total production of 1333 million tonnes (FAO 2003) and productivity of 65.3 t/ha. India by contributing 21.0% area and 22.0% production ranks second among sugarcane growing countries of the world for both area and production of sugarcane. The crop is grown all over the country except cold hilly areas. Besides, sugar, jaggery and *khandsari* production, sugarcane also provides useful raw material as co-products (Table 1) to over 25 industries which utilize its different components to produce a range of agro-industrial products viz., alcohol, paper, boards, chemicals, cattle feed, electricity, etc.

Table 1. Sugarcane co-products (million tonnes) availability in India (2007-08)

Bagasse	Molasses	PMC	Green tops	Trash
102	11.3	8.8	68	34

Indian sugar industry

During 2007-08, Indian sugar industry utilized 73.5% of the total cane produced (340 million tonnes) and produced 26.4 million tonnes sugar. The industry is the second largest agro-based industry (next to cotton) textile. It is instrumental in generating sizeable employment in rural sector directly and through ancillary industries. It is estimated that about 50 million farmers and their dependents are engaged in the cultivation of sugarcane and another 0.5 million skilled and unskilled workers including highly qualified and trained technologists are engaged in the manufacturing of sugar.

Industrial utilization of sugarcane co-products

1. Bagasse

Bagasse is a fibrous residue of the cane stalk left after the extraction of cane juice. Sugarcane stalk contains 12-15% fibre. Fresh bagasse is known as mill wet bagasse and has about 50% moisture. Thus the quantity of bagasse is about 30% of the quantity of cane crushed.

Composition of fresh bagasse

Moisture	50%
Fibre	47%
Soluble Solid	3% (Mainly sugars)

Composition of bone dry bagasse

Cellulose	45%
Lignin	20%
Pentosans	28%
Ash	2%
Sugar	5%

Ninety per cent of the bagasse produced in sugar mills of India is used to meet their own fuel requirement. Rest amount is available for other uses.

Cellulose content of bagasse is useful as fibrous raw material for all the fibre based industries, whereas the pentosan content is used in the manufacture of furfural. The rind portion of cane consists of high quality cellulose. The internal portion of cane, known as pith, contributes 25-30% portion of the bone dry bagasse is not of use in the manufacture of paper boards and other cellulosic products. Pith is made of soft parenchymatus tissues and has undesirable properties. Therefore, it has to be removed from the bagasse. Lack of suitable equipment for complete removal of pith from bagasse was mainly responsible for unsuccessful attempts in the early years to use bagasse as fibrous raw material for manufacture of papers, boards etc. The pith portion can be recycled to produce briquettes or pellets to be used as fuel in the sugar factory or could be used in the preparation of cattle feed.

A. Fibrous products from bagasse

(i) Bagasse based paper industry

Bagasse has been successfully used for the production of different grades of paper which are normally produced from other fibrous material like wood, bamboo straw etc. India has banned the setting up of any more pulp and paper factories based on forest raw materials. In view of this, pulp, paper and industries are on the look out for alternate fibrous raw materials to replace wood and bamboo. Bagasse holds a big promise in this context.

Tamil Nadu Newsprint Ltd. at Pugalur (Trichy) was the first mill in the world to produce newsprint using 100% bagasse pulp. However, the growth of bagasse based paper mills has been tardy in this country due to non-availability of adequate bagasse since a large portion of bagasse is used by sugar mills themselves as a fuel. Recently large scale paper mills based on bagasse were established at Dalmianagar in Bihar, Raniganj in west Bengal, Yamunanagar in Haryana and Erode in Tamil Nadu. The numbers of bagassed based paper mills are gradually increasing in India.

(ii) Particle and fibre boards

Bagasse has been found to an excellent raw material for particle and fibre boards. In India, technology of bagasse based particle board manufacture has been introduced recently. The first particle board plant based on bagasse has been established at Gajraula in U.P. Besides this, many co-operative sugar factories in Maharashtra are going to start particle board plants based on bagasse.

(iii) Bagasse – cement board

These types of boards are generally obtained from the mixture of particles of lingo-cellulosic fibre with portland cement, calcinated magnesite or gypsum.

(iv) Moulded bagasse products

Any lingo-cellulosic material like bagasse admixed with thermosetting synthetic resins can be hot pressed into any desired shape, by the use of an appropriate mould. Some of the articles that can be produced are boxes for packing fragile instruments, ceiling and roofing tiles, trays, party plates, doors and window frames etc. The process of manufacture of moulded products is exactly the same as of particle boards. For small articles like food trays, fine particles are necessary and for large size article like boxes, large particles are used. In India, a firm known as Sitapur Ply Wood Products Ltd. at Sitapur (U.P.) started bagasse based moulded products like door and window frames from 1985.

(v) Rayon grade pulp

Research work have indicated that it would be possible to produce rayon grade pulp from depithed bagasse. Rayon grade pulp is mainly used for making yarn (rayon) for fabrics and absorbent sanitary products. There is no commercial scale plant at present to produce such pulp based on bagasse.

B. Bagasse products as fuel and energy

(i) Bagasse briquettes or pellets and logs

These are prepared by densifying the bagasse. The production of bagasse briquettes or pellets and logs and their use as domestic fuel is common in some countries.

(ii) Bagasse charcoal

The preparation of charcoal from bagasse involves carbonization. Such charcoal is used as fuel for home and industry.

(iii) Methane and producer gas from bagasse

Methane (CH_4) is produced in nature by anaerobic fermentation of all organic matter in the presence of methogenic bacteria. Methane and producer gas (a mixture of CH_4, CO_2 CO, H_2 and N_2) are produced from bagasse.

(iv) Ethanol

Conversion of cellulosic material of bagasse into ethanol by simultaneous saccharification-cum-fermentation method has been developed at IIT, New Delhi. However, at present, this technology is not economically viable due to high cost.

(v) Co-generation based on bagasse

The term 'co-generation' means, the combined production of motive, electric and thermal power by the sequential use of energy from a common source of fuel (bagasse). Co-generation facility uses significantly less fuel to produce many forms of energy together than would be needed to produce different forms of energy, separately. Sugar factories are now designed to have furnaces to burn the bagasse to generate high pressure steam required to drive the prime-movers (milling plant and power generator) and make use of the low pressure exhaust steam obtained from the prime movers for heating and concentration of cane juice. Thus, the steam generated by burning of bagasse as fuel is used for simultaneous generation of motive, electric and thermal powers.

C. Chemicals from bagasse

(i) Furfural and furfuryl alcohol

Pentosan content of bagasse is used in the manufacture of furfural. Furfuryl alcohol is produced by furfural. It is used in petroleum refinery as a selective solvent for the purification of lubricating oil. In India, Southern Agrifurane Industries Ltd. with an annual production of 3000 tonnes of furfural went into operation in 1981.

(ii) Xylitol

Bagasse contains pentosans, which are mainly composed of xylan. Xylitol is obtained by hydrogenation of xylose. Xylitol is a non-sugar sweetner so it is used in diabetic food. At present, the biggest use of xylitol is in the manufacture of the chewing gums. It is also used in tooth paste, mouthwashes, cough syrup etc.

(iii) Carboxy methyl cellulose (CMC)

CMC is an ether of cellulose and monochloroacetic acid. It is used in painting and ceramic industry.

(iv) Activated charcoal

It is produced when bagasse is pyrolised for 20 hours at 450°C in the absence of atmospheric air. It is used as decolarating agent for sugar, oils, fats, beverages, spirits etc.

D. Other uses of bagasse

(i) Cattle feed

Experiments conducted in many countries to feed raw bagasse to the animals have failed as it cannot be digested by animals in raw form. However, digested bagasse may be fed to the cattle. In order to increase the digestibility, the bond between the fibres and lignin has to be broken. This can be achieved by treating raw bagasse with steam or with small quantity of alkali at high pressure and temperature for a short duration. The Surat District Co-operative Milk Producers' Union Ltd. in Gujarat has set-up a plant to produce pre-digested bagasse to be used as low cost fodder for the cattle.

(ii) Cultivation of edible mushrooms on bagasse

Bagasse, being a rich source of cellulose and hemicellulose, is one of the best substrate for the cultivation of edible mushroom. Among the various cultivated mushrooms, Oyster mushroom (*Pleurotus)* can be produced successfully on bagasse.

2. Molasses

Molasses is the mother liquor left over after the crystallization of sucrose from which further quantity of sucrose cannot be recovered economically. The quantity of molasses produced is about 4% of the cane crushed. In India, 501 sugar factories produce about 11.3 million tonnes of molasses per annum out of which 80-90% is used for the manufacture of ethanol (ethyl alcohol) and rest is used for other purposes. Molasses contains about 40-50% total sugar content. The industrial value of molasses is on account of its fermentable sugar content (usually 42-48%) which is being used to produce many products of great economic value. Brazil is a leading country in the world, which uses ethyl alcohol produced from molasses as fuel in internal combustion engines. About 20% alcohol and 80% gasoline (petrol) known as gasohol may be used in all the existing motor cars without requiring any modifications in the engines. Pure ethyl alcohol cannot be directly used in engines because of very high self ignition temperature and propensity to knock. The industrial utilizations of cane and beet molasses are more or less the same, as the total sugar content of both these are almost same. However, cane molasses contains more invert sugar as compared to beet molasses.

(i) Fermentation products from molasses

Various industrial products of great commercial value are obtained by fermentation of sugar in molasses. These are mentioned in Table 2.

Table 2. Fermentation products form molasses

Product	Fermented by	Used in	Manufacturing units
Ethanol (C_2H_5OH)	*Saccharomyces cerevisiae*	Potable purposes (50%), industrial purposes (40%)rest as fuel and others	246 distilleries in India producing 1158 million litres of ethanol
Lactic acid	*Lactobacillus delbruckii* *L. bulgaricus* **L. leichmanii**	Food products-jams, jellies, dairy products, soft drinks, pickles, canned vegetables, fish products Non-food uses- dyeing of textiles and woods, adhesives, cleaning and polishing formulations, pesticides, resins	Munirabad Chemicals, Karanataka Tamil Nadu Dadha Pharma Ltd., Madras
Oxalic and Tartaric acids	*Aspergillus niger* **Penicillium**	Tanning of leather and textile industry	10 Small scale plants
Malic and fumeric acids	*Rhizopus nigricans*	Foods and pharmaceutical preparations	
Citric acid	*Aspergillus niger*	Food, pharmaceutical , leather and textile industry	Citurgia Biochemicals Ltd., Surat Punjab Khand Udyog Ltd.
Mono-sodium glutamate (MSG)	*Micrococcus glutamicus*	Food industry as flavor intensifying agent	
Glycerol (glycerine)	*Saccharomyces cerevisiae*	Cosmetics, toothpaste, mouthwashes, adhesives, photographic products	Hindustan Live Ltd., Bombay
Butanol –acetone	*Clostridium acetobutylicum*	Solvent for lubricants, oils, waxes, resins, gums	A plant for defence purposes
Dextran	*Leuconostoc dextranicum*	As a stabilizer for ice cream and confectionery	
Lysine	*Corynebacterium glutamicus*	Fortification of foods and feeds, pharmaceuticals	
Acornitic and itaconic acids	*Aspergillus terreus*	Plastilizers, wetting and flavouring agent, antioxidant	
Ephedrine	*Saccharomyces* Sp.	Ingredient in all cough syrups	Two plants operating in India
Xanthan gum	*Xantomonas compestris*	Textile, paint, petroleum and ceramic glass industries	
Yeast	*Saccharomyces cerevisiae* *Torulopsis utilis* *Candida arborea* *Oidium lactis*	Bakery and distillery	Indian Yeast Co., Dhampur Yeast

(ii) Live-stock feeds

Animals also like sweet feeds. The carbohydrate (sugars), minerals and organic content of molasses are very nutritious for all live-stock. National Dairy Development Board (NDDB), Anand has recently developed Molasses-Urea Licks or blocks (animal chocolates) for keeping before the animals for licking. NDDB has so far set up 10 plants in different states to produce Molasses-Urea blocks.

(iii) Molasses for curing of tobacco

Tobacco leaves are dipped in a thin solution of molasses, before they are cured to attain a golden colour which is very valuable for the manufacture of high class cigarettes.

(iv) Mixing of molasses with hookah tobacco

Hookah tobacco is mixed with small quantity of molasses to make the tobacco burn slowly and emit a particular flavour.

3. Press mud cake (PMC)

The PMC is one of the major co-products of sugar industry and constitute about 3-7% of the total weight of cane crushed depending upon the clarification process employed. In the sulphitation factories, it amounts to about 3% and in the carbonation factories about 7%. An estimated amount of about 8-9 million tonnes of PMC is being released by sugar industry in India.

The PMC is the residue obtained by filtration which settles out in the process of clarification of juice for sugar manufacture. This material is also known as filter cake. It is soft, spongy light weight material of dark brown or black colour normally containing 50-70% moisture (when it comes from sguar factory). It consists of a mixture of sugarcane fibres, sucrose, coagulated collide including cane wax, albuminoids, inorganic salts and soil particles.

(i) PMC as organic manure

PMC is a good source of organic matter and plant nutrients. Riga Sugar Factory in Bihar is producing a bio-fertiizers using PMC known as *Hara Bhara*. For making *hara bhāra*, PMC is air dried and screened and this powder is incorporated to the extent of 12-15% in a mixture of urea, DAP, SSP and MOP, so that mixture contains NPK in the proportion of 18:18:6. The mixture is fed to revolving drums, into which a small quantity of water is sprayed. The mixture is dried and packed. It is quite popular among the cane growers of the factory area.

(ii) Wax from PMC

Waxy coating found on the rind of sugarcane stalks constitute about 0.125% of the weight of the cane. During the processing of sugarcane in the sugar factory a part of the wax goes with bagasse and is burnt. The balance of wax enter the cane juice and precipitate during clarification of juice. Thus, PMC is the raw material for extraction of wax. Ravalgaon Sugar Farm Ltd., in India was reported to producing wax from sugarcane. Cane wax can be used in manufacture of shoe polish, carbon paper etc.

(iii) Bio-gas from PMC

It is very well known that an aerobic digestion of organic matter in the presence of methogenic bacteria results in production of bio-gas, which is generally a mixture of methane (60-65%) and CO_2 (35-40%). In India Pravara SSK Ltd. (Maharashtra) has set up a commercial scale bio-gas plant based on PMC. The bio-gas generated is being supplied to the employees of the sugar factory for its use as domestic fuel.

4. Cane tops and green leaves (CTGL)

On average CTGL constitute 20% of the weight of cane stalks and is a valuable cattle feed. In India, about 68 million tonnes of CTGL are produced annually.

The CTGL may also be ensiled for off-season use. Dilute solution of molasses and ammonium sulphate are sprinkled on a chopped tops and dumped on a platform and covered with a tarpaulin. Molasses solution undergoes fermentation producing lactic acid, which maintains the keeping quality of the ensiled material.

5. Sugarcane trash

Dried leaves of sugarcane are known as trash. It is 8-10% of the weight of cane stalks. Trash availability in India, ranges from 30 to 34 million tonnes per annum. Generally, trash is burnt away in the field soon after the cane harvest. Low bulk density of trash has limited its commercial and industrial use and therefore, its cellulosic value (42-46%) could not be fully exploited. It is quite likely that in future baler could be used to increase bulk density which will greatly reduce transportation problems.

(i) Bio-gas and bio-manure from trash

NSI, Kanpur has developed a technology for production of bio-manure and bio-gas from cane trash. The bio-gas can be used for variety of purposes such as for generation of electricity, operating engines, lighting gas mantles and for cooking.

(ii) Mushroom cultivation on cane trash

Cultivation of edible mushroom *Pleurotus* sp. is possible on trash singly or in combination with wheat straw.

A concept of 'sugarcane agro-industrial complex' is emerging which means cluster of industries based on sugarcane co-products (bagasse, molasses, PMC, green tops and trash), in addition to sugar, at one place. An example of this is the Jiangmen Sugar Complex in South China which produces 28 products, making use of sugarcane as the main raw material.

Sweet sorghum

Research work has proved the potentiality of sweet sorghum as a supplementary sugar crop. The crop matures in 100 days and yields about 30 t/ha millable stalks. The juice has a brix of about 19%, pol of 15% and a purity of 79% with a sugar recovery of about 9%. The only disadvantage is that besides sugar, the juice contains starch, aconitic acid etc. which interferes with the process of crystallization of sugar in the juice. However, the production of sweet syrup and alcohol from the juice is successful. More studies are needed to explore the possibilities of commercial exploitation of this crop.

Sweet maize

Certain varieties of maize are found to contain fairly good quantities of sugar in the juice extracted from their stalks. However, to extract sugar from maize stalks is not economical at present.

□□□

Socio-Economic and Climatic Facets

sector evolves from traditional subsistence level farming to commercial agriculture producing high value and processed products. Strong macro-economic fundamentals, sustained economic growth rate of about 8% during last three-four years and the changing socio-economic scene are driving what was once a traditional, small-scale processed food production system into a modern industry aimed at catering to the evolving tastes and needs of discerning consumers. There is high potential demand in the domestic as well as export market. On the supply side, with vast arable land and agro-climatic diversity, India has the [illegible]. With the interaction of demand and supply forces the food processing sector has the potential to be [illegible]. Food processing industry is widely recognized as a sunrise industry with a vision to make India the food basket of the world.

Socio-Economic and Climatic Facets

India is a one of the leading producer of agricultural and livestock commodities-grains, fruits, vegetables, milk etc. however, the share of these products going for processing is very small, Also, India's share in world trade of processed food in 1.6% and value addition in 20% (Ministry of Food Processing, Annual Report 2005-06)

Table 1. Size of processing relative to production

Commodity	Production (m t)	World position	% Processed
Milk	91	I	35
Fruits & Vegetable	150	II	2
Livestock	485	I	21
Cereals	204	III	
Fish	6.3	III	

Source: Ministry of Food Processions Industries, Annual Report 2005-06

Key issues in food processing in India are the low level of processing, consumer perception regarding processed food, and unorganized nature of processing sector. Lower level of industry level processing does not mean that Indians do not consume processed foods. Instead, most of the processing is done at the household level-jams, pickles, jellies, and other traditional items are usually made at home and preserved through traditional means such as storing with salt, sun drying etc. Another important aspect of the processing sector relates to traditional habits and perception that "processed food in not fresh". However, this notion is changing fast because of exposure to a variety of processed foods in the organized food retailing outlets. Currently most of the processing is in the unorganised sector. Technically, organised sector means the units registered under Section 2m of Factories Act, 1948,

39

Socio-economic and Policy Issues in Relation to Commercial Agriculture

JAGDISH KUMAR

This paper provides an overview of commercial agriculture pertaining to the socio-economic and Indian agricultural policy issues. After presenting a brief overview of the policies adopted in India for agricultural development over the planning period, the important issues of the National Agricultural Policy are discussed.

Important policy measures introduced in the agriculture sector in India during the planning periods

1. *Technological measures*: These include creation of irrigation potential on a large scale (by undertaking major and medium irrigation projects and rural electrification project for minor irrigation) and popularizing new agricultural strategy in the form of package programme for the selected regions of the country introduced in mid 1960s. Steps were taken to increase the production of high-yielding varieties of seeds, fertilizers and pesticides within the country and supplement domestic production by imports whenever necessary.
2. *Land reforms*: Land reform measures undertaken includes: (i) Abolition of intermediaries, (ii) Tenancy reforms to (a) regulate rent paid by tenants

to landlords, (b) provide security of tenure to tenants, and (c) confer ownership rights on tenants; and (iii) imposition of ceilings on size of holdings in a bid to procure surplus land for distribution among landless labourers and marginal farmers.

3. *Consolidation of holdings*: This programme aimed at consolidating all plots of land owned by a particular farmer in different places of the village by sanctioning him land at one place equal in area (or value) to his plots of land.

4. *Institutional credit*: It includes expansion of institutional credit to farmers, especially through cooperatives and commercial banks. After nationalization of banks in 1969, commercial banks have paid increasing attention to the needs of agriculture. Further to facilitate access to short-term credit by farmers, a Kisan Credit Card (KCC) scheme was introduced in 1998-99.

5. *Procurement and support prices*: It includes announcement of procurement and support prices to ensure fair returns to the farmers so that even in years of surplus the prices do not tumble down and farmers do not suffer losses. The price policy also intended to facilitate adoption of improved farm technology which in turn gave boost to the farm sector.

6. *Input subsidies to agriculture*: The government has provided massive subsidies to farmers on agricultural inputs like irrigation, fertilizers and power to promote commercialization of agriculture in the country.

7. *Other measures*: In addition to the measures mentioned above, the Indian agricultural policy contained a number of other elements, some of which are outlined below:

- Improving the system of agricultural marketing through the establishment of regulated markets and introducing a variety of measures like standardization of weights and measures, grading and standardization of farm output, providing information regarding market prices to farmers, etc. efforts have also been made to strengthen the cooperative marketing structure.
- Provision and expansion of storage and ware-housing facilities.
- Promotion of agricultural research and training to discover new high-yielding varieties of seeds, avoid wastage of grains in storage, successfully counter the attacks of pests, insects and rodents, etc.
- In order to increase the production of pulses, a centrally sponsored National Pulses Development Programme was launched in 1986-87.

- The Small Farmers' Agribusiness Consortium (SFAC) was set up in January 1994 to generate agribusiness activities to expand employment opportunities and raise income levels in the rural areas through effective support to various types of agribusiness.
- In recent years a number of policy changes have been introduced to make agricultural exports more viable.

National agricultural policy

The first ever National Agricultural Policy was announced on 28th July, 2000. The Policy seeks to actualize the vast untapped growth potential of Indian agriculture, strengthen rural infrastructure to support faster agricultural development, promote value addition, accelerate the growth of agro business, create employment in rural areas, secure a fair standard of living for the farmers and agricultural workers and their families, discourage migration to urban areas and face the challenges arising out of economic liberalization and globalization. Over the next two decades, it **aims** to attain:

- A growth rate in excess of 4 per cent per annum in the agriculture sector;
- Growth that is based on efficient use of resources and conserves our soil, water and bio-diversity;
- Growth with equity, i.e., growth which is widespread across regions and farmers;
- Growth that is demand driven and caters to domestic markets and maximizes benefits from exports of agricultural products in the face of the challenges arising from economic liberalization and globalization;
- Growth that is sustainable technologically, environmentally and economically.

New policy initiatives

Reduction in rate of interest on agricultural loan: From kharif 2006-07, it was decided by the government that the farmers would receive crop loans up to a principal amount of Rs. 3 lakhs at 7 percent rate of interest and the government of India would provide necessary interest subvention to NABARD and banks for the purpose.

Revamping of cooperative credit structure: In January 2006, the Government announced a package for revival of short-term Rural Cooperative Credit structure and designated NABARD as the implementing agency.

National Agricultural Insurance Scheme (NAIS): This scheme has been implemented from rabi 1999-2000 season with the objective of providing insurance coverage in the event of failure of any of the notified crops as a result of natural calamities, pests and diseases on 'Area Approach' basis to all the farmers. There is provision of 10 per cent subsidy in premium to small and marginal farmers.

Agricultural Marketing–Reforms through APMC Act: To provide dynamism and efficiency into the marketing system, large investments are required for the development of post-harvest and cold-chain infrastructure nearer to the fields. The Government is taking steps to motivate corporate sector to invest in agricultural marketing system by making necessary amendments in APMC Act.

Notional Food Security Mission (NFSM): The mission aims at increasing food grains productions of rice (10 million tonnes), wheat (8 million tonnes) and pulses (2 million tonnes) through area expansion and productivity enhancement; restoring soil fertility and productivity; creating employment opportunities; and enhancing farm level economy to restore confidence of farmers of targeted districts.

Rashtriya Krishi Vikas Yojana (RKVY): The RKVY aims at achieving the 4 percent annual growth in the agriculture sector during the eleventh five year plan period by ensuring a holistic development of agriculture and allied sectors. The main objectives of the scheme are:

- To incentives the states to increase public investment in agriculture and allied sectors.
- To provide flexibility and autonomy to the states in planning and executing agriculture and allied sector schemes.
- To ensure the preparation of plans for the districts and the states based on agro-climatic conditions, availability of technology and natural resources.
- To ensure that the local needs/crops/priorities are better reflected.
- To achieve the goal of reducing the yield gaps in important crops, through focused interventions.
- To maximize returns to the farmers.

National Policy for Farmers, 2007: The national policy for farmers, among other things, has provided for a holistic approach to development of the farm sector. The broad areas of its coverage include:

- Focus on the economic well-being of the farmers
- Asset reforms

- Water use efficiency
- Encouraging adoption of new technologies
- Establishment of National Agricultural Bio-security System.
- Seeds and Soil Health
- Support services for women
- Credit & Insurance
- Gyan chaupals at village level with the help of ICT and farm school
- Putting in place an appropriate social security scheme
- Effective implementation of Minimum Support Price (MSP) mechanism.
- Enlarge Food Security basket to include nutritious millets.

□□□

40

Agroforestry for Socio-economic and Environmental Sustainability

O.P.S. BANA

Over the years, farmers have been practicing the integrated land-use (tree+crop+animal components) for sustenance. With the passage of time the agricultural component received priority over woody elements for sustaining self-sufficiency in food-grains. Subsequently, the population pressure resulted in smaller holdings and lesser number of trees on farms providing food, fodder, fibre, fuel, fertilizers, fruits etc. The introduction of high yielding crops and varieties with adoption of improved production technologies has no doubt resulted in almost 450 per cent increase in food-grain production but not without misgivings in terms of degradation of natural resources (including land/soils, plants/biodiversity, environment etc.) and increased cost of production and resource management. In order to bring a balance among food-sufficiency, socio-economic well-being of farmers and environmental stability its important to follow nature and eschew the false dichotomy between agriculture, forestry and/or animal husbandry by integrating them on the same land-management unit either simultaneously or sequentially. These interdependent and inseperable interprises being land based could be counter productive when delinked either at field, farm or village level, or even in the government functioning.

Agroforestry systems, sub-systems and practices with their element of intercropping are in essence more productive, ecologically sound and socio-economically acceptable models similar to natural systems and have, therefore, an element of sustainability associated with them. The agroforestry can serve the community socially (reducing drudgery, providing employment and better health), functionally (i.e. supporting farm income, heading from subsistence to commercial) and ecologically (i.e. conserving unique resource base).

Present day challenges

In the post-independent era agricultural production witnessed a radical boost of more than four times with an annual compound growth rate of nearly 2.5 per cent. Despite this impressive achievement keeping pace with surging multifarious social demands is facing tough time with following challenges :

- *Resource scarcity* : In global terms, India has only 2.0 per cent of geographical area, 1.5 per cent forest area and growing stock and 0.5 per cent of grazing land which is to support about 17 per cent human population and 15 per cent of global livestock population. Estimates indicate that Indian population will grow to 130 billion by 2020 AD requiring a production of 325 m tones of food grain. This may be achieved by producing 5-6 million tones additional food grain every year as against 3.1 million tones per year over the last 40 years (Priyanka, 2000). The total forest cover of the country is 67.71 million ha (i.e. 20.60% of geographical area). Of this, very dense and moderately dense forest constitute 38.72 million ha (i.e. 11.78% of geographical area) only (FSI, 2005), which is not enough to maintain the ecological balance and to meet the needs of timber, fuel, fodder, fibre etc. However, at present about 42 per cent (5160 million) of total tree population (12,252 million) in our forests grow out side the forest area mostly in the form of agroforestry that constitute almost 35 per cent of growing stock.
- *Degradation of natural resource base* : The degradation of soils (physically, chemically and biologically), water, loss of biodiversity and environmental imbalance are other problems threatening the sustainability of agriculture.

  Sustainable agriculture is the successful management of the resources to satisfy changing human needs while maintaining or enhancing the quality of the environment, conserving the natural resources and also encompass economic viability. The comparative economic advantage has been the basis of most agricultural systems in post-green revolution era which led to intensification in agriculture involving HIV, intensive use of agro-chemical and other inputs, expansion of area and deforestation, adding the component of unsustainability to Indian agriculture. As a result decline in agricultural productivity and uncertain climatic conditions (as rain fall distribution in time and space) have been noticed in many parts of our country.

Like India, many other developing countries are also facing several agricultural and ecological problems viz. over exploitation of forest resources, soil erosion and degradation, uncertain weather and natural calamities, environmental pollutions etc.

- *Socio-economic insecurity* : Increased cost of cultivation i.e. lower profitability of common agriculture systems, unemployment in farming families, greater health hazards etc. are some less talked but important features of today's non-sustainable agriculture.

Premises of agroforestry land-use

The best test for the soundness and validity of the concept of any land-use system is certainly its field performance, based on the results of which the concepts can be modified and the system suitably refined. However, the premises on which the concept is based and the evidence for its possibilities can be taken as strong indications of its validity. The premises on which the concepts of agroforestry are based, are both bio-ecological and socio-economic;

a. Bio-ecological premises

It may include all the advantages of the forests on the productivity of soil and the environment, such as :

- *Higher yields* : Increase in both the total biological (Org, et.al. 1991) and in economic terms (Pant and Bana, 1998; Dutta and Thakur, 2004) outputs have been observed under agroforestry compared with monocultures (sole cropping). But for such biological and economic gains, the agroforestry models have to be planned with great care in selection of component species, their proportions, arrangement and post planting management as per the agro-ecological region. The higher productivity of agroforestry systems can be attributed to; a larger amount of solar energy captured, better utilization of vertical space, mitigation of temperature extremes for crops by trees, reduced impact of strong winds and rain drops on crops, smoothering effect of trees on weeds, a greater diversity of fauna, the species diversity and their spatial arrangement may deter insect proliferation etc.

- *Closed and efficient nutrient cycling and maintenance of organic matter* : Tree roots capture nutrients and water that move through the soil profile beyond the reach of annual or perennial crop plants and recycle a large amount of biomass returned to the soil as organic matter through leaf fall, fruits, flowers, twigs, bark and fine roots. Trees and their roots also tend to improve the soil structure by producing a higher amount of stable aggregates and avoiding/dismantling various types of hard pans. The improvement in physico-chemical properties of soils has been reported by several workers (Pant and Bana, 2001; Swamy, *et.al.*2008). However, the magnitude of improvement will depend on tree species, their age and spatial arrangement, litter quality and management etc.

- *Soil and water conservation* : Trees in agroforestry are known to prevent surface run-off and erosion through increased interception of rain, improved soil physical properties (more infiltration and storage), litter provide mulch to reduce water evaporation from soil and adds considerable organic matter (Wiersum, 1991).

- *Regulation of micro-climate* : The temperature, light, humidity and wind conditions are substantially modified in micro-climate of agroforestry. Temperature extremes are mitigated (Lower maxima and higher minima) for the benefit of crops and animals close to the ground in agroforestry systems. The lower maxima reduce the speed of decomposition of organic matter.

  It was noticed that 25-29 per cent reduction in net radiation (Rn) improved the microclimate in *Dalbergia* : Salix (50:50) mixture during resproductive periods of wheat crop and caused no significant reduction in yield (Saini and Singh, 2001). They also observed that late sown wheat under *Eucalyptus tereticornis* can be grown upto 40 per cent reduction in Rn. Whereas, direct seeded rice with 54 per cent Rn under *Dalbergia sissoo* and 40 per cent Rn under *Albizia lebbak* can provide 90 per cent yield compared to sole crop. However, micro-climate under agroforestry systems can greatly be modified by tree row spacing and orientation and tree management options (pruning). Other changes in micro-climate may be associated directly or indirectly to the host-pest relationship and ET requirement of the components of agroforestry systems.

- *Suitability of agroforestry to marginal lands* : The adaptability of trees to the soils that are incapable of sustaining annual agricultural crops also favours the adaptation of agroforestry in marginal and degraded lands to make them productive.

b. Socio-economic premises

The socio-economic factors that substantiate the potential of agroforestry are that the poor farmers in developing countries, existing in an environment of mounting population pressures and lack of resources, are forced to utilize inherently unproductive lànds for food production and practice systems that have disastrous consequences, such as deforestation, desertification, soil degradation, flood and droughts. These farmers need alternative land management system that provide food and wood without causing deterioration of the ecosystem. From agroforestry, farmer obtain atleast a part, direct economic benefits from trees that satisfy their needs for firewood, post, poles, sawn timber, fruits, fodder, flowers for honey, medicinal products etc. and do not need to buy them or transport. Trees that produce salable wood constitute "standing capital", as an insurance to meet emergencies in case of immediate cash needs. Dependency and catastrophies associated with a single crop are overcome or mitigated, particularly in case of irregular rainfall patterns, market fluctuations, pest out breaks, difficulty in acquiring imported products, such as pesticides, fertilizers, machinery, concentrates for cattle etc.

The presence of trees usually reduces weeding costs. Trees can be used to mark properly the field boundaries and constitute a safeguard against land usurpation. Longer employment to family labour with greater flexibility to distribute the work load throughout year is ensured through agroforestry. The employment generation off-farm i.e. agroforestry based industries and processing units is also possible and more than one lakh people were found employed in various agroforestry based industries in single district (Yamunanagar) of Haryana (Bisht and Panse, 2008). There is obviously considerable scope to allow a gradual change from distructive land-use practices towards more stable systems without diminishing productivity and to improve an existing stable systems by designing new systems with higher production and productivity by associating the most desirable plant species (and/or enterprises), in space and/or time.

Agroforestry is here to stay and grow. When teak was selling for Rs. 300/ton grown with irrigation, manuring etc. did not make sense. Now that teak is selling for Rs. 30,000/ton it does not make sense not to apply irrigation, plant nutrition and good management. Top quality poplar wood sells at the same prices per ton as wheat and rice in India but produces much more wood per area per year than wheat and paddy combined. Agroforestry has the same strengths as natural forestry; providing pollution control and environmental protection etc.

References

Bisht, N.S. and Panse, S.S. 2008. Employment Potential of Agroforestry ; A case study from Yamunanagar, Haryana. *Indian J. Agroforestry* 10(1) : 19-23.

Dutta, V. and Thakur, P.S. 2004. Bio-economics of cropping systems combining medicinal and aromatic herbs with commercial timber tree species. *Indian J. Agroforestry* 6 (1) : 1-7.

Forest Survery of India (FSI). 2005. State of Forest Report-2005. F.S.I. MOEF, GOI, Dehradun 172 pp.

Org, C.K., Corlett, J.E., Singh, R.P. and Black, C.K. 1991. Above and below ground interactions in agroforestry system. *For. Ecol.* and *Management* 45:45-57.

Pant, A.K. and Bana, O.P.S. 1998. Crop Productivity of Wheat (*Triticum aestivam*) and Urd (*Phaseolus mungo*) under plantation crops. *Indian Journal Forestry* 21(3): 238-243.

Pant, A.K. and Bana, O.P.S. 2001. Soil Fertility improvement under various tree species in mixed agroforestry systems. *J.Tree Sci.* 20 (1&2):86-92.

Priyanka, K. 2000. Natural resource management : Towards sustainable agriculture in 21st century. *Employment News* 24(46):1-3.

Saini, B.C. and Singh, R.V. 2001. Performance of *Salix* based agroforestry systems under shallow water table conditions of *Tarai* of Uttarakhand. *J. Indian Water Resource Soc.*18;15-19.

Swamy, S.L., Bhartiya, J.K. and Misra, A. 2008. Nutrient storage and crop productivity under different tree spacings of *Gmelina arborea* in Agrisilivicultural system. *Indian J. Agroforestry* 10(2):3-9.

Wiersum, K.F. 1991. Soil erosion and conservation in agroforestry systems. *In* : Biophysical Research for Asian Agroforestry (Avery, M.E., Cannell, M.G.R. and Org, C.K. eds.), Winrock Intern., USA and South Asia Books USA Publ. 208 p.

□□□

41

Climate Change and Agriculture

A.S. NAIN

Indian economy is expanding at a very healthy rate, which has surpassed 9 % in the fiscal year 2006-07. The booming has been observed mainly in IT, manufacturing, service and banking sectors, while agriculture sector, which remained backbone of Indian economy since independence, is lingering on far behind the GDP growth rate. Rather, agriculture sector witnessed the ever time low growth rate in the current past. The population of India is increasing at an alarming rate, which has already surpassed the landmark of 1 billion and is expected to traverse the 1.5 billion by the year 2050. The situation will worsen by the compounding effect of climate change. It is expected that the atmosphere will become warmer by several degrees, which will exert the negative impact on crops. The duration of the crop will decrease and also the availability of the water. The regions, which are suitable of raising different crops, will become unsuitable due to the northward shifting of the agroecological zones. The food demand of the country is increasing with every passing day, while growing conditions are worsening and the availability of the land is decreasing. Therefore, it is the right time to think, plan and act accordingly to minimize the future threats.

The climate of the globe is slowly but steadily changing, which exhibits accelerated trends in the recent past. Climate change is not only a major global environmental problem, but is also an issue of great concern to a developing country like India. The earth's climate has demonstrably changed on both global and regional scales since the pre-industrial era, with some of these changes attributable to human activities. The changes observed in the regional climate have already affected many of the physical and biological systems and there are indications that social and economic systems have also been affected. Climate change is likely to threaten food production, increase water stress and decrease its availability, result in sea-level rise that could flood crop fields and coastal settlements, and increase the occurrence of pests and diseases. Given the lack of resources, and access to technology and finances, developing countries such as India have limited capacity to develop and adopt strategies to reduce their vulnerability to changes in climate.

Global climate change scenario

Prediction of long-term vegetation response to climatic changes induced by increasing greenhouse gas (GHG) concentrations is an important scientific problem connected with evaluation of possible effects of mitigation and adaptation activity (IPCC, 1992). Climate affects the society directly (water, food, habitat, health, economic infrastructure such as energy, transport and industry) and also through the environment (precipitation, land and sea processes, extreme events, ecosystems). Globally the Earth's climate is warmer today than at any time during the last one hundred and forty years. An annual mean global warming of 0.4 to 0.8°C has been reported since the late Nineteenth Century. Surface temperature records indicate that the 1990s has been the warmest decade of the past millennium in the Northern Hemisphere. Concomitant with this temperature increase, sea level has risen by 10 – 25 cm and there has been a general retreat of glaciers worldwide. There is evidence that El Niño episodes since the mid 1970s have been relatively more frequent than the opposite La Niña episodes. Even with the recent rate of warming, the 1997-98 El Niño event stands out in both surface and tropospheric temperature as a very unlikely event. There has been an increase in heavy and extreme precipitation in areas where the total precipitation has increased. On the other hand, across most of the globe there has been a decrease in the frequency of much below normal seasonal temperatures.

There is now stronger evidence for a human influence on the global climate. The global warming threat is real and the consequences of the climate change phenomena are many, and alarming. Climate change will add to the current stresses on vulnerable natural ecosystems. Agricultural productivity is projected to decrease in many countries in the tropics and sub-tropics. The impact of climatic change will be felt more severely in developing countries such as India whose economy is largely dependent on agriculture and is already under stress due to current population increase and associated demands for energy, fresh water and food. In spite of the uncertainties about the precise magnitude of climate change and its possible impacts particularly on regional scales, precautionary measures must be taken to anticipate, prevent or minimize the causes of climate change and mitigate its adverse effects.

Indian scenario

In India, the analysis of seasonal and annual surface air temperatures has shown a significant warming trend of 0.57°C per hundred years. The warming is found to be mainly contributed by the post-monsoon and winter seasons. The monsoon temperatures do not show a significant trend in any major part of the country except for a significant negative trend over Northwest India. In order to predict the plausible changes in the seasonal as well as annual variability of the monsoons in response to increases in radiative forcing of the atmosphere, detailed climate change scenarios over Indian subcontinent under the SRES 'Marker' scenarios have been developed. Over land regions of the Indian subcontinent, the area-averaged annual mean surface temperature rise is projected to range between 3.5°C and 5.5°C (least in B1 scenario and maximum in A2 scenario) by the end of this century. The projected surface warming is more pronounced during winter than during summer monsoon season. The spatial distribution of surface warming as a consequence of increase in anthropogenic radiative forcing (with respect to 1961-90) suggests that north India may experience an annual mean surface warming of 3°C or more by 2050s, depending upon the future trajectory of anthropogenic forcing. The spatial pattern of temperature change has a large seasonal dependency. The model simulates peak warming of about 3°C over north and central India in winter.

Causes of climate change

Earth's climate is determined by complex interactions between the sun, oceans, atmosphere, land, and living things. The composition of the atmosphere is particularly important because certain gases (including water vapor, carbon dioxide, methane, halocarbons, ozone, and nitrous oxide) absorb heat radiated from the Earth's surface. As the atmosphere warms, it in turn radiates heat back to the surface, to create what is commonly called the "greenhouse effect" (Figure 1). Changes in the composition of the atmosphere alter the intensity of the greenhouse effect. Such changes, which have occurred many times in the planet's history, have helped determine past climates (Figure 2) and will affect the future climate as well. Humans are exerting a major and growing influence on some of the key factors that govern climate by changing the composition of the atmosphere and by modifying the land surface. The human impact on these factors is clear. The concentration of carbon dioxide (CO_2) has risen about 30% since the late 1800s (Figure 3). The concentration of CO_2 is now higher than it has been in at least the last 400,000 years. This increase has resulted from the burning of coal, oil, and natural gas, and the destruction of forests around the world to provide space for agriculture and other human activities. Rising concentrations of CO_2 and other greenhouse gases are intensifying Earth's natural greenhouse effect. Global projections of population growth and assumptions about energy use indicate that the CO_2 concentration will continue to rise, likely reaching between two and three times its late-19th-century level by 2100. This dramatic doubling or tripling will occur in the space of about 200 years, a short span in geological history. Other important gas responsible for greenhouse effect is methane, which is about 60 times more efficient than CO_2. As we add more CO_2 and other heat-trapping gases to the atmosphere, the world is becoming warmer (which changes other aspects of climate as well). Historical records of temperature and precipitation have been extensively analyzed in many scientific studies. These studies demonstrate that the global average surface temperature has increased by over 1°F (0.6°C) during the 20th century. About half this rise has occurred since the late 1970s. Seventeen of the eighteen warmest years in the 20th century occurred since 1980. In 1998, the global temperature set a new record by a wide margin, exceeding that of the previous record year, 1997, by about 0.3°F (0.2°C). Higher latitudes have warmed more than equatorial regions, and nighttime temperatures have risen more than daytime temperatures. As the Earth warms, more water evaporates from the oceans and lakes, eventually to fall as rain or snow. During the 20th century, annual precipitation has increased about 10% in the mid- and high-latitudes. The warming is also causing permafrost to thaw, and is melting sea ice, snow cover, and mountain glaciers. Global sea level rose 4 to 8 inches (10-20cm) during the 20th century because ocean water expands as it warms and because melting glaciers are adding water to the oceans.

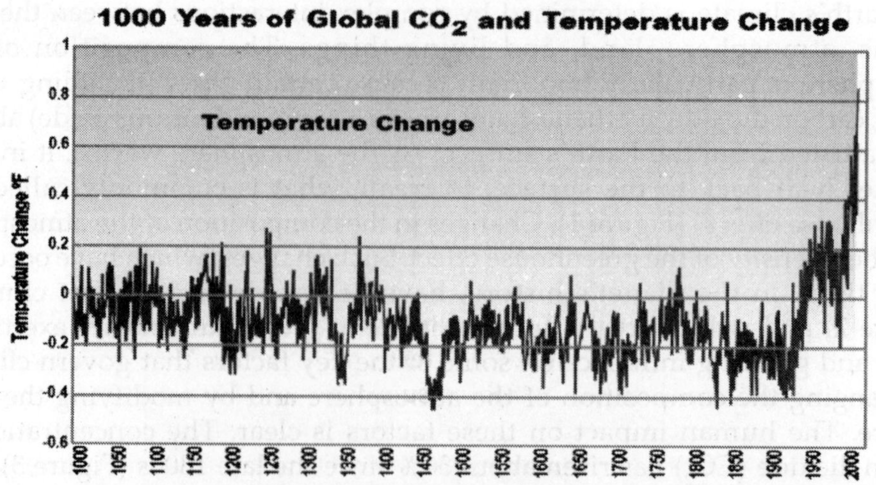

Figure 1. The greenhouse effect of earth atmosphere

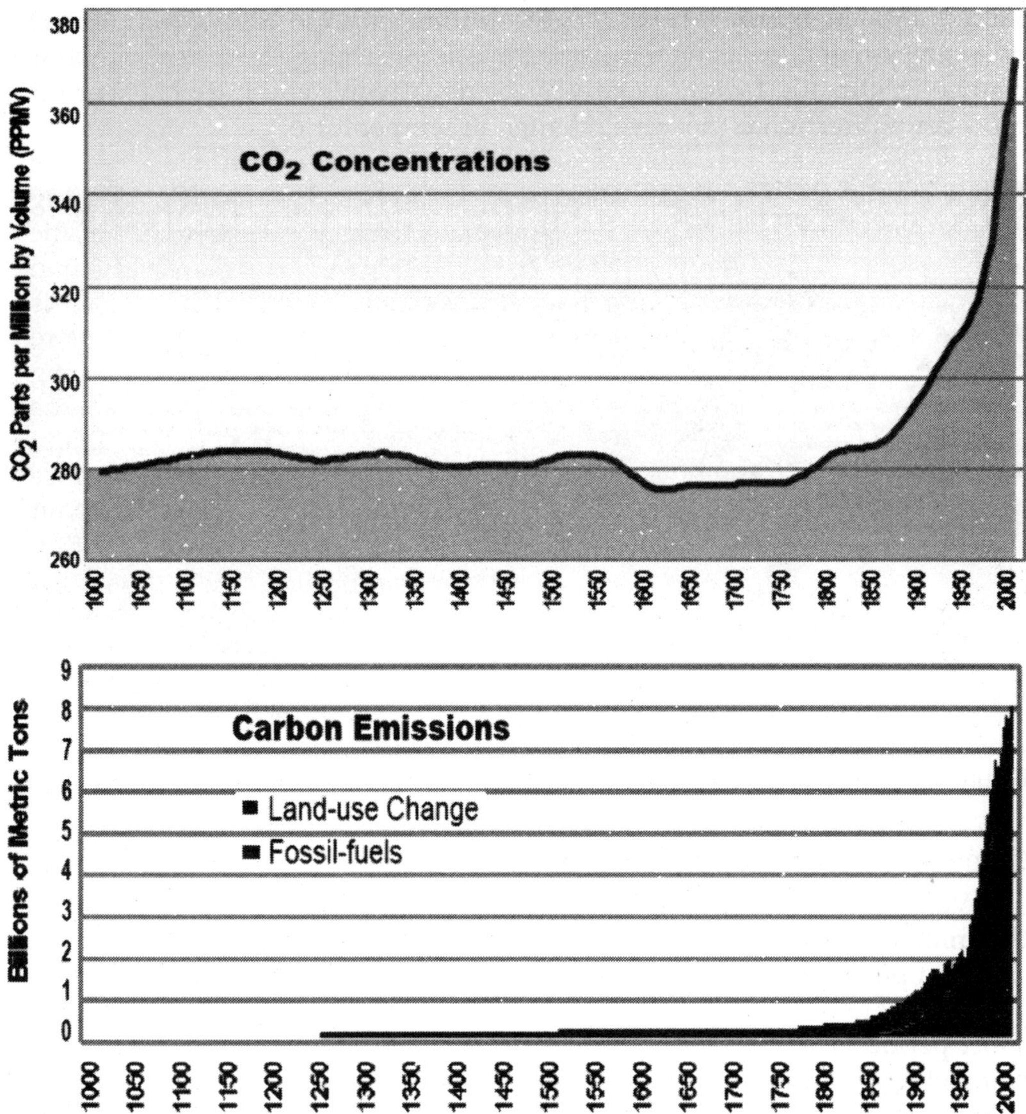

Figure 2. The estimated temperature change during last 1000 years

Global warming vs climate change

People very often use global warming and climate change interchangeably or synonymously, however these two terms are different and convey separate meanings. Climate change is the broad term, which considers change from the normal in all the climatic variables such as temperature, precipitation, wind circulation, atmospheric humidity, clouds and atmospheric pressure etc. While global warming is relatively a narrow term and conveys the message of rising temperature globally. Global warming suggests the gradual increments in temperature, while climate change also considers the phenomenon of decreasing temperature. Climate change term accommodates change in climatic factors positively or negatively, while global warming reflects unidirectional (positive) change in temperature.

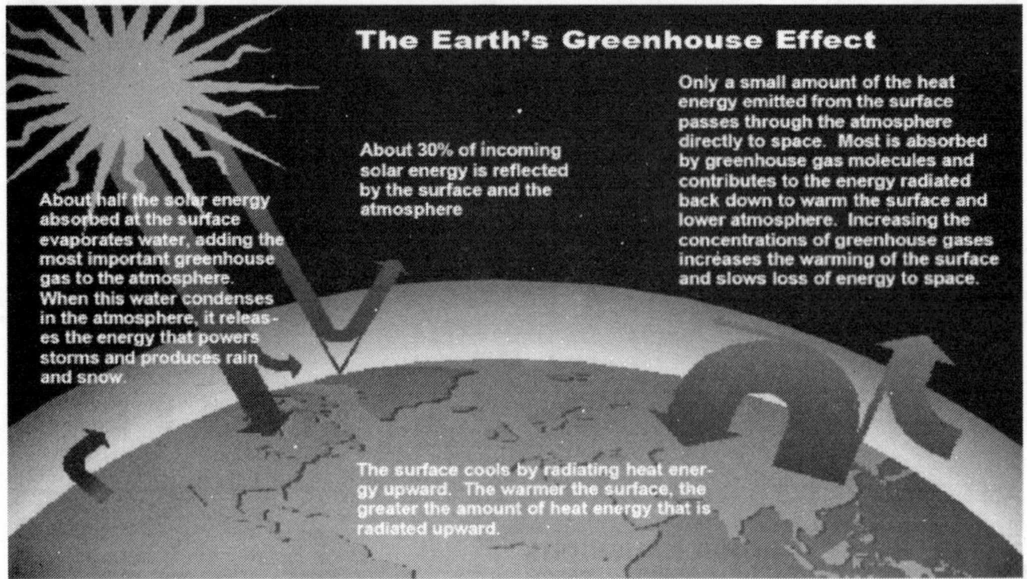

Figure 3. The CO_2 emission and concentration in the atmosphere during last 1000 years

Climate variability vs climate change

Variability in climate and climate change some time create confusion in the minds of the readers, however, these two terms are used for entirely different purposes. Climatic variability is used to denote the random changes in the climate of the region, which tends to attain the normal values after a short period. The climatic variability is a cyclic phenomenon, which after completing the cycle will again come at the initial point / stage. While climate change refers to the systematic change in climatic factors in a particular direction, which keep on happening for longer period. Climate change do not exhibits the reversing tendency at least for considerably longer periods. In addition to this, climate change also includes the extreme weather events, their frequencies and values.

Climate change and agriculture: a complex scenario

Climate constitutes a complex of inter-related variables (Figure 4). On average, through a set of regulatory mechanisms, a smooth change in one variable triggers smooth changes in most others. With the exception of possible qualitative and abrupt variations, which will be mentioned below, such interrelations are independent of atmospheric carbon dioxide (CO_2). The latter and other greenhouse gases play a part largely through their effect on the radiation balance of the atmosphere. There is only a weak link between such factors as cloudiness and wind. Temperature, evaporation and rain are strongly correlated, which illustrates the likely intensification of the hydrological cycle. Combined with the projected pressure on land and water use, competition for land and water will certainly become a key social and political issue.

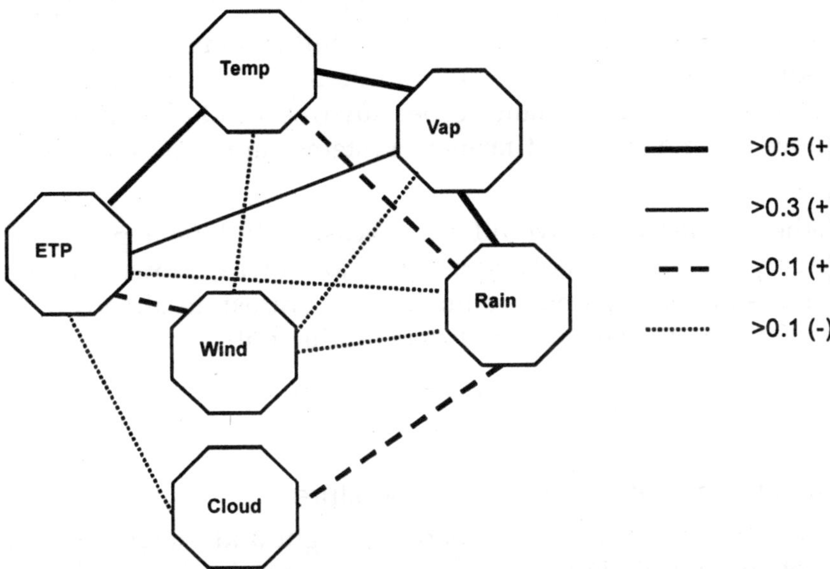

Figure 4. Some relationships between major climate variables. Solid and dotted lines indicate positive and negative correlations, respectively. The strength of the correlation decreases from double heavy lines to thin single lines. Computed from annual averages of 3263 land stations, mostly from developing countries (FAO, 1995)

Climate variability is likely to increase under global warming (Katz and Brown, 1992), both in absolute and in relative terms. This is linked with thresholds which affect the occurrence of many meteorological phenomena. For instance, tropical cyclones are 'fed' by water vapour evaporating from oceans at a temperature above 26 or 27°C. Therefore, higher average sea surface temperatures are bound to result in a higher frequency of tropical cyclones.

Indirect effect of climate change on agriculture

In qualitative terms, many indirect effects of climate change on agriculture can be conjectured. Most of them are estimated to be negative and they catch most of the attention of the media. These effects include:

- The overall predictability of weather and climate would decrease, making the day-to-day and medium-term planning of farm operations more difficult;
- Loss of biodiversity from some of the most fragile environments, such as tropical forests and mangroves;
- Sea-level rise (40 cm in the coming 100 years) would submerge some valuable coastal agricultural land;
- The incidence of diseases and pests, especially alien ones, could increase;
- Present (agro) ecological zones could shift in some cases over hundreds of kilometres horizontally, and hundreds of meters altitudinally, with the hazard that some plants, especially trees, and animal species cannot follow in time, and that farming systems cannot adjust themselves in time;
- Higher temperatures would allow seasonally longer plant growth for the crop growing in cool and mountainous areas, allowing in some cases increased cropping and production. In contrast, in already warm areas climate change can cause reduced productivity;
- The current imbalance of food production between cool and temperate regions and tropical and subtropical regions could worsen.

Direct effect of climate change on agriculture

The weather conditions - temperature, light and water - determine the capacity of the populations to cultivate sufficient food for themselves and their animals. Any short-term or long-term fluctuations of the climate can have dramatic effects on the agricultural productivity. Thus, the climate has a direct incidence on food safety. The risks associated with climate change lie in the interaction of several systems with many variables that must be collectively considered. Agriculture (including crop agriculture, animal husbandry, forestry and fisheries) can be defined as one of the systems, and climate the other. If these systems are treated independently, this would lead to an approach which is too fragmentary. The issue is more global. It is now held as likely that human activities can affect climate, one of the components of the environment. Climate in turn affects agriculture, the source of all food consumed by human beings and domestic animals. It must be further considered that not only climate may be changing, but that human societies and agriculture develop trends and constraints of their own which climate change impact studies must take into consideration.

The greenhouse gases CH_4, N_2O and chlorofluorocarbons (CFCs) have no known direct effects on plant physiological processes. They only change global temperature and are therefore not discussed further. Instead, concentration should be on the effects of increased CO_2 tropospheric O_3, increased UV-B through depleted stratospheric ozone, increased temperatures and the associated intensification of the hydrological cycle.

The fertilization effect of CO_2

CO_2 is an essential plant 'nutrient', in addition to light, suitable temperature, water and chemical elements such as N, P and K, and it is currently in short supply. Higher concentrations of atmospheric CO_2 due to increased use of fossil fuels, deforestation and biomass burning, can have a positive influence on photosynthesis (Figure 5). Under optimal growing conditions of light, temperature, nutrient and moisture supply, biomass production can increase, especially of plants with C_3 photo-synthetic metabolism, above and even more below ground. A total of 10 to 20% of the approximate doubling of crop productivity over the past 100 years could be due to this effect (Tans *et al.*, 1990) and forest growth or regrowth may have been stimulated as well. Further productivity increases may occur in the coming century, in the order of 30% or more where plant nutrients and moisture are adequate. Higher CO_2 values would also mitigate the plant growth damage caused by pollutants such as NO_x and SO_2 because of smaller stomatal openings. Higher percentages of starch in grasses improves their feeding quality, implying less need for feed mixes when silaging.

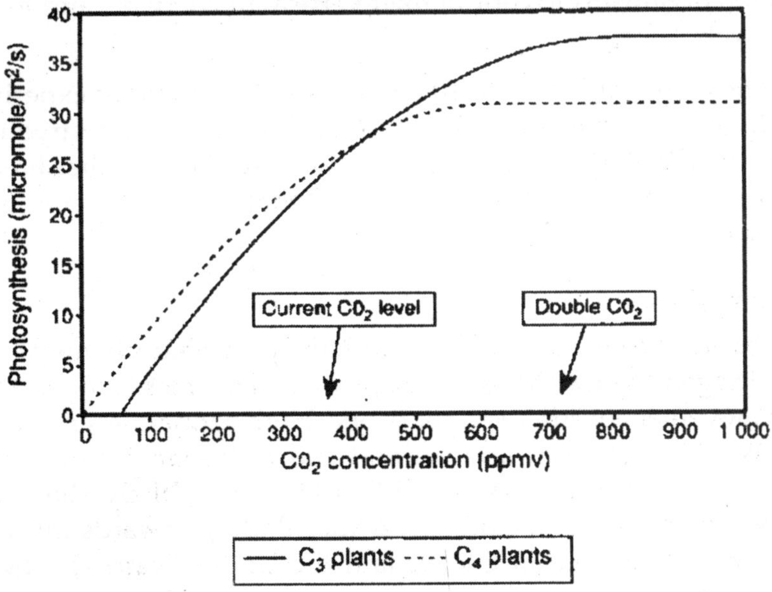

Figure 5. Schematic effect of CO_2 concentrations on C_3 and C_4 plants (after Wolfe and Erickson, 1993). The main mechanism of CO_2 fertilization is that it depresses photo-respiration, more so in C_3 than in C_4 plants

The CO_2 anti-transpirant effect (improved water-use efficiency WUE)

With increased atmospheric CO_2 the consumptive use of water becomes more efficient because of reduced transpiration. This is induced by a contraction of plant stomata and/or a decrease in the number of stomata per unit leaf area. This restricts the escape of water vapour from the leaf more than it restricts photosynthesis (Wolfe and Erickson, 1993). With the same amount of available water, there could be more leaf area and biomass production by crops and natural vegetation. Plants could survive in areas hitherto too dry for their growth.

Ultraviolet radiation

Increased ultraviolet radiation (UV-B, between 280 and 320 nanometres), due to depletion of the stratospheric ozone layer, mainly in the Antarctic region, may negatively affect terrestrial and aquatic photosynthesis and animal health. Over the last decade, a decrease of stratospheric ozone was observed at all latitudes (about 10% in winter, 0% during summer and intermediate values during spring and autumn). However, the 'Biological Action Factor' of UV-B can vary over several orders of magnitude with even slight changes in the amount and wavelength of UV-B. In particular it can be noted that:

- There are damaging effects of increasing UV-B on crops, animals and plankton growth. It has been reported that UV-B affects the ability of plankton organisms to control their vertical movements and to adjust to light levels;

- Reductions in yield of up to 10% have been measured at experimentally very high UV-B values, and would be particularly effective in plants where the CO_2 fertilization effect is strongest. On the other hand, UV-B increase could increase the amount of plant internal compounds that act against pests.

Tropospheric ozone

Tropospheric ozone originates about half from photochemical reactions involving nitrogen oxides (NO_x), methane or carbon monoxide, and half by downward movement of stratospheric ozone. High ozone concentrations have toxic effects on both plant and animal life (German Bundestag, 1991; MacKenzie and El-Ashry, 1988; UNEP, 1993). It is likely that ozone, in conjunction with other photo-oxidants, is contributing towards the 'new type of forest damage' observed in Europe and the United States. In the tropics, tropospheric ozone concentrations are generally lower than at northern mid-latitudes. However, this does not apply to periods when biomass burning releases precursor substances for the photochemical formation of ozone.

Temperature

In general, higher temperatures are associated with higher radiation and higher water use. It is relatively difficult to separate the physiological effects (at the level of plants and plant organs) of temperatures from the ecological ones (at the level of the field or of the region). There are both positive and negative impacts at the two levels, and only crop- and site-specific simulation can assess the global 'net' effect of temperature increases. It is generally agreed that:

- Rising temperatures - with smallest increases in the tropics (IPCC, 1992) - would diminish the yields of some crops, especially if night temperatures are increased (the temperature increase since the mid-1940s is mainly due to increasing night-time temperatures, while CO_2-induced warming would result in an almost equally large rise in minimum and maximum temperatures (Kukla and Karl, 1993);
- Higher temperatures could have a positive effect on growth of plants of the CAM type. They would also strengthen the CO_2 fertilization effect and the CO_2 anti-transpirant effect of C_3 and C_4 plants unless plants get overheated;
- Higher night temperature may increase dark respiration of plants, diminishing net biomass production;
- Higher cold-season temperatures may lead to earlier ripening of annual crops, diminishing yield per crop, but would allow locally for the growth of more crops per year due to lengthening of the growing season. Winter kill of pests is likely to be reduced at high latitudes, resulting in greater crop losses and higher need for pest control;
- Higher temperatures will allow for more plant growth at high latitudes and altitudes.

The hydrological cycle and soils

Even a slight increase in surface temperatures will affect evaporation, atmospheric moisture and precipitation. While it is generally agreed that rainfall will increase (by an estimated 10 to 15%), two aspects have to be elucidated: how will rainfall intensities be affected, and what are the details of spatial changes. This is still largely a matter of discussion among experts. If the increase in precipitation should be associated with increased rainfall intensities, then the quality and quantity of soil and water resources would decline, for instance through increased runoff and erosion, increased land degradation processes, and a higher frequency of floods and possibly droughts. However:

- The extra precipitation on land, if indeed including present subhumid to semi-arid areas, will increase plant growth in these areas, leading to an improved protection of the land surface and increased rainfed agricultural production; in already humid areas the extra rainfall may, however, impair adequate crop drying and storage;
- The extra precipitation predicted to occur in some regions provides possibilities for off-site extra storage in rivers, lakes and artificial reservoirs (on-farm or at subcatchment level) for the benefit of improved rural water supply and expanded or more intensive irrigated agriculture and inland fisheries:
- The effects on water resources and water apportioning of international river and lake basins can be very substantial, with political overtones.

Reversing climate change trends

The trends in climate change could be reversed by adopting the certain measures in order to decrease the GHG gasses in atmosphere. These measures are:

- Carbon dioxide is supposed to be the major factor responsible for climate change. The CO_2 concentration in Atmosphere can be reduced by fixing carbon in soil and increasing forest cover (Afforestation).
- Reducing CO_2 and other Green House Gasses emission from Industrial sector by establishing pollution control / treatment plants.
- Reducing the emission of CFC gasses from refrigerating industry.
- Reducing soil erosion to restore the soil productivity as well as to check the organic carbon mixing in water which leads to emission of methane, CO_2 and other GHG gasses in Atmosphere.
- Promoting the low carbon emission fuels such as biofuels.
- Promoting renewable energy such as wind energy, solar energy, hydropower energy, geothermal energy etc.

Coping strategies / mitigation strategies

It has been established from the instrumental records that cliamate change is taking place. Thus, there is an urgent need to mitigate the adverse effect of climate change by adopting the different measures at 1) farm / field level, 2) Institutional level and 3) Government level.

Farm / field intervention

- Substituting crops best adapted to the changed climate.
- Diversification of Agriculture so that at least some production is achieved in the eventuality of extreme weather condition for specific crop.
- Improving agricultural management.
- Improving tillage practices to combat the loss of organic matter and to improve the soil quality.
- Adjustment of sowing dates and other agricultural operations to cope with extreme climatic conditions.
- Judicious and integrated use of agricultural inputs such as fertilizers, manures, bio-pesticides and chemicals
- Improving water use efficiency (increase agriculture productivity per drop of water).
- Adaptation of multistoried, relay cropping etc. for efficient use of soil moisture and prevailing weather conditions.
- Efficient rainwater harvesting by water storage in ponds/dams/ reservoirs etc.
- Adaptation of micro-irrigation techniques such as drip irrigation, sprinkler irrigation etc.
- Adaptation of Farm pond based approach.

Institutional intervention

- Agro-ecological conditions based research, technological development, and package of practices.
- Evolving new high yielding but short duration varieties so that crop is exposed to extreme weather for shorter period.
- Developing and promoting use of hybrids, especially for rainfed agro-ecosystems.
- Improving productivity of crops, existing plantations and livestock.
- Small farm mechanism.
- Development and adaptation of the measures/ procedures for micro-climate modification.

Governmental / policy Intervention

- Agro-ecological Zones based approach for agricultural planning and research.

- Watershed based approach / micro-planning to use rainwater within watershed boundary.
- Expansion of irrigated area for assured irrigation to save the crop from droughts / dry spells.
- Improving availability of seed/planting material of high yielding varieties.
- Reducing soil erosion to restore the soil productivity.
- Controlling / reversing the Desertification process by increasing water availability and governmental aids.
- Developing infrastructure for post-harvest management, marketing and agribusiness,
- Establishing Monitoring and Early warning system for growth and development of crops.
- Strengthening Agricultural insurance schemes.
- Improving surface water quality.
- Increasing ability to shift water within and between sectors.
- Use pricing and market mechanisms proactively to increase efficiency of water use.
- Establishing climate change centers at regional level to study the regional climate change and their effects.
- Strategic food reserve for bad harvest year.
- Genuinely Free World Food Trade.

Conclusion

There are certain evidences which suggest that climate is changing globally. However the degree of change may differ from one place to other. Even climatic factors will exhibit uncertainty in their behavior. One climatic factor may show changing trend at one place, while other factor may show changing trend at other places. There are views that climate change is only a myth and it is a dissemination of filtered information. However, all the scientists are unanimous on the fact that concentration of CO_2 is increasing in atmosphere. Therefore global temperature rise is inevitable because of close proximity of CO_2 and temperature. Climate change, if it occurs, will definitely affect agriculture. Most mechanisms, and two-way interactions between agriculture and climate, are known, even if not always well understood. It is evident that the relationship between climate change and agriculture is still very much a matter of conjecture with many uncertainties. Major uncertainties affect both the Global Circulation Models (GCMs) and the response of agriculture, especially at the national and subregional levels. In addition, many of the models do not take into consideration CO_2 fertilization and improved water-use efficiency, the effect of cloud cover (on both climate and photosynthesis), or the transient nature of climate change.

References

FAO, 1995. FAOCLIM 1.2, Worldwide agroclimatic data. *FAO Agrometeorology Working Paper Series No. 11*. FAO, Rome. 1 CD-ROM and 66 p.

German Bundestag. (ed.). 1991. *Protecting the Earth*. A status report with recommendations for a new energy policy. Bonn, Deutscher Bundestag, Referat Oeffentlichkeitsarbeit. Vol. 1, 672 p.

IPCC. 1992. *Climate Change 1992*. The supplementary report to the IPCC scientific assessment. J.T. Houghton, B.A. Callandar and S.K. Varney (eds.). Cambridge University Press. Cambridge.

Katz, R.W. and Brown, B.G. 1992. Extreme events in a changing climate: variability is more important than averages. *Clim. Change* **21:** 289-302.

Kukla, G. and Karl, T.R. 1993. Nighttime warming and the greenhouse effect. *Envir. Sci. Technol.* **27** (8): 1468-1474.

MacKenzie, J.J. and El-Ashry, M.T. 1988. Links between air pollution and damage to crops. In: *Ill Winds: Airborne Pollution's Toll on Trees and Crops*. World Res. Institute. pp. 25-33.

Tans, P.P., Fung, I.Y. and Takahashi, T. 1990. Observational constraints on the global atmospheric CO_2 budget. *Science* **247**: 1431-1438.

UNEP. 1993. *Action on Ozone*. UNEP, Nairobi. 24 p.

Wolfe, D.W. and Erickson, J.D. 1993. Carbon dioxide effects on plants: uncertainties and implications for modelling crop response to climate change. In: *Agricultural Dimensions of Global Climate Change*. H.M. Kaiser and T.E. Drennen (eds.). pp. 153-178.

□□□

ABOUT EDITORS

Dr. Surendra Kumar Saini, did his graduation in agriculture, master and doctorate in agronomy from G.B. Pant University of Agriculture & Technology, Pantnagar, Udham Singh Nagar. He has 35 year experience of teaching, research and extension in the field of crop management. He has published more than 100 research papers in the national as well as international journals of repute. He is the national project investigator of the All India Coordinated Research Project on Sugarcane (Agronomy), Head Department of Agronomy-cum-Director, Center of Advanced Studies at G.B. Pant University of Agriculture & Technology, Pantnagar, Udham Singh Nagar. He has guided 25 M.Sc and 07 Ph.D students.

Dr. Subhash Chandra, obtained graduate degree in agriculture and doctorate degree in agronomy from G.B. Pant University of Agriculture & Technology, Pantnagar and the masters' in agronomy from PAU, Ludhiana. He served ICAR from October, 1989, till Feb, 2006 as ARS scientist and Senior scientist. Presently, he is actively engaged in teaching, research and extension programme especially in the field of water management at G.B. Pant University of Agriculture & Technology, Pantnagar, Udham Singh Nagar as Chief scientist of water management. He has more than 50 publications in reputed journals to his credit. He was conferred ICAR 'Team Research' award for the biennium 2006-07.